Introduction to
Process Technology

Introduction to
Process Technology

by

Charles E. Thomas

Uhai Publishing, Inc.
1731 Helderberg Trail
Berne, NY 12023
Toll free 1-877-842-4782

NOTICE TO THE READER

This book is based on research and the author's experience in the chemical processing industry. Neither the publisher nor the author make any further claims as to the fitness of the products or procedures represented herein. The reader is cautioned to consider the safety hazards involved in the procedures and process contained in this book and to adopt such safety precautions as are appropriate. The publisher disclaims any responsibility to obtain information from the manufactures of products other than that provided by the manufactures and contained herein.

Cover Design:
   Bi-County Graphics

Cover Photograph:
   PhotoDisc

Illustrations by:
   C. Thomas/ Bi-County Graphics

ISBN: 1-930528-00-0

COPYRIGHT © 2001
UHAI Publishing, Inc.

Printed in the United States of America

For more information, contact:
Uhai Publishing, Inc.
1731 Helderberg Trail
Berne, NY 12023
Toll free 1-877-842-4782

All rights reserved. No part of this work covered by the copyright hereon may be reproduced or used in any form or by any means—graphic, electronic, or mechanical, including photocopying, recording, taping, or information storage and retrieval systems—without the written permission of the publisher.

1 2 3 4 5 6 7 8 9 10   XXX   04 03 02 01 00

Preface

The chemical processing industry (CPI) is anticipating severe shortages in skilled technicians to operate their plants. As the large baby boomer group quickly approaches retirement age, the CPI braces for a seventy to eighty percent employee turnover. The next ten to fifteen years will bring massive changes as education levels in the United States continue to drop. The CPI is painfully aware of the changing requirements for a process technician. New technology, rightsizing, and redistribution of technical skills have created a new profile for this group. The term "gold collar" is being applied to the field of process technology that can command incomes in the six-digit area.

The process technician of the future will have a specialized degree in process technology that will include instruction in engineering principles and physics, chemistry, maintaining unit operations, safety, checking equipment, sampling, taking readings, making rounds, troubleshooting, filling out quality charts, and operating new computer systems. These new apprentice technicians will need strong technical and problem solving skills, the ability to assimilate cutting edge technologies quickly, and the ability to apply innovative ideas. In addition to these skills a process technician will need to be able to handle conflict, look at a complex situation and see the overall picture, and communicate effectively.

Over the past six years, process technology has become one of the most popular programs in community colleges and universities located in heavily industrialized areas. The program appeared virtually overnight in response to industry and community needs. Process technology is defined as the study and application of the scientific principles associated with the operation and maintenance of the chemical processing industry. Process technicians can be found in petrochemical and refinery operations, food processing, paper and pulp, and many other areas. This group represents the fourth largest U.S. manufacturing industry.

Introduction to Process Technology is the first class a new student takes in a college process technology program. The course is designed to be an overview of the entire curriculum and provides the apprentice technician with the foundation that future classes will build upon. This text breaks into chapters each of these classes and provides key objectives that instructors can use to develop lesson plans and enhance their instruction. Each chapter includes objectives, key terms, photographs and line drawings, lecture material, summaries, and review questions. An instructor guide is available; however, the author strongly encourages each teacher to develop his or her own tests and learning activities. These activities can be linked to instructional videos, lab exercises, or field trips. Key topics covered in this text include:

- Introduction to Process Technology
- Process Equipment
- Process Systems
- Safety, Health, and Environment
- Process Instrumentation
- Chemistry
- Physics
- Quality Control
- Troubleshooting
- Process Operations

The author would like to express his thanks to those individuals who have been involved in the development of the process technology program.

Charles E. Thomas Ph.D.

Table of Contents

Chapter 1   History of the Chemical Processing Industry **1**
    1.1   Key Terms **2**
    1.2   History of the Chemical Processing Industry **3**
    1.3   Working in the Chemical Processing Industry **11**
    1.4   College Programs for Process Technology **14**
    1.5   Your Career as a Process Technician **19**
    1.6   Careers in the Chemical Processing Industry **22**

Chapter 2   Introduction to Process Technology **27**
    2.1   Key Terms **28**
    2.2   Roles and Responsibilities of a Process Technician **29**
    2.3   Safety, Health, and Environment **34**
    2.4   Quality Control **35**
    2.5   Process Equipment, Systems, and Operations **37**
    2.6   Instrumentation and Process Control **39**
    2.7   Applied Chemistry and Physics **39**
    2.8   Industrial Processes **40**
    2.9   Troubleshooting **40**

Chapter 3   Safety, Health, and Environment **43**
    3.1   Key Terms **44**
    3.2   Basic Safety **45**
    3.3   OSHA **46**
    3.4   The PSM Standard **47**
    3.5   The Hazard Communication Program **48**
    3.6   Safe Handling, Storage, and Moving of Hazardous Chemicals **50**
    3.7   Physical Hazards Associated with Chemicals **50**
    3.8   Health Hazards Associated with Chemicals **50**
    3.9   The Material Safety Data Sheet (MSDS) **52**
    3.10   Toxicology **52**

Introduction to Process Technology • Contents

3.11	Respiratory Protection Programs	52
3.12	Personal Protective Equipment (PPE)	53
3.13	Emergency Response	53
3.14	Plant Permit System	54
3.15	Classification of Fires and Fire Extinguishers	54
3.16	HAZWOPER	56
3.17	Hearing Conservation and Industrial Noise	56
3.18	Department of Transportation	56

Chapter 4 Basic Process Principles **61**
4.1	Key Terms	62
4.2	Basic Principles of Pressure	63
4.3	Heat, Heat Transfer, and Temperature	69
4.4	Fluid Flow	70
4.5	Basic Math for Process Technicians	73

Chapter 5 Equipment One **85**
5.1	Key Terms	86
5.2	Basic Hand Tools	87
5.3	Valves	88
5.4	Piping and Storage Tanks	93
5.5	Pumps	96
5.6	Compressors	99
5.7	Steam Turbines	100
5.8	Steam Traps	102
5.9	Electricity and Motors	102
5.10	Equipment Lubrication, Bearings, and Seals	105

Chapter 6 Equipment Two **109**
6.1	Key Terms	110
6.2	Heat Exchangers	111
6.3	Cooling Towers	112
6.4	Boilers	113
6.5	Furnaces	115
6.6	Reactors	116
6.7	Distillation	116
6.8	Plastics Plant Equipment	120
6.9	Pressure Relief Equipment	122

Chapter 7 Process Instrumentation ... 125
- 7.1 Key Terms ... 126
- 7.2 PFDs and P&IDs ... 127
- 7.3 Process Instruments ... 139
- 7.4 Basic Elements of a Control Loop ... 139
- 7.5 Process Variables and Control Loops ... 140
- 7.6 Primary Elements and Sensors ... 142
- 7.7 Transmitters and Control Loops ... 142
- 7.8 Controllers and Control Modes ... 143
- 7.9 Final Control Elements and Control Loops ... 145
- 7.10 Interlocks and Permissives ... 146
- 7.11 P&ID Components ... 146

Chapter 8 Systems ... 153
- 8.1 Key Terms ... 154
- 8.2 Flow Diagrams and Equipment Relationships ... 155
- 8.3 Compressor System ... 155
- 8.4 Heat Exchanger and Cooling Tower System ... 156
- 8.5 Lubrication System ... 156
- 8.6 Electrical System ... 156
- 8.7 Furnace System ... 158
- 8.8 Plastics System ... 158
- 8.9 Reactor System ... 160
- 8.10 Steam Generation System ... 160
- 8.11 Distillation System ... 161
- 8.12 Refrigeration System ... 162
- 8.13 Water Treatment System ... 164
- 8.14 Hydraulics ... 165
- 8.15 Utilities ... 166

Chapter 9 Industrial Processes ... 169
- 9.1 Key Terms ... 170
- 9.2 Common Industrial Processes ... 171
- 9.3 Petrochemical Processes ... 173
- 9.4 Benzene ... 173
- 9.5 BTX Aromatics ... 173
- 9.6 Ethylbenzene ... 175
- 9.7 Ethylene Glycols ... 175

Introduction to Process Technology • Contents

9.8	Mixed Xylenes	175
9.9	Olefins	177
9.10	Paraxylene	177
9.11	Polyethylene	177
9.12	Xylene Isomerization	177
9.13	Ethylene	178
9.14	Refining Processes	178
9.15	Alkylation	179
9.16	Fluid Catalytic Cracking	179
9.17	Hydrodesulfurization	180
9.18	Hydrocracking	181
9.19	Fluid Coking	183
9.20	Catalytic Reforming	183
9.21	Crude Distillation	183

Chapter 10 Operations ... 187
10.1	Key Terms	188
10.2	Team Skills	189
10.3	Quality Tools	189
10.4	Statistical Process Control (SPC)	190
10.5	Troubleshooting	193
10.6	Benchtop Operations	195
10.7	Pilot Plant Operation	195

Chapter 11 Process Chemistry ... 201
11.1	Key Terms	202
11.2	Fundamental Principles of Chemistry	205
11.3	Chemical Equations and the Periodic Table	207
11.4	Chemical Reactions	213
11.5	Material Balance	216
11.6	Percent-by-Weight Solutions	217
11.7	pH Measurements	218
11.8	Hydrocarbons	219
11.9	Applied Concepts to Chemical Processing	221

Chapter 12 Applied Physics ... 229
12.1	Key Terms	230
12.2	Fundamental Concepts	231
12.3	Density and Specific Gravity	233

12.4	Pressure in Fluids	237
12.5	Complex and Simple Machines	239

Chapter 13   Environmental Control 249

13.1	Key terms	250
13.2	Air Pollution Control	251
13.3	Water Pollution Control	252
13.4	Solid Waste Control	253
13.5	Toxic Substances Control	254
13.6	Emergency Response	254
13.7	Community Right-to-Know	254

Chapter 14   Quality Control 259

14.1	Key Terms	260
14.2	Principles of Continuous Quality Improvement	261
14.3	Quality Improvement Cycle	261
14.4	Supplier-Customer Relationship	262
14.5	Quality Tools	262
14.6	Statistical Process Control	263
14.7	Flow Charts	267
14.8	Run Charts	267
14.9	Cause-and-Effect Diagrams (Fishbone)	267
14.10	Pareto Charts	268
14.11	Planned Experimentation	269
14.12	Histograms or Frequency Plots	269
14.13	Forms for Collecting Data	270
14.14	Scatter Plots	271

Chapter 15   Self-Directed Job Search 273

15.1	Key Terms	274
15.2	The Job Search	275
15.3	Pre-Employment Testing	278
15.4	Sample Cover Letter and Resume	279
15.5	Work Experience	280

Glossary 283

Index . . . . .B

Chapter 1

History of the Chemical Processing Industry

OBJECTIVES

After studying this chapter, the student will be able to:
- *Define key terms used in process technology.*
- *Explain the history and development of the chemical processing industry.*
- *Describe batch operations, thermal cracking, catalytic cracking, and fractional distillation.*
- *Contrast the development of the hydrocarbon industry with advances in modern society.*
- *List the skills required to work in the chemical industry.*
- *Discuss the key elements of working in a diverse work force.*
- *Describe future trends in the area of process technology.*
- *Identify behaviors that lead toward sexual harassment.*
- *Explain techniques used by successful college students.*
- *Describe college programs in process technology.*

History of the Chemical Processing Industry • Chapter 1

KEY TERMS

Batch process—adds all ingredients to the process up front.

Catalytic cracking—uses a catalyst to separate hydrocarbons.
catalyst is not consumed

Diversity training—identifies and reduces hidden biases between people with differences.

Faculty expectations—college faculty expect process students to be responsible for their own learning, to set goals, to manage their time, to participate in class activities, and to attend scheduled class meetings.

Fractionating column—the central piece of equipment in a distillation system. Fractionating columns separate hydrocarbons by their individual boiling points.

Goal setting—reasonable, specific, measurable objectives that lead toward the successful completion of a goal.

Lifelong learning—process technicians will spend a lifetime learning new technologies and equipment. Global competition requires companies to adopt new and innovative techniques. Process technicians will come into contact with learning opportunities that can not be found anywhere else.

Process technician—a person who operates and maintains the complex equipment, systems and technologies found in the chemical processing industry. Since these people work closely with specific pieces of equipment or processes, they are commonly called boiler operators, compressor technicians, distillation technicians, refinery technicians, or waste water operators.

Process technology— the study and application of the scientific principles associated with the operation and maintenance of the chemical processing industry.

Sexual harassment—behavior that constitutes unwelcome sexual advances. This behavior could come in the form of verbal or physical abuse or unwelcome requests for sexual favors. The behavior may involve people of the opposite sex, same sex, supervisor to employee, student to student, employee to employee, teacher to student, etc. (For further information on sexual harassment, see Title VII of the Civil Rights Act of 1964.)

Thermal cracking—uses heat and pressure to separate small hydrocarbons from large ones.

Time management—a structured system that incorporates an individual's study needs with time management principles.

Cracking - Separating Hydrocarbons

2

1.2 History of the Chemical Processing Industry

The lifeblood of modern society is found in petroleum products. Cars, planes, trains, ships, and farm equipment all require petroleum products to operate. It is difficult to look around our world and not see the results of modern petroleum manufacturing. Before 1800, few people recognized the value or potential of hydrocarbon processing.

Petroleum

The term petroleum combines two Latin words, *petra* (rock) and *oleum* (oil). Petroleum is a natural resource that took many years to develop. The most dominant theory is that crude oil is made up of the remains of small ocean animals and plants that died, dropped to the bottom of the shallow ocean floor, and were covered by sediment. Over a long period the tremendous weight of the sediment, combined with a low oxygen content and sustained temperatures around 150 degrees, formed the oil. Under these conditions a chemical reaction occurs as carbohydrates, proteins, and other compounds are converted to crude oil. Natural gas forms under these same conditions if the temperature is maintained near 200 degrees. As the land masses shifted, the oil was forced by water into cracks, openings, and porous rocks.

Crude oil is a mixture of hydrocarbons that vary in molecular structure and weight. Modern manufacturers separate these components through the distillation process. Distillation separates the various components in a mixture by boiling point.

Petroleum Products

Some examples of petroleum products are asphalt for roads, gasoline, kerosene, plastic products, carpet material, baby diapers, aspirin, lubricating oils, butane, propane, detergents, cosmetics, insecticides, fertilizers, wax, milk cartons, and toothpaste. It is difficult to see our culture at its present level of technology without petroleum; however, most experts agree that the Earth's entire oil reserves are around 700 billion barrels. Present global consumption is around 27.4 billion barrels per year. The United States produces 3 billion barrels per year and refines over 5.5 billion barrels per year. These reserves can not be replaced once they are used, and some projections indicate that, at our present rate of consumption, our oil reserves will be depleted during the next 25 years.

Process technicians will find themselves operating processes that use alternate fuel sources such as coal, oil shale, and tar sands, along with operating new technologies. Huge reserves have been located in Canada, Utah, Wyoming, and Colorado. New conservation strategies, better oil reserve projections, alternate fuel sources, and new technologies can help to extend our supply of energy.

Over the last ten years, advances in technology and massive oil finds in Russia, Columbia, and Africa have added to global reserves. New offshore drilling techniques allow the oil industry to drill at depths previously considered impossible. An offshore platform in the Gulf of Mexico called the Genesis extends 2,600 feet to the sea floor. The surface rig extends over two and one half football fields. The Genesis produces over 55,000 barrels of oil and 72 million standard cubic feet of natural gas. Although this is impressive, a consortium of oil companies led by Chevron recently set a well in the Gulf of Mexico in waters 7,718 feet deep.

Technological advances in converting natural gas into oil could add 1.6 trillion barrels to our reserves. This figure represents more oil than we could use in sixty years. At the present time, natural gas is used for home heating, cooking, and generating electricity. The technology exists to convert natural gas to gasoline, kerosene, diesel, and lubricating oils; however, it is still impossible to produce heavy bottoms products like asphalt. Modern natural gas plants can be constructed for $10 billion and produce a barrel of oil for under $20. In 2000, the cost of a barrel of oil is between $20 and $30.

The procedure to use natural gas to create gasoline starts by passing methane and oxygen over a heated catalyst. This releases the hydrogen from the carbon atom and allows it to bond with the oxygen. This reaction produces carbon monoxide and hydrogen called syn-gas, the building blocks for the conversion process. In step two, chains of eight or more carbons are combined to form gasoline. Products produced from natural gas burn cleaner because they do not contain sulfur, nitrogen, or molecular carbon ring arrangements.

Early Contributions

The history of the chemical processing industry can be traced back thousands of years. The Bible reports that Noah used pitch as a building material for the Ark. The ancient Chinese connected bamboo poles to pipe natural gas deposits into containers where the gas was burned in order to separate brine water and natural gas from salt. The ancient Egyptians coated mummies with pitch. Pitch was also used to build the streets and walls of ancient Babylon. Before the first European stepped foot on the North or South American continents, Indians used crude oil for medicine and fuel. Around 600 AD, temples built near Baku, Azerbaijan had eternal flames that burned continuously and were a source of awe for worshippers.

Jan Baptista van Helmont and John Clayton– Manufactured gas was first discovered in 1609 by Jan Baptista van Helmont, a Belgian physician and chemist. Helmont noticed that when coal is heated, it produced fumes he called "gas." Almost a century later, an Englishman named John Clayton captured the escaping gas from heated coals in an animal bladder. Clayton continued his experiment by sealing the bladder and puncturing a small hole in its side. The escaping gas was ignited, demonstrating a variety of new applications that natural gas could be used for.

William Murdock– In 1792, a British engineer named William Murdock used the gases from heated coal to light his home. From 1802 to 1804, Murdock installed over 900 gaslights in local cotton mills. This gave him the title: "the father of the gas industry." Large-scale operations adopted Murdock's process and began to expand. The United States did not adopt this technology until 1817 when Baltimore, Maryland decided to light up their streets.

William Aaron Hart– In 1821, the first natural gas well in the United States was drilled in Fredonia, New York by a gunsmith named William Aaron Hart. Hart piped the gas from a twenty-seven foot well to nearby buildings for use as a fuel in lighting. Between 1821 and 1865, over three hundred natural gas companies sprang up. In 1859, crude oil was discovered in Titusville, Pennsylvania, and with this discovery, natural gas research and production took a serious downturn that would not rebound until 1920. Natural gas is frequently used today for cooking, industrial and residential heating, and as an alternative fuel source.

Abraham Gesner– One of the most significant technological improvements occurred in 1840 when Abraham Gesner, a Canadian geologist, discovered how to produce kerosene from crude oil or coal. Kerosene provided a cheap fuel source for heating and lighting and laid the foundation for the beginning of the chemical processing industry. Fortunately, because communications and documentation were very crude, Samuel Kier and J.C. Booth would repeat this experiment in 1851.

James Young and Samuel Kier– By the mid 1800's, a number of chemists, educators, and inventors were working on useful applications for coal, shale, and crude distillation. In 1847, James Young of Scotland found a way to distill coal oil from coal and shale. Around 1851, Samuel M. Kier, a Pittsburgh pharmacist, enlisted the support of J. C. Booth, a chemist, to see if kerosene or coal oil could be distilled from crude oil. The experiments were a success and found immediate application in the kerosene market. Kier believed oil was a cure for many sicknesses.

Benjamin Sillman Jr.– In 1854, a Yale University professor named Benjamin Sillman Jr. was asked to analyze a barrel of salt-skimmed crude oil. Sillman had a theory that the various components of the crude mixture could be separated by distilling at different temperatures. He suspected that each component in the mixture had a different boiling point. During the experiment, Professor Sillman distilled gasoline, kerosene, and a thick, dark, waxy oil.

Edwin Drake– In 1859, Colonel Edwin L. Drake adapted an old steam engine to fit a drill. Drake selected a spot near Titusville, Pennsylvania to drill for oil. Almost immediately, Drake's well produced oil and with this success, other oil drillers set down wells. The beautiful Pennsylvania landscape was transformed into an industrial community of wooden derricks, roughnecks, carpenters, and unskilled labor. Oil was shipped out on wagons to waiting river barges for transportation to a handful of east coast refineries. Figure 1.2-1 shows an early wooden storage tank and piping.

Production was immediately limited because of product transportation problems and the limited number of refineries. The railroad attempted to offset the transportation problem by laying track down to a point within five miles of the oil fields; however, wagons were still used to transport the product from the derricks to the railroad. The transport bottleneck was not relieved until 1865 when the first oil pipeline was built between the oil fields and the railroad station.

The Batch Process
Refinery operation developed overnight as new oil wells were discovered. In 1860, the first refinery was built by William Abbott and William Barnsdall at Oil Creek. Over the next ten years, a hundred refineries would spring up.

Figure 1.2-1 *Early Chemical Processing*

The basic process could be described as batch operation, Figure 1.2-2. Process technicians charged crude oil to a vessel, and the temperature was raised in steps, from 180 degrees F to 1000 degrees F. The products produced from this process included gasoline, naphtha, kerosene, and bottom residuum. It was a common practice to treat the kerosene with caustic soda, sulfuric acid, and a water bath. The gasoline and naphtha were discarded, and the bottoms product was treated and used as a lubricant. Process technicians treated the residuum with acid and naphtha, blended it with steam-refined feedstock, and then ran it through the distillation process again. This final product was blended with brightstock and chilled. The chilled product was stored in canvas bags so that the lighter fractions could escape. Heavy petroleum greases were made by combining the chilled bottoms product with fatty oils and wax.

Early refiners were able to produce eleven barrels of gas from every one hundred barrels of crude oil. Because of this low 11% yield, the entire industry began to look for ways to increase gasoline production without increasing the reserves of less profitable products. Over the next fifty years, refiners increased yields to 20%. Modern refiners are able to convert 45% of a barrel of crude oil into gasoline using cracking processes and combining processes. Cracking processes fall into two categories: thermal and catalytic. Combining processes include alkylation, polymerization, and reforming.

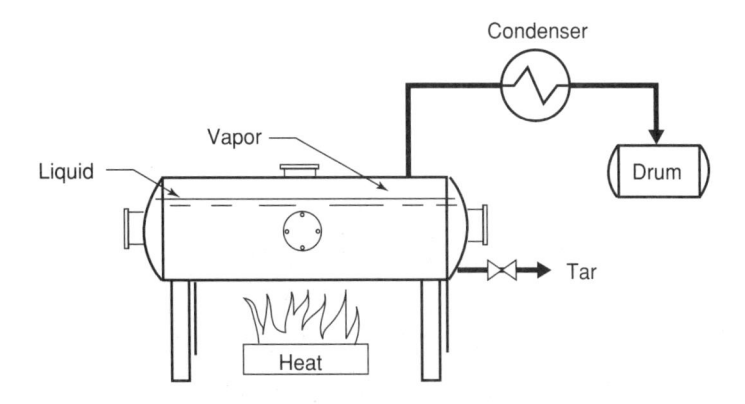

Figure 1.2-2 *Simple Batch Process- 1860*

Near the turn of the century, technology took a large step forward. Two inventions were about to change the world we live in forever: the invention of the automobile and the light bulb. In 1879, Thomas A. Edison invented the electric light bulb which slowly replaced the kerosene lamp and natural gas. Natural gas found a market in cooking and heating while kerosene found a market in the infant aviation field. The second invention was the automobile. As the automobile industry exploded, the need for gasoline increased. At the turn of the century gasoline was considered a worthless by-product of kerosene production and was often dumped on the ground or in local rivers and streams.

During this time period a useful application was found for the residuum or bottom product of crude distillation. New automobiles required bigger and better roads to travel on. Residuum could be used to produce a new product called asphalt. Asphalt in large quantities was being produced as crude oil production increased. Soon, government sponsored road building projects were springing up in every state.

On January 10, 1901, the chemical processing industry struck the first oil gusher in North America. Located near Beaumont, Texas, the Spindletop oil field instantly gave the CPI an unlimited oil supply. Other wells were soon discovered in Louisiana and Oklahoma.

Thermal Cracking
The Burton Process: 1913-1920– Two of the early problems with the batch process were the poor yield of gasoline (8.4 gallons) from a forty-two gallon barrel of crude oil and the residuum that was left over after each run. Early technicians were required to climb into the vessel and chip it out by hand. This procedure was dangerous, inefficient, difficult, costly, and time consuming.

Dr. William Burton was a Standard Oil of Indiana chemist who developed a process for cracking hydrocarbons using high pressure. The "cracking process" is a term used to describe how lighter hydrocarbons are separated (cracked) from heavier hydrocarbons using conventional methods and higher pressure. Dr. Burton was aware of some experimental studies in England that had

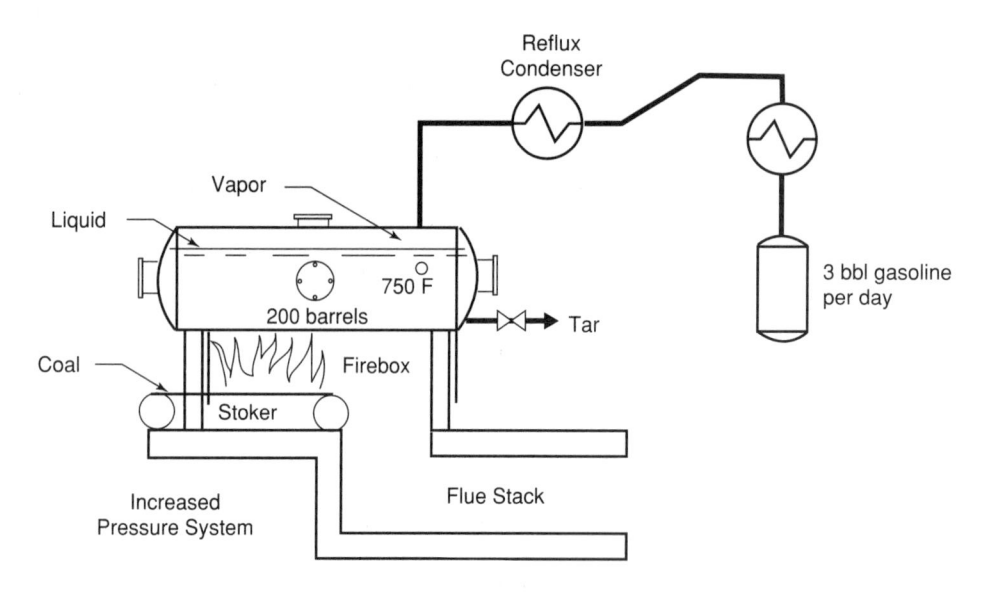

Figure 1.2-3 *Thermal Cracking- 1913*

produced good results using higher operating pressures. Unfortunately, the process had been conducted in a lab and not on a large commercial level. Boilermakers did not have the modern welding technology we use today. The seams were filled with molten metal and beat into place. Finding a large vessel that would withstand the higher pressures was a difficult task. Burton's process produced greater yields: 70% distillates, half of which was gasoline (14.7 gallons). Although the yields improved, the vessel still needed to be cleaned out after each run.

In the Burton Process, Figure 1.2-3, process technicians charged the vessel with 200 barrels of crude oil and slowly heated it to 700 degrees F.

Fractionating Columns
The first fractionating column was introduced in 1917 and featured a "still upon a still" design, Figure 1.2-4. As the heated crude oil flowed into the column, a fraction of the feed stock would vaporize and rise up through the upper stills. The heavier components would flow through the lower stills to the bottom of the column. A liquid seal that allowed the hot vapors to pass through it was established on the bottom of each still. This process allowed each component in the crude oil mixture to find its place in the column, where it could then be removed from the liquid seal and stored. Early fractionating columns were linked together in groups of nine, with a common feed line.

Catalytic Cracking
The Houdry Process: 1936– Eugene J. Houdry was the heir apparent to a French structural steel firm. During World War I, he distinguished himself as a hero. If a catalyst could be found to enhance the cracking process, a higher yield could be obtained from a barrel of crude oil. As the impending war closed in, Houdry experimented with a variety of catalysts. A catalyst is designed to speed up a reaction without becoming part of the reaction.

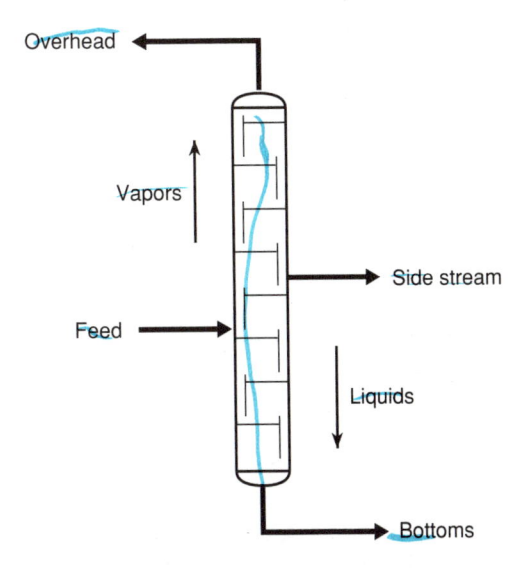

Figure 1.2-4 *Fractioning Column- 1917*

Using a series of bench-top units Houdry attempted to find a catalyst that would enhance the cracking process. He would also need to develop a procedure to burn off the carbon that formed on the catalyst during the reaction, Figure 1.2–5. Three years after the experiment started, Houdry found one of his reactors operating within design specifications. The reactor was filled with aluminum silicate.

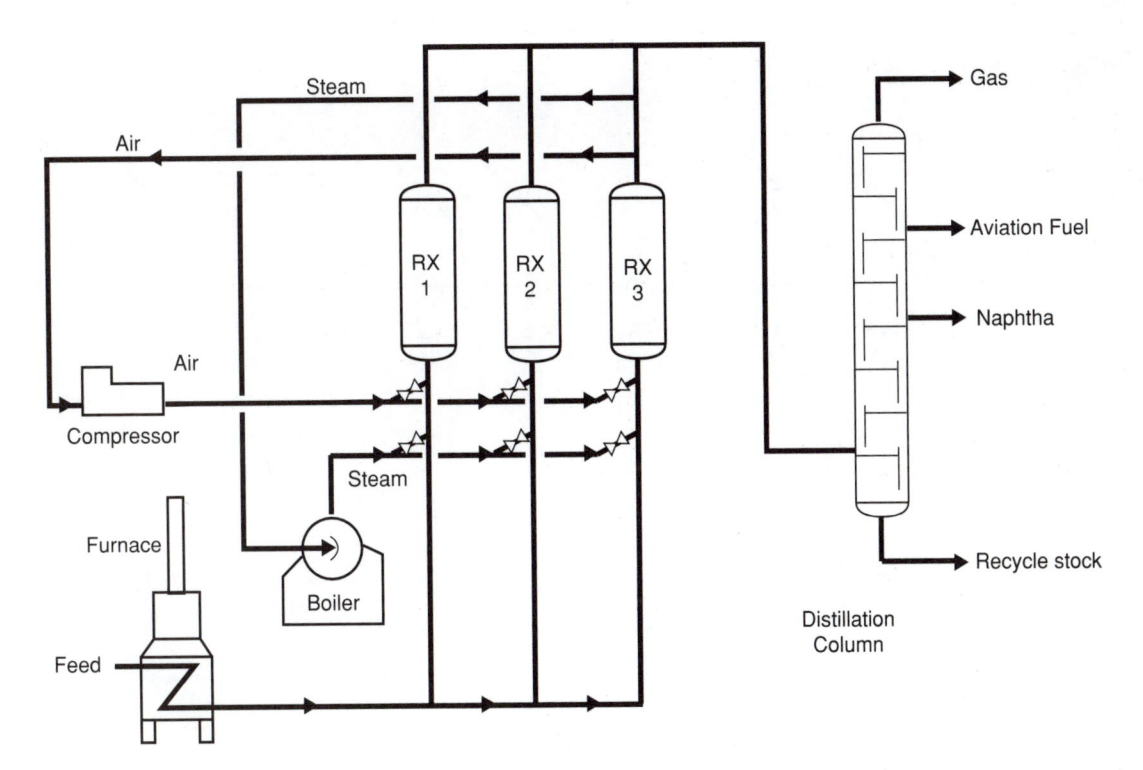

Figure 1.2-5 *Catalytic Cracking- 1936 (The Houdry Process)*

History of the Chemical Processing Industry • Chapter 1

Modern Fractional Distillation

Modern refineries and chemical plants are a lot more efficient than their counterparts one hundred years ago. Today, the process goes through: the separation process, the conversion process, and the treatment process, Figure 1.2–6.

Separation Process. When crude oil is pumped out of the ground, it is desalted, treated, and sent on for additional processing. This material is heated in a large industrial furnace to 385 degrees C (725 degrees F) and pumped to a fractional distillation column. Hot vapors rise in the column and condense on the various trays while hot liquids drop down the column until they gain enough energy to vaporize or separate from lighter components and congregate on their designated tray. This step is referred to as the separation process.

Conversion Process. The conversion process includes vapor recovery and alkylation on the overhead light gases and gasoline lines. Reforming and aromatic recovery are used on the kerosene line. The industrial fuels midsection of the column is still sent to the catalytic cracking section to squeeze out every drop of light product. The bottom lines used in the production of lubricating oils, greases, and asphalt traditionally go through solvent recovery and the crystallization process.

Treatment Process. During the treatment process, each product stream is treated and blended for product purity. The modern distillation column produces high octane gasoline, gasoline, jet fuel, kerosene, heating oil, diesel oil, industrial fuels, waxes, lubricating oils, greases, and asphalt.

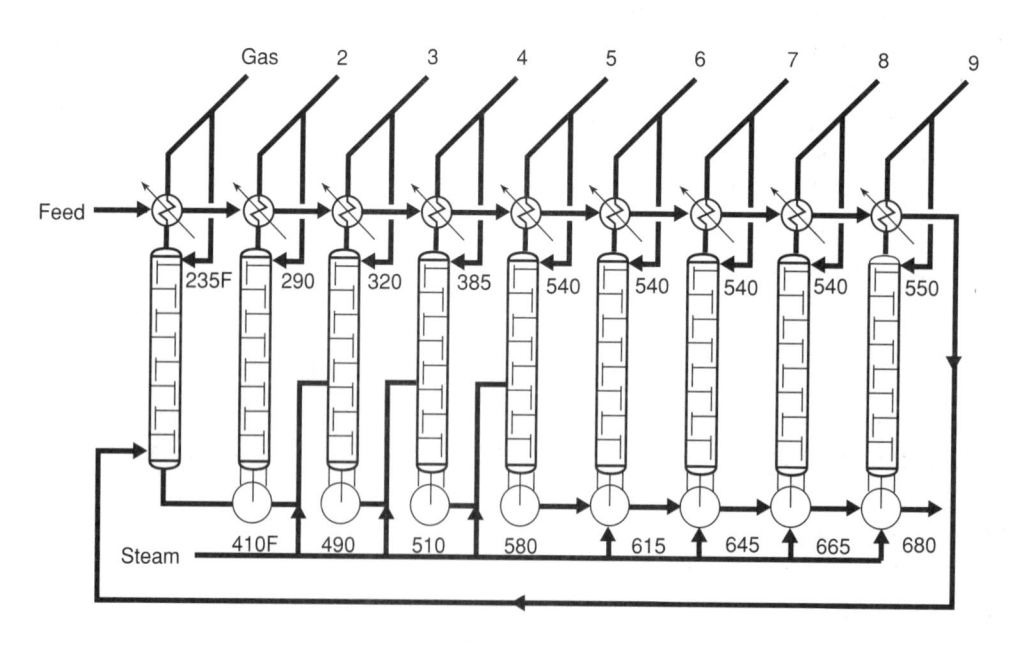

Figure 1.2-6 *Fractional Distillation*

Chapter 1 • *History of the Chemical Processing Industry*

Important Events

1859	Colonel Edwin L. Drake took an old steam engine and adapted it to fit a drill. Drake selected a spot near Titusville, Pennsylvania to drill for oil.
1860	Batch operation. The first refinery was built by William Abbott and William Barnsdall at Oil Creek. Crude oil was charged to a vessel, and the temperature was raised in steps, from 180 degrees F to 1000 degrees F. The products produced from this process included gasoline, naphtha, kerosene, and bottom residuum.
1870	John D. Rockefeller consolidated control of the oil industry and founded Standard Oil Company.
1879	Thomas A. Edison invented the electric light bulb.
1896	Henry Ford designs a gasoline engine.
1901	First oil gusher. Spindletop Texas oil gusher draws thousands on January 10.
1908	Middle East finds large oil reserves in Masjed Soleyman, Persia.
1913	Thermal cracking. Dr. William Burton was a Standard Oil of Indiana chemist who developed a process for cracking hydrocarbons using high pressure.
1917	The first fractionating column was introduced.
1920	Gas stations open in the United States.
1936	Catalytic cracking. Eugene J. Houdry found a catalyst, "alumina silicate", that enhanced the cracking process and gave higher yields from a barrel of crude oil.
1941	Oil embargo placed on Japan by US, Britain, and the Netherlands. Japan bombs Pearl Harbor in December.
1944	Germans create new technology to convert natural gas into oil.
1969	Santa Barbara, California oil spill sparked environmental movement.
1973	Arab oil embargo. US faces gas lines for the first time since World War II.
1977	Alaskan pipeline opens.
1979	Iranian revolution. Gasoline price tops $1.00 per gallon. More gas lines.
1984	Bhopal, India vapor release. Kills and injures thousands.
1989	Exxon Valdez oil release at Prince William Sound.
1989	Phillips explosion in Houston, Texas kills 23 technicians.
1990	ARCO explosion in Houston, Texas kills 17.
1990	Kuwait is invaded by Iraq, starts Gulf war, oil fields burned.
1999	Gas prices plummet below eighty cents then rebound to over $2.00 per gallon.

1.3 Working in the Chemical Processing Industry

Preparation and Basic Skills

Preparation for work in the chemical processing industry starts early for a process technician. Students should take classes in high school that will prepare them for the fast paced processes and technologies they will face in industry. A solid core curriculum would include microcomputers, communications, math, and science. Some high schools have programs that offer dual credit for process technology classes. These classes give graduating seniors a jump on other students entering two-year community college programs.

Jobs in the chemical processing industry are usually high paying with full benefit packages. Because of advances in process control, fewer positions are available for job seekers. These rapid changes in technology have been integrated into the competitive global structure of the chemical processing industry. Job descriptions for process technicians include a two-year degree in process technology, good scores on pre-employment tests and interviews, and passing a medical examination.

Reading. To do well on most plant entry exams, above-average reading skills are needed. Operators must read and interpret operating procedures, training procedures, quality and environmental guidelines, customer requests, and many other technical documents.

Writing. Process technicians are required to document most of their activities on the job. These documents include logbook entries, lock-out, tag-out, process samples, training procedures, operational procedures, permits, shift relief, work orders, and quality control charts.

Listening. Effective listening skills are helpful to process technicians during equipment malfunctions, troubleshooting, shift relief, training, and team meetings.

Interpersonal Skills: Communication. Interpersonal skills can be enhanced with proper coaching and study inside a normal process technology program. Most people develop basic skills years before entering their occupation and may find a need for improvement.

Computer Technology. Process technicians interface with their equipment through advanced instrumentation and electronic networks. The computer console, which is a window to the process, is becoming the central focus of the control room. Technicians need to know how to use personal computers. Skills required in this area include word processing, spreadsheets, data bases, graphic art, networking with other sites, electronic mail, accessing operating procedures, accessing material safety data sheets (MSDS), computer architecture, and applying new technology being developed.

Math. In order to pass a typical pre-employment exam, a process technician needs a sound understanding of addition, subtraction, division, multiplication, fractions, percentages, decimals, and measurement metrics. The technician of the future will need a much stronger understanding of applied mathematics, including: basic math, algebra, geometry, applied college algebra, trigonometry, physics, and calculus. These foundational courses can enhance a technician's ability to perform chemical calculations, to control and troubleshoot unit operations, and to interface with unit chemists and engineers.

Science. Process technology programs utilize the principles of general science, chemistry, and physics. Raw materials are combined with advanced technology to produce useful end products.

The science behind this technology is impressive but is usually transparent to the technician. The depth of the technology provides a lifelong learning opportunity for the operator.

Industrial manufacturers usually upgrade technology frequently in order to compete in the global economy. Most technicians are exposed to cutting-edge technology throughout their careers.

It is important to understand what is happening as raw materials are combined to form new products. Operators do not open and close valves blindly. They carefully study and prepare prior to operating the unit.

Successful plant operation requires theoretical knowledge and observational knowledge. Typically, the engineering staff is trained in the theoretical area whereas operators control the observational area. An operator who possesses both theoretical and observational skills will be a valuable asset to the company. Corporate rightsizing and restructuring should require technicians of the future to perform more challenging and technical job functions.

Key scientific principles used by technicians include:
- Fundamentals of chemistry—atoms, elements, atomic structure, hydrocarbons, states of matter, gases, solutions
- Physics—fluids, temperature, pressure, heat transfer, work, and energy
- Math and statistics
- Basic equipment and technology
- Computer literacy skills
- Communication skills
- On-the-job skills

Punctuality. Most companies terminate trainees after several unexcused tardies. Punctuality is considered very important to shift workers. Other key areas the chemical processing industry looks at are fighting, lack of teamwork, drug abuse, safety violations, and excessive absences and tardies. Studies indicate that job satisfaction is linked directly to these infractions.

Multitasking. Process technicians typically have many things going on at the same time. Being able to control several work tasks at once is important. Process operators commonly carry small notebooks around the unit with them to simultaneously document and keep up with a variety of tasks.

Problem Solving. Problem solving is a technique that improves as you become more familiar with the equipment. The trick is to know your equipment and process. It will help you to identify the symptom, the problem, and the solution.

Safety Awareness. Safety awareness is taught from the first moment you step into a plant. Statistics indicate you are safer in the plant than at home, yet every year, over 14,000 work-related fatalities and 2.5 million disabling injuries occur in the United States. Evidence indicates that

Figure 1.3-1 *Spill Releases*

a well managed safety program drastically reduces occupational illnesses and injuries. Safety statistics are important to an industrial manufacturer, and extreme pressure is applied to each employee to work safely.

Quality Awareness. *meeting customer needs* The new global economy has introduced a competitive way of doing business. Industrial manufacturers use advanced quality techniques to stay ahead of the competition. These techniques are taught openly and used by the entire company. Technicians should be aware of these quality techniques, which include flow charts, control charts, statistical process control, scatter plots, histograms, pareto, run charts, ISO-9000, and training.

Environmental Awareness. Technicians should be aware of the impact they can have on the environment. Industry refers to these programs as air pollution, water pollution, solid waste disposal, toxic waste disposal, emergency response, community-right-to-know, and spill release guidelines, Figure 1.3–1.

1.4 College Programs for Process Technology

Process technology is the study and application of the scientific principles associated with the operation and maintenance of chemical processing plants. Process technology students are required to study the equipment and technology common to most industrial processes and the relationship they share. For example, piping, valves, pumps, and tanks share a unique relationship common between processes.

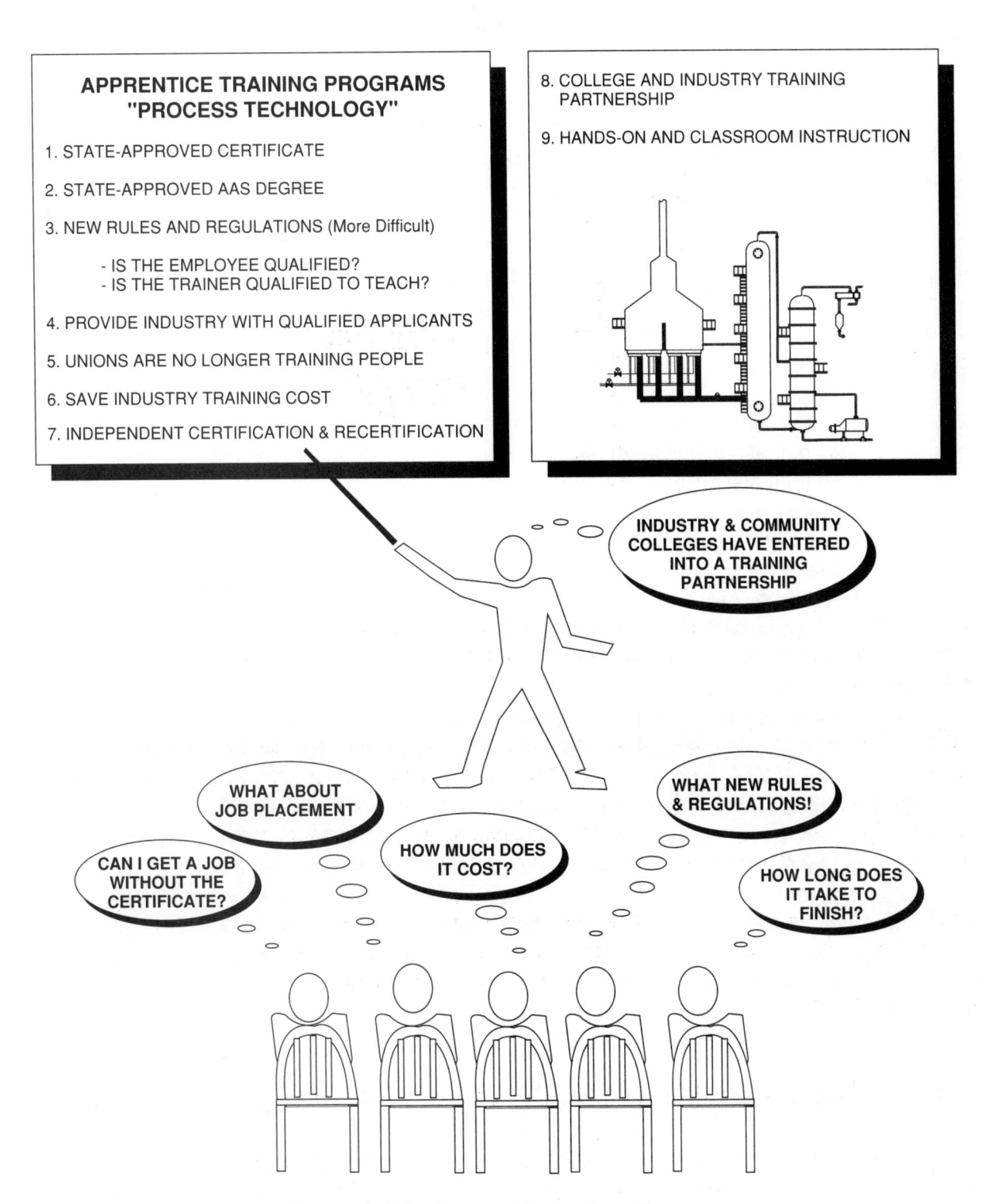

**APPRENTICE TRAINING PROGRAMS
"PROCESS TECHNOLOGY"**

1. STATE-APPROVED CERTIFICATE
2. STATE-APPROVED AAS DEGREE
3. NEW RULES AND REGULATIONS (More Difficult)
    - IS THE EMPLOYEE QUALIFIED?
    - IS THE TRAINER QUALIFIED TO TEACH?
4. PROVIDE INDUSTRY WITH QUALIFIED APPLICANTS
5. UNIONS ARE NO LONGER TRAINING PEOPLE
6. SAVE INDUSTRY TRAINING COST
7. INDEPENDENT CERTIFICATION & RECERTIFICATION
8. COLLEGE AND INDUSTRY TRAINING PARTNERSHIP
9. HANDS-ON AND CLASSROOM INSTRUCTION

INDUSTRY & COMMUNITY COLLEGES HAVE ENTERED INTO A TRAINING PARTNERSHIP

CAN I GET A JOB WITHOUT THE CERTIFICATE?

WHAT ABOUT JOB PLACEMENT

HOW MUCH DOES IT COST?

WHAT NEW RULES & REGULATIONS!

HOW LONG DOES IT TAKE TO FINISH?

Figure 1.4-1 *Process Technology Programs*

History of the Chemical Processing Industry • Chapter 1

In a process technology program, a student will learn the principles of modern process control and troubleshooting. Most programs start this process using the five elements of a control loop as a guide. Since new governmental guidelines require process technicians to understand the chemistry of the processes they are operating, a solid foundation is required in math, physics, and chemistry. Calculating product transfers, mixing raw materials to form new products, and dealing with pressure, level, flow and temperature problems are all common areas to which the math/science foundation is applied.

In order to complete the program, a student will need to be exposed to advanced quality control techniques, safety training from a process technician's view, and human relations. The knowledge and skills learned in the process technology degree program can be directly applied to a number of hands-on learning activities at the educational institution before being applied on-the-job.

When students enroll in the process technology program, they are expected to complete a regimented curriculum that has has been developed by industry, education, and governmental agencies. Successful completion of a regionally accredited program requires a student to demonstrate the following skills:
- self-directed study habits—attendance, participation, critical thinking, troubleshooting, goal setting, time management, motivation, reading and study, homework, self-directed worker ethic.
- interpersonal skills—listening, communication, diversity awareness, using quality tools, honesty, integrity, work with supervision.
- safety awareness—safety, health, and environmental.
- basic understanding of the equipment and technology—computer savvy, basic math and science, mechanical aptitude, assimilation of skills, hands-on operation.

High School to College Transition

The transition between high school and college can be a difficult one for many students. Process technology students come from a wide array of backgrounds and experiences. College classes are typically diverse and composed of women and men between the ages of sixteen (16) and sixty (60). A significant number of these students have college degrees or have completed college classes; however, the largest block of students have never enrolled in a college course. Making the adjustment between high school and college is easier if a student is aware of the differences and given the tools to succeed. Table 1.4–1 illustrates the differences between college and high school.

High school is vastly different from college. Perhaps the biggest difference between high school and college is in the area of freedom. Most high school programs are structured with rules that dictate how personal time is spent. College students are considered to be adults who are allowed to establish their own set of rules. In high school the teacher was primarily responsible for selecting and presenting the material. College students are given the opportunity to decide what is important to them and when they will take it. Because of this freedom, the method of learning is shifted from the high school instructor to the college student. College instructors place the

Table 1.4–1 Differences Between College and High School

Topic	College	High School
Freedom	Controlled by student	Controlled by administration
Cost	Paid by student	Paid by parents (taxes)
Learning	Student responsibility	Teacher responsibility
Resources	Vast and confusing	Limited
Job	High % work	Low % work
Married	High % married	Low % married
Reward	High paying career	High school diploma
Good student	Sometimes struggle because they do not know how to apply what they have learned	As & Bs, good GPA
Curriculum	Selected by student Designed by industry, education & government Direct job application	Selected by administration Designed by education Not directly applied to career tasks e.g.: history, english

responsibility for learning on the student. Emphasis should be on learning application and not memorization! Technical instructors use a hands-on approach to learning which is similar to the simple practice exams used by high school teachers.

Unlike high school, the college student makes a significant financial investment in his or her education. This sacrifice buys a specific product and a huge educational responsibility. College instructors cover a much larger volume of material than high school instructors. Tests are taken from class lectures, reading assignments, structured experiments, bench-top labs, pilot units, videos, computer programs, and standardized tests.

Tools to Succeed in College

When a new process technology student enters college for the first time, a number of tools can be used that have been proven to enhance college performance. Table 1.4–2 provides a list of mental tools used by successful college students.

Table 1.4–2

Tools to Succeed in College
Understanding the college system
Goal setting
Time management
Applied learning
Attitude and Participation

History of the Chemical Processing Industry ● Chapter 1

The educational and administrable methodology that exists in a college can be quickly decoded and understood by a new process technology student. The first step is to get the college catalog and review the rules, procedures, policies, course descriptions, degree programs, and faculty. The second step is to set a college course schedule. Fall, spring, and summer course schedules will provide you with a detailed listing of classes, locations, instructors, and times. In order to start school you will need to complete step three; register with the college, provide identification, agree to have your high school send transcripts, and take a series of minor tests for placement purposes. The process technology degree program will require a student to take between eighteen (18) to twenty-two (22) classes. Full time students will take five or six classes over a sixteen week period and spend an average of 21 hours per week in the classroom or laboratory.

College instructors usually provide students with a syllabus that contains the following information: course description, performance objectives, standards, grading policy, attendance policy, textbooks and supplies, disability assistance, and scheduled exams. Process instructors typically provide students with class outlines. Outlines can be used to prepare for upcoming tests and applied learning activities. The degree program provided by your school will list the required courses to complete the program. Do not get off the path and take classes that will not help you graduate.

College process technology programs fall into the following areas: one year certificates and two-year degrees. Certificates require a minimum of thirty semester hours and two-year degrees require a minimum of sixty hours. Typical course topics include:
- Introduction to Process Technology
- Process Technology 1—Equipment
- Process Technology 2—Systems
- Process Instrumentation
- Safety, Health, and Environment
- Process Technology 3—Operations
- Quality Control
- Troubleshooting
- Chemistry
- Physics
- Academic classes

During the educational process, a number of snares and traps can damage a student's ability to progress. Be prepared to drop a class before the scheduled deadline if any of the following situations exist:
- You are hopelessly lost and have a D or an F
- Instructor-student problems
- Work schedule conflict
- Family tragedy

Goal Setting

Goal setting is a college level activity used by successful students. Goals should be specific, measurable, and realistic. Short-term goals should be distinguished from long-term goals. The process technology degree program should be broken down into bite size pieces and attached to weekly, monthly and yearly goals. Job search activities are typically more effective using this structured approach.

Time Management

Time management combines a student's knowledge of their personal study needs with a structured system.
The typical time management system includes:
- To Do List & Weekly Schedule
- Specific study times
- Self discipline
- Adequate sleep periods—Don't burn the candle at both ends and don't sleep too long.
- Move to next action item when time is expired
- Relax mechanism

Applied Learning

Typical instructional techniques use a simple-to-more-complex approach for learning. The process technology program presents the theory of process technology in modular blocks prior to introducing applied techniques. As the program builds, the learner is exposed to laboratory equipment that uses hands-on activities. In the classroom, a student may be introduced to the theoretical concepts of pressure, heat transfer, fluid flow and distillation; however, in the laboratory, they are asked to apply these concepts. Transitioning between the book and the lab is a fundamental requirement for success in the program. Students who perform well in the classroom may struggle on the bench-top, computer simulator, or pilot plant. Students may perform well in the lab and poorly in the classroom.

Attendance and Participation

Students who decide early to attend every class and participate in classroom discussion have a tremendous edge over those who do not. Instructors learn the names of these students faster and identify their individual needs quicker. Participation and attendance are essential elements in the applied learning process.

1.5 Your Career as a Process Technician

Successful job applicants are notified by phone or mail and given a starting date. The first few weeks include orientation, paperwork, safety training, tours, and apprentice training. Key individuals in the organization are given the opportunity to speak to the new process technicians. New employees also spend time getting sized for flame retardant clothing, safety glasses, work boots, as well as being introduced to co-workers.

Industrial training programs vary from one company to another. Some are certified by the US Department of Labor. This certification requires a specific number of on-the-job training hours and scheduled classroom hours. These programs can run from one to five years and usually are attached to pay increases.

Key elements of apprentice training programs include:
- Orientation, followed by 1 to 8 weeks industrial classroom training
- Mandatory safety training before being allowed to go to the unit
- Meeting with the supervisor and training coordinator, and planning work and training activities
- Meeting with the trainer and supervisor to plan on-the-job training and a new job assignment

After the introductory period, the process technician is assigned to a unit and a trainer. During this time period, the new technician reports through the formal chain of command: the trainer and the unit supervisor. After meeting with the supervisor, the trainer knows the specific area and responsibilities of the trainee. Trainees are typically watched very closely during the first twelve months of employment. Training on the unit includes tracing lines, catching samples, filling out paperwork, housekeeping, checking equipment, making line-ups, starting and stopping equipment, etc. This process continues until the trainer feels comfortable with the trainee's progress. During this time frame, the new technician works shift work. This is a very difficult transition for an individual that has never worked rotating shifts.

Diversity and Sexual Harassment
Handling stress, conflict, cultural diversity, and sexual harassment are all important aspects of a process technician's job. The people who make up the work force within a plant are typically diverse and well educated. Diversity training identifies and reduces hidden biases between people with differences. Sexual harassment is defined as behavior that constitutes unwelcome sexual advances. The behavior could come in the form of verbal or physical abuse or unwelcome requests for sexual favors. This behavior may involve people of the opposite sex, same sex, supervisor to employee, or employee to employee. Since work relationships can be expected to go as long as thirty-five to forty years, it is important to fit in on your unit. Understanding your assignments, multiple roles, and responsibilities and contributing to the overall team effort is important to a successful work career.

New technicians describe entering a chemical processing plant as an unusual experience similar to being transplanted into a foreign environment with pipes, tanks, strange equipment, noises, smells, and advanced computer technology. This initial experience is very stressful for the new technician. Each plant has a variety of techniques for reducing the stress on a new technician. Some of these techniques include:
- Systematic, competency-based training
- Trainer-trainee on-the-job training
- Job shadowing on-the-job training

Stress levels will drop as the new technician qualifies on a job post and becomes more familiar with the environment.

Conflicts will naturally occur during the work career of most technicians. How these conflicts are handled can be used to determine retention rates, evaluations, promotions, absenteeism, and job satisfaction. Conflicts must be handled professionally and through the proper channels. New technicians sometimes feel they are being singled out and asked to do the most routine and dirty jobs. Training is typically structured from simple to more complex. As you learn and qualify on additional job posts, further responsibilities and the respect of your peers will increase.

It is important to remember that during the first twelve months of employment the apprentice technician should not:
- Miss work or come in late
- Sleep on the job
- Use illegal drugs or alcohol
- Fight on the job
- Be caught in the control room with his/her feet up

During the first year of employment, new technicians are expected to be on their best behavior. A proven track record has not yet been established.

Chemical plants and refineries are divided into major sections or divisions that make the best sense for the overall operation of the plant. Each section appears to run independent of the other with a designated section head. Process section heads report directly to the plant manager. Plant managers, section heads and second line supervisors typically have engineering degrees. The chemical processing industry follows a pyramid type management structure that includes the plant manager, section head, second line supervisors, first line supervisors, and process technicians. A variety of management structures are available, however, large pyramid type organizations rarely escape the original design. Work teams vary in size from five to twenty technicians.

Inside and Outside Operators

Process technicians can be classified as inside or outside operators. Inside operators are typically experienced technicians who are familiar with the outside functions of their unit. As the name implies, inside operators spend most of their time inside a control room monitoring and controlling process variables, filling out unit log books, and working with the outside operator. The majority of process technicians are outside operators who inspect equipment, perform unit start-ups and shutdowns, troubleshoot problems, perform routine housekeeping, catch readings, and collect samples.

Most companies provide formal apprentice training for new employees regardless of what type of experience or training they have received previously. Portable credentials (an AAS degree or a certificate) are needed to address the CPI's apprentice training and experienced technician retraining problems. These programs provide prospective employers with a list of qualified candidates who already have completed key elements of government required training. Portable credentials could save companies as much as 700 classroom hours.

1.6 Careers in the Chemical Processing Industry

Electricians, instrument technicians, lab/research technicians, machinists, mechanical craftsmen and process technicians work as a team to control the operations of a plant. They work with chemists, engineers, secretarial and clerical staff, attorneys, legal assistants, computer specialists, industrial hygienists, and human resources. Each of these occupations starts at different pay rates. The primary financial difference among the four craft occupations and process is shift differential and overtime. Most operating facilities run between 20% and 25% overtime for process technicians. In the gulf coast area (Texas and Louisiana), in 2000, a typical starting rate is $16.00 to $17.00 per hour for an eight-hour shift with a top-out rate around $28.00. Depending on the amount of overtime worked, new technicians will earn between $33,000 and $50,000 during their first year. (1999-2000 Gulf Coast Area) Top-out rates in the chemical processing industry are presently between $22.00 and $28.00 per hour. Time and a half can add up to as much as $42.00 per hour for a technician working overtime. A typical year will include 2080 scheduled work hours plus about 25% overtime.

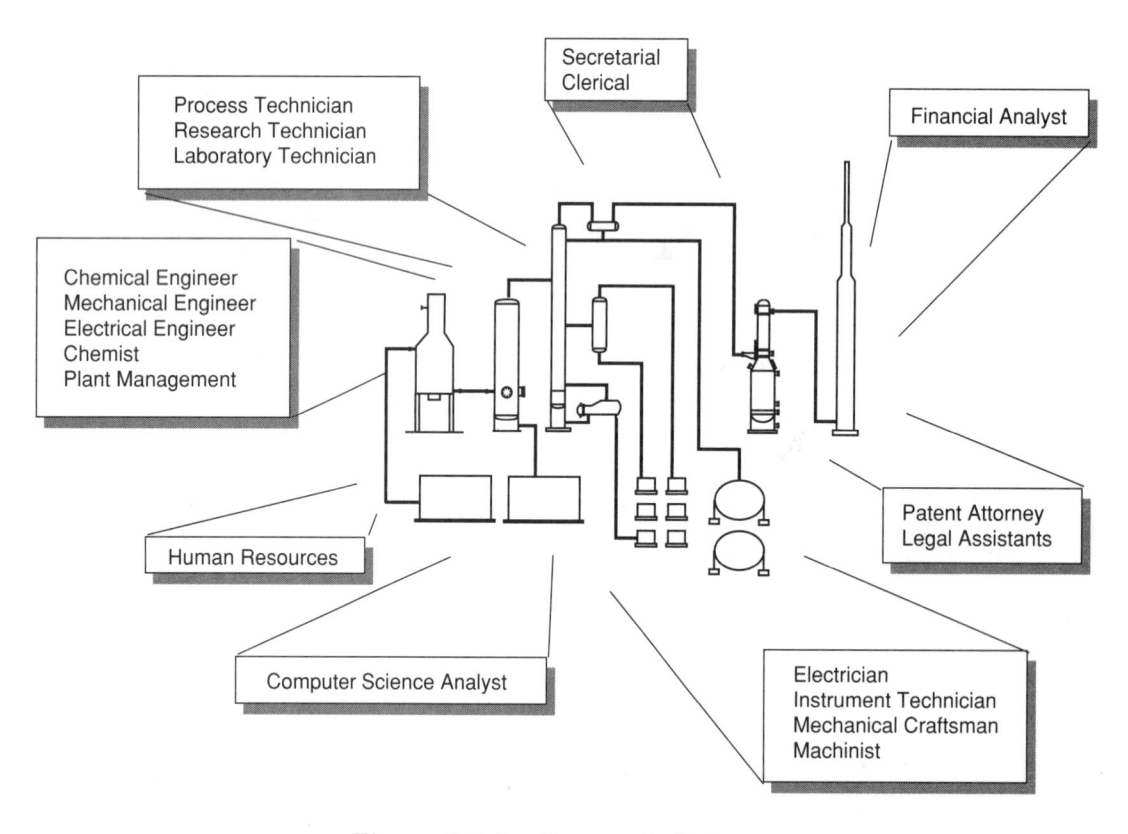

Figure 1.6-1 *Careers in Industry*

Process, Research, and Chemical Technicians

Process Technician.
> Start: $16.00 to $28.00 per hour, $33,000 to $50,000 per year to start. Certificate or AAS degree. Maintain unit operations; check equipment, catch samples, take readings, make rounds, troubleshoot, fill out quality charts, operate computer systems, and do housekeeping. Must have strong technical and problem solving skills, ability to assimilate cutting edge technologies quickly, and ability to apply innovative ideas. In addition to these skills a process technician needs to be able to handle conflict, look at a complex situation and see the overall picture, and communicate effectively.

Research Technician.
> Start: $16.00 to $28.00 per hour, $33,000 per year to start. AAS-BS degree. Same as process technician plus; operate bench-top units and pilot plants. Special emphasis on technical and problem solving skills, ability to assimilate cutting edge technologies quickly, and ability to apply innovative ideas.

Lab Technician.
> Start: $16.00 to $28.00 per hour, $33,000 per year to start. AAS-BS degree. Perform quality control tests.

Mechanical Crafts

Electrician.
> Start: $16.00 to $28.00 per hour, $33,000 per year to start. AC/DC voltage hook-ups, circuit testing, troubleshooting, electrical controls.

Instrument Technician.
> Start: $16.00 per hour to $28.00 per hour, $33,000 per year to start. Work on level, fluid flow, pressure and temperature instruments and control loops, troubleshoot, maintain operations.

Machinist.
> Start: $16.00 to $28.00 per hour, $33,000 per year to start. Maintain mechanical equipment, check rotating equipment alignments.

Mechanical Craftsman.
> Start: $16.00 to $28.00 per hour, $33,000 per year to start. Includes pipe fitting and welding, equipment maintenance, troubleshooting.

Engineering and Chemists

Contact Engineer.
> Start: $24.23 per hour, $50,400+ per year. Bachelor of Science in Chemical Engineering (BSCE). Assigned to operations unit for technical support.

Design Engineer.
> Start: $22.98 per hour, $47,800+ per year. Bachelor of Science in Mechanical Engineering (BSME). Troubleshoot equipment and machinery problems.

Chemist.
> Start: $32.21 per hour, $67,000 per year. Ph.D. Chemistry. High grade point average and experience.

Administrative Support Staff
Secretarial, Clerical, and Legal Assistant.
> Start: $12.60 per hour, $26,208 per year. Word processing, computer literacy, communication skills, type reports, memos, letters, legal research, analysis, special services for attorneys.

Computers
Computer Science Analyst.
> Start: $21.92 per hour, $45,600+ per year. Bachelor of Science in Computer Science. Maintain plant computer systems.

Personnel
Human Resources Analyst.
> Start: $26.44 per hour, $55,000+ per year. Master's Degree in Industrial Relations. Recruiting, labor relations, training, EEO.

Safety
Industrial Hygienist.
> Start: $22.98 per hour, $47,800+ per year. Bachelor Degree in Environmental Engineering. Ensure compliance with OSHA; help employees to recognize, control, and evaluate occupational hazards.

Other
Financial Analyst.
> Start: $26.44 per hour, $55,000+ per year. Master's Degree in Business Administration (MBA). Develop budgets, analyze costs, monitor expenses.

Patent Attorney.
> Start: $26.44 per hour, $55,000+ per year. BS degree, Law degree (Juris Doctor). Protect company inventions, patents, contracts, license agreements.

Summary

The lifeblood of modern society is found in petroleum products. Cars, planes, trains, ships, and farm equipment all require petroleum products to operate. The term "petroleum" combines two Latin words, *petra* (rock) and *oleum* (oil). Crude oil is a mixture of hydrocarbons that vary in molecular structure and weight. Modern manufacturers separate these components through the distillation process. Distillation separates the various components in a mixture by boiling point.

In 1859, Edwin L. Drake adapted an old steam engine to fit a drill. Drake selected a spot near Titusville, Pennsylvania to drill for oil. Almost immediately, Drake's well produced oil, and with this success, other oil drillers set down wells. In 1860, the first refinery was built by William Abbott and William Barnsdall at Oil Creek. The basic process could be described as a batch operation. Process technicians charged crude oil to a vessel and the temperature was raised in steps, from

Chapter 1 • *History of the Chemical Processing Industry*

180 degrees F to 1000 degrees F. The products produced from this process included gasoline, naphtha, kerosene, and bottom residuum.

Early refiners were able to produce eleven barrels of gas from every one hundred barrels of crude oil. Because of this low 11% yield, the entire industry began to look for ways to increase gasoline production without increasing the reserves of less profitable products. Modern refiners are able to convert 45% of a barrel of crude oil into gasoline, using thermal and catalytic methods. Combining processes include alkylation, polymerization, and reforming.

The first fractionating column was introduced in 1917, featuring a "still upon a still" design. As the heated crude oil flowed into the column, a fraction of the feed stock vaporized and rose up through the upper stills. The heavier components flowed through the lower stills to the bottom of the column.

Preparation for work in the chemical processing industry starts in high school with interpersonal skills, microcomputers, communications, math, and science. College students are given an opportunity to complete courses developed by education and industry, structured upon a set of objectives that are nationally accepted and taught by people with years of industrial experience.

College programs have three classes that focus on the equipment found in the chemical processing industry. The initial course goes into some depth about the various areas. The second course presents the various systems that the equipment can be arranged in. Since hundreds of systems are possible only the more common ones are used. The last core course allows a process technician to operate one or more of the systems found in the chemical processing industry.

Chapter 1

Review Questions

1. Describe process technology training programs found at local colleges.

2. What skills do technicians need in order to succeed?

3. Describe a typical apprentice training program.

4. List three differences between high school and college.

5. What is your motivation for pursuing a career in process technology?

6. What is time management?

7. What are the benefits of diversity for a company?

8. List positive and negative aspects of the sexual harassment law.

9. List the five most significant events in the history and development of the chemical processing industry.

10. What tools must a student acquire and use in order to be successful in a process technology program?

11. What is process technology?

12. List all process technology classes offered at your school.

13. Calculate the gross income of a first-year technician.

14. Calculate the gross income of a senior technician at top rate. Use the standard of 2080 hours plus 1000 overtime hours at time and a half.

15. Describe batch operation.

16. Contrast thermal and catalytic cracking.

17. Identify two inventions that revolutionized the chemical processing industry.

18. Who was given the title, "the father of the gas industry"?

19. What significant event took place in 1859 that changed the chemical processing industry?

20. Describe fractional distillation.

Chapter 2

Introduction to Process Technology

OBJECTIVES

After studying this chapter, the student will be able to:
- *Identify the roles and responsibilities of a process technician.*
- *Describe the chemical processing industry and future trends.*
- *Explain the basic principles of safety, health, and environment.*
- *List the basic principles of quality control.*
- *Identify the different types of process equipment.*
- *Describe industrial processes and systems.*
- *Explain the principles of instrumentation and modern process control.*
- *Apply the principles of chemistry and physics to the chemical processing industry.*
- *Identify the characteristics of a self-directed work team.*

Introduction to Process Technology • Chapter 2

KEY TERMS

Chemical Processing Industry—is composed of refinery, petrochemical, paper and pulp, power generation, and food processing technicians. *know the difference*

Gold collar—a new term used to describe process technicians.

HAZCOM—hazard communication; known as "workers' right to know." Provides technicians with chemical lists, Material Safety Data Sheets (MSDS), targets critical operations, physical and health hazards, Personal Protective Equipment (PPE), release detection, and regulations for chemical manufacturers.

OSHA—Occupational Safety and Health Act. Composed of three agencies: the Occupational Safety and Health Administration; the National Institute for Occupational Safety and Health; and the Occupational Safety and Health Review Commission.

PFDs and P&IDs—process flow drawings and piping and instrumentation drawings. Simple flow diagrams are used to describe the path a process travels as it moves through an operating unit. Process and instrumentation drawings are advanced schematics that include instrumentation, equipment, piping, and tanks.

Process—a collection of equipment systems that work together to produce products. Example: crude distillation.

PSM—Process Safety Management; designed to prevent the catastrophic release of toxic, hazardous, or flammable materials that could lead to a fire, explosion, or asphyxiation. *gov. program for safety*

System—a collection of equipment designed to perform a specific function. Example: refrigeration system.

— *Feedstock is a raw materials that goes in process*
— *Petrochemical — a plant that makes finished products that came from refinery*

The standards developed by the ACS identify the knowledge and skills that process technicians need when they begin work in a manufacturing environment. This identification of standards is part of a much larger grass-roots movement toward the development of two structured professions: laboratory and process technician. These two professions have developed in response to the technology revolution.

Training Programs
In the past, very little formal training was required prior to taking a job in the chemical process industry. Industrial manufacturers relied on pre-employment screening and in-house training programs to educate and recertify their employees. Nationally, this method for training is changing. Because of intense competition in the global community, the chemical process industry is evaluating whether the focus is on training or producing products. When a company identifies a part of its day-to-day operation that could better be operated by an outside organization and hires or uses this organization, this is called outsourcing. Outsourcing is becoming a very popular option for industrial manufacturers.

Formal training programs have been established in local community colleges and universities nationwide. At the present time, these programs are limited to three or four geographic regions across the United States. Students can attend these institutions and receive state-approved certification and two-year degrees in laboratory or process technology. These programs relieve the burden of typical apprentice training and allow industrial trainers to focus on higher level, site-specific training. Graduates from these types of programs provide a much larger pool of qualified applicants from which the chemical processing industry (CPI) can screen and choose. In time, pre-employment tests will evolve from the typical math and mechanical aptitude tests into a more comprehensive exam covering entry level skills discussed in this text. Another popular option being discussed is to waive the pre-employment test and use the candidate's college transcripts. This method appears to work well with other occupations such as engineering, law, medicine, and chemistry.

New Hiring Standards
Employers are requiring prospective employees to have one or more of the following: (1) formal training; (2) state-approved certification; (3) a technical degree; (4) experience; (5) satisfactory scores on a pre-employment test; or (6) a combination of these attributes. The chemical process industry and various educational institutions have entered into formal partnerships to facilitate the technical training of employees.

Program Justification
The key reasons driving the development of these technical programs are: (1) rapid advances in technology; (2) desire to eliminate accidents in the workplace; (3) potential catastrophic risks; and (4) new regulations and guidelines from the government. The Occupational Safety and Health Administration (OSHA) recently enacted the process safety management (PSM) standard that requires employers to train their employees on process fundamentals. This standard applies to initial certification and recertification of employees.

According to the ACS, over 240,000 chemical laboratory technicians and 500,000 plant technical operators are employed in the United States. This group makes up the fourth largest U.S. manufacturing industry.

Introduction to Process Technology • Chapter 2

Figure 2.2-3 *Chemical Processing Industry*

Work Force Development

Studies of work force development indicate that much of the existing work force is comprised of the "baby boomer" generation and is mature. In the near future, this large baby boomer group will "boom out" or retire, leaving a significant number of vacancies. Measures must be taken soon to stop the loss of technical expertise from this generation, to capture it in the form of technical programs, and to assimilate it into the modern US work force. Figure 2.2–3 shows the development of the chemical processing industry as illustrated by this large chemical processing plant.

The ACS has taken significant steps toward the development of a practical, technical foundation. The ACS standards need to be used in the development of future technology programs. Many of the objectives listed in the ACS's major categories are found in the body of this text. A report on the project "Foundations for Excellence in the Chemical Process Industries" can be obtained by writing to the American Chemical Society, 1155 Sixteenth Street, NW, Washington, DC 20036, (202) 872-8734. The standards identified by the group for laboratory and process technicians follow.

Employability Performance-Based Skill
- Math and statistics   (22 lab objectives)   (13 process objectives)
- Computer literacy skills   (19L)   (12 P)
- Communication skills   (31L)   (14 P)
- Workplace skills   (25L)   (19 P)
- General plant and lab skills (32L)   (16 P)

Chapter 2 • *Introduction to Process Technology*

Critical Job Functions: Laboratory
- Maintain a safe and clean laboratory adhering to environmental/health and safety regulations (34L).
- Sample and handle chemical materials (31L).
- Conduct physical tests (20L).
- Perform chemical analysis (37L).
- Perform instrumental analysis (38L).
- Plan and design experiments; synthesize compounds (53L).

Critical Job Functions: Process
- Maintain safety, health, and environmental standards in the plant (30P).
- Handle, store, and transport chemical materials (33P).
- Operate, monitor, and control continuous processes (27P).
- Operate, monitor, and control batch processes (33P).
- Provide routine and preventative maintenance and service to processes, equipment, and instrumentation (32P).
- Analyze plant materials (36P).

Basic process equipment and technology standards are covered at the beginning of each chapter in this book. The subject matter covered in this text is designed to closely resemble current information found in a typical apprentice training program.

Industrial manufacturers spend millions of dollars on equipment and technology to produce their products. These same manufacturers employ process technicians to operate and maintain their plants. Taking care of the equipment and operation is the primary responsibility of a process technician. Process technicians maintain and operate the equipment twenty-four hours a day, seven days a week. Because of this unique relationship, process technicians become the "hub" of everyday operations.

Operators are responsible for:
- Knowing the basic equipment, design, and operation
- Equipment operation and specific maintenance procedures
    - making relief
    - performing shift tasks
    - making rounds
    - troubleshooting unit problems
    - filling out control documentation
    - maintaining and monitoring equipment
    - inspecting equipment
    - placing equipment in service
    - removing equipment from service
    - responding to emergency situations
- Safety, health, and environment
- Quality control
- Strong technical and problem solving skills, with the ability to assimilate cutting edge technologies quickly and apply innovative ideas

33

- Ability to handle conflict, look at a complex situation and see the overall picture, communicate effectively, and use and understand modern process control
- Ability to understand basic chemistry, physics, and math

Modern manufacturing plants are comprised of complex networks that work closely with each other.

The people who operate and maintain these networks include:
- Process, research, and laboratory technicians
- Maintenance technicians: instrument technicians, electricians, mechanics, and machinists
- Engineers and chemists
- Administrative, human resources, attorneys, and financial analysts
- Computer analysts
- Safety and industrial hygienists
- Janitorial technicians
- Construction: brick, carpentry, structural steel, concrete, and rigging workers

2.3 Safety, Health, and Environment

Over the past 15 years a number of incidents have occurred that have quietly changed the CPI forever. Incidents like Bhopal, Phillips, Arco, and the Exxon Valdez have made us aware of the potential for catastrophic events that exist in our modern manufacturing environment. Technology advancements are so rapid in the chemical processing industry that many of these technologies are outdated a few months after they are installed. Process technicians use this technology to control many of their processes. A process technician can remotely control a company's vast equipment resources from a single control room.

In 1992, OSHA and the Environmental Protection Agency (EPA) released the Process Safety Management standard. The PSM standard was developed in response to a number of incidents that had alarmed the chemical processing industry, community, and government. After years of research and investigation into the causes of industrial explosions, fires, and vapor releases the government issued the Process Safety Management standard. Key elements of the standard include employee participation, process safety information, operations procedures, process hazard analysis, employee training, emergency response, and hot work permit.

Both OSHA and the EPA believe that the key to preventing catastrophic emergencies inside of the chemical processing industry is adequate employee training. This was the conclusion of the governmental groups that investigated the Phillips Chemical Company and ARCO vapor release and explosions. The employee training aspect of the PSM standard includes seven sections:
- Process overview
- Training records and method used to administer training. You must document attendance and competency achieved.
- Identify chemicals used in the process
- Control access to and from the process unit
- Training materials must reflect current work practices
- Provide refresher training
- Contract labor must inform, train, and document that training

Within months of the release of the PSM standard, industry joined with education to form a number of industrial partnerships. These early partnerships initiated the development of a new two-year degree program, Process Technology. The key reasons driving the development of this program were: (1) rapid advances in technology; (2) desire to eliminate accidents in the workplace; (3) potential catastrophic risks; (4) new regulations and guidelines from the government; and (5) loss of the baby boomer work force.

Safety programs have a rich tradition inside the chemical processing industry. The CPI has been very receptive to adopting sound safety principles and government regulations. Process technicians have a wide variety of government mandated training and regulations. The following list contains ten of the most common safety training issues:
- Process safety management 29 CFR 1910.119, OSHA
- Hazard communication 29 CFR 1910.1200
- HAZWOPER 29 CFR 1910.120
- Fire fighting 29 CFR 1910.157
- Permit system
- Environmental awareness
- Department of transportation
- Respiratory protection 29 CFR 1910.134
- Personal protective equipment 29 CFR 1910.133 &135

The hazard communication (HAZCOM) standard is a central feature in the safe operation of the chemical processing industry. HAZCOM ensures that process technicians can safely handle, transport, and store chemicals. The standard provides chemical lists, material safety data sheets, personal protective equipment, information on physical and health hazards, toxicology, hazardous chemicals and operations, manufacturers' information, and warning labels.

Permit systems are designed to protect workers from hazardous energy, hot work, opening and blinding, confined space entry, and cold work. A good permit system can easily integrate into normal operations and protect employees, equipment, and the environment.

Fire protection, prevention, and control are principles that provide protection from industrial fires. Process technicians are required to participate in yearly training using fire extinguishers, monitors, and hoses. A variety of fires are aggressively attacked and extinguished by every member of the team. Fire prevention educates technicians about fire hazards and the steps to take to eliminate them.

2.4 Quality Control

During the early 1980's, US industry was taught a valuable lesson in the area of quality improvement. Using advanced quality practices, the Japanese captured major economic markets from US counterparts. A decade earlier, American business had refused to listen to several leading gurus in quality. This lack of vision cost the stockholders of these companies dearly as Dr. W. Edward Deming, Joseph M. Juran, and others took their message to the Japanese. By 1985, all of the leading oil and automotive giants were listening very closely to what these people had to say.

Introduction to Process Technology • Chapter 2

A basic principle of quality control states that each process has its own range that it naturally moves through. For example, the normal temperature range for your home may fluctuate between 70 and 80 degrees. Your desired setpoint may be 75 degrees. Before statistical process control (SPC), an adjustment was made each time the process variable rose above or fell below the setpoint, Figure 2.4–1. If natural variation is not taken into consideration, the process could find itself completely out of control. SPC allows a process to operate within its own variation by making adjustments only after a number of samples have been caught.

Quality - understanding the customers needs and meeting those needs

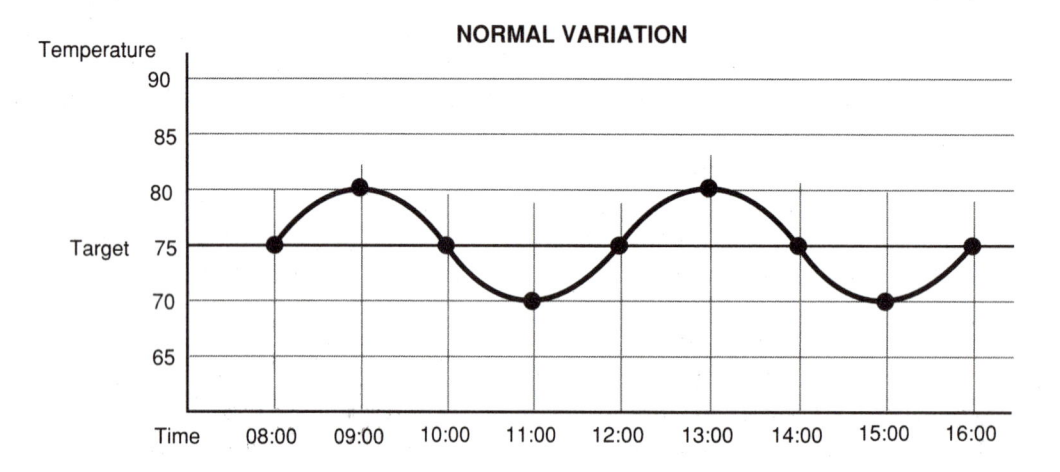

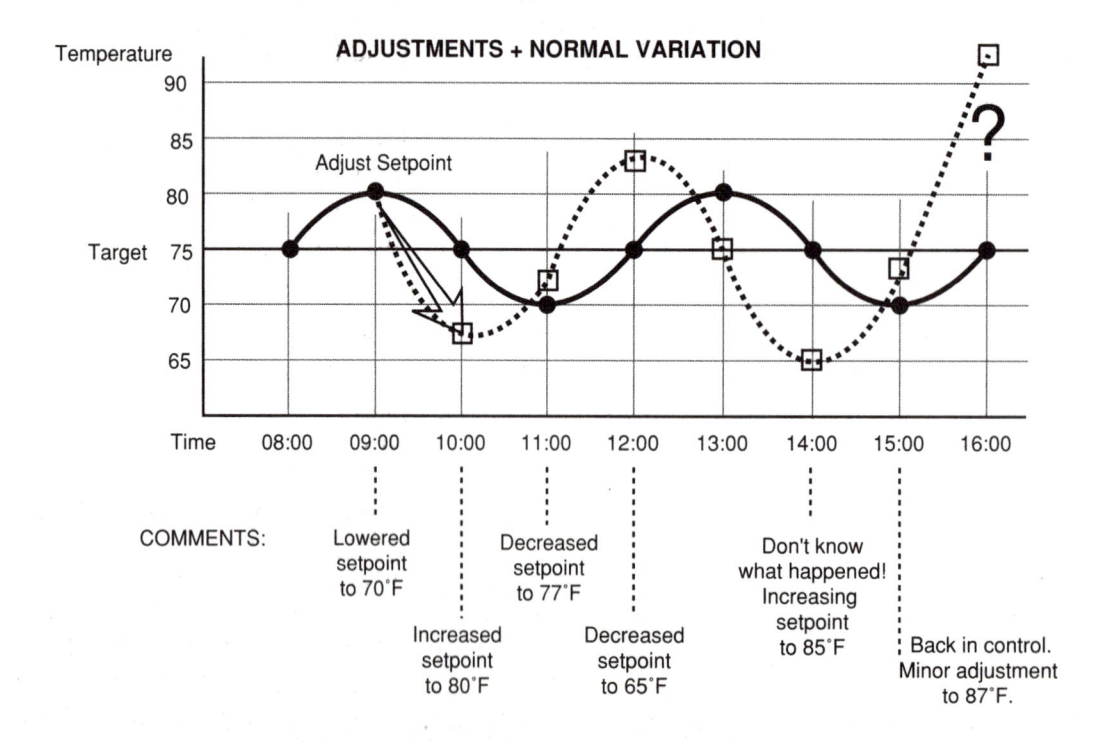

Figure 2.4-1 *Temperature Control Before SPC*

2.5 Process Equipment, Systems, and Operations

Equipment

Process training for operators includes an in-depth study of the basic equipment found in the chemical processing industry, Figure 2.5–1. This knowledge forms a basis for future site-specific training activities. While all of the equipment reviewed in the training program will not be found on the unit you will be assigned to, the odds indicate that at some time in your work career you will come in contact with all of the equipment that is initially studied, and much more.

Equipment training focuses on five basic skills: (1) familiarity with the equipment and basic components, (2) understanding the operation of the device (scientific principles and technology), (3) equipment relationships within a system, (4) preventive maintenance and troubleshooting, and (5) operating the equipment. Process technicians are not required to become mechanical, instrument, or electrical technicians, however they are required to have a sound understanding of the equipment that makes up their process. Understanding the five basic skills allows a technician to understand the process and communicate effectively with maintenance and engineering.

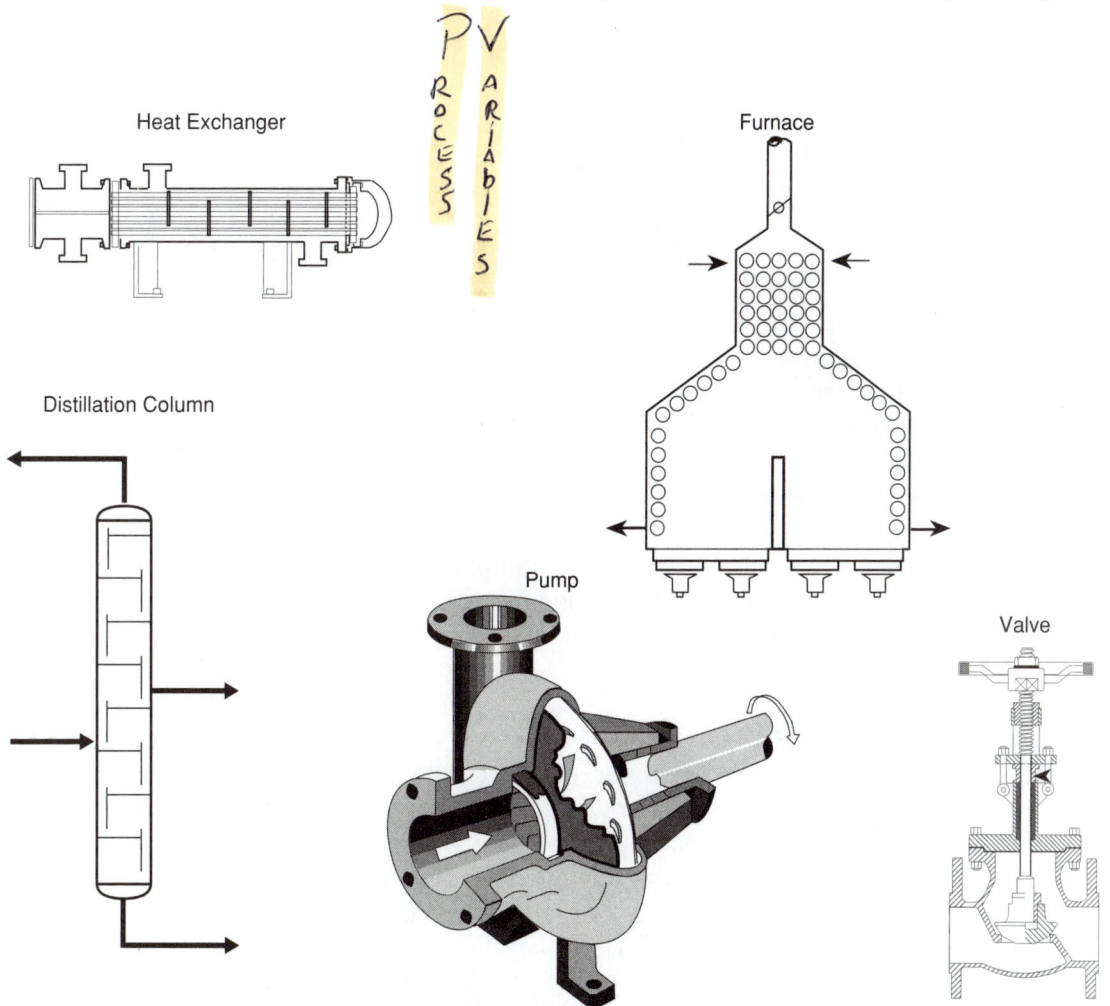

Figure 2.5-1 *Equipment*

Most entry level training programs cover the following types of equipment:
- Valves, piping, and vessels
- Pumps, compressors, fans, and blowers
- Steam turbines and motors
- Heat exchangers and cooling towers
- Boilers and furnaces
- Reactors and distillation columns
- Instrumentation
- Basic hand tools
- Lubrication, bearings, and seals
- Flares, mixers, and steam traps

Systems

Until recently, systems training was left up to site-specific, on-the-job training. An average, full-time, working process technician would spend several years studying the different systems in his/her plant. A system is a collection of equipment that works together to produce a product. Process systems take their specific characteristics from the equipment that makes up the process unit. Some of the basic systems found in the chemical processing industry include:
- Pump-around system
- Compressor system
- Heat exchanger and cooling tower system
- Lubrication system
- Electrical system
- Furnace system
- Plastics system
- Reactor system
- Steam generation system
- Distillation system
- Refrigeration system
- Water treatment system
- Process control system

Modern manufacturing plants are comprised of complex networks that work closely with each other. The people who operate and maintain these networks include process technicians, maintenance technicians, instrument technicians, electricians, laboratory technicians, chemists, and engineers.

Operations

Operations is the capstone experience for most college training programs. This course is designed to allow students to apply the knowledge they have learned. The operations course is ideally attached to an operating unit either at the college or local industry. The class is designed to closely represent the first three months of working in the chemical processing industry.

In order to operate a process unit a number of steps must be taken. These steps include:
- Orientation
- Overview of unit
- Safety, health, and environment review
- On-the-job training

- Reading and using operational procedures
- Complete qualification process
- Operate the unit

2.6 Instrumentation and Process Control

Process instrumentation is a core class designed to teach the process technology student the basic principles for reading process blueprints, the primary function of instruments, and how they work together to automatically control a process. Process instruments fall into five different groups: (1) primary elements and sensors, (2) transmitters, (3) controllers, (4) transducers, and (5) final control elements.

Each part found in a plant has an equivalent symbol or diagram and a specific relationship to other pieces of equipment. Examples of this include:
- Transmitters and controllers
- Piping, tanks and valves
- Pumps and compressors
- Motors and steam turbines
- Heat exchangers and cooling towers
- Fired heaters and boilers
- Distillation columns and reactors

2.7 Applied Chemistry and Physics

Applied chemistry and physics are two fundamental courses that have been recommended by industry to be included in a process technology program. It is clear that information contained in modern chemistry and physics courses are not addressing key topics required by the occupation. The Process Safety Management standard requires that process technicians have an understanding of the chemistry associated with the processes they are operating.

Process technicians frequently mix chemicals together under a variety of conditions to produce new products. These chemical mixtures may be heated, cooled, blended, passed over a catalyst, or distilled. The chemistry associated with these processes can be simple or complex. Documentation associated with these mixtures should be designed so a new technician will be able to understand the basic chemistry.

Common chemistry and physics topics include:
- Pressure
- Heat transfer
- Viscosity
- Atoms and elements
- Compounds
- Hydrocarbons
- Temperature
- Heat and energy
- Specific gravity
- Bonding
- Solutions
- Distillation
- Fluid flow
- Density
- Reactors
- Molecules
- Mixtures
- Matter

2.8 Industrial Processes

Industrial processes are categorized as petrochemical, refinery, environmental and gas processes. At the present time, there are hundreds of different processes. In recent years, the petrochemical and environmental areas have significantly added to this overall total. The more common petrochemical processes include ethylene, olefins, benzene, ammonia, and aromatics. Popular refinery operations include the traditional crude distillation, reforming, cracking, isomerization coking, and alkylation. Since the early 1980's, environmental issues have been heating up. The more popular environmental systems are applied to water treatment, air pollution, solid waste, and toxic waste.

2.9 Troubleshooting

Process troubleshooting is an important part of a process technician's job description. Troubleshooting incorporates three basic components: knowledge of the equipment, instrumentation, and technology; understanding of the scientific principles associated with your unit; and an understanding of a basic troubleshooting system.

Experience is another important factor in troubleshooting; however, studies indicate that when new technicians are exposed to a specific troubleshooting model they tend to compete well with experienced technicians under controlled situations. If troubleshooting is not included in the company's corporate training culture, then experience is the best teacher.

The basic tools used in modern troubleshooting processes are:
- Checklists
- Run charts
- Control loops
- Data collection
- Scatter plots
- Variable vs. setpoint
- Interlocks, permissives
- Equipment knowledge
- SPC charts
- Cause and effect diagrams
- Flow charts
- Histograms-frequency plots
- Brainstorming
- First out, alarms
- System(s) knowledge
- Scientific principles

Troubleshooting systems vary depending on the type of automation used on the process unit. Systems can be classified as fully automated, partially automated, and no automation.

Summary

The standard roles and responsibilities of process technicians include mastering an understanding of basic equipment, design, operation, and maintenance. Process technicians are also responsible for making relief with other technicians and working a job post, performing shift tasks, making rounds, catching samples, taking readings, troubleshooting unit problems, filling out control documentation, maintaining and monitoring equipment, inspecting equipment, placing equipment in service, removing equipment from service, and responding to emergency situations.

Safety, health, and environment training includes initial and continuous training and the employment of safety systems that are carefully integrated into everyday operation. Some of these systems include permits, personal protective equipment, fire fighting, hazard communication, HAZWOPER, and process safety management.

Using quality techniques like statistical process control and control charting, a process technician can operate equipment more productively.

The basic equipment used by process technicians includes valves, pumps, compressors, steam turbines, heat exchangers, cooling towers, reactors, distillation columns, boilers, furnaces, and instruments.

A system is defined as a collection of equipment that works together to produce a product. Process systems take their specific characteristics from the equipment that makes up the process unit. Some of the basic systems found in the chemical processing industry include the pump-around system, compressor system, and heat exchanger and cooling tower system.

Process instruments fall into five different groups: (1) primary elements and sensors; (2) transmitters; (3) controllers; (4) transducers; and (5) final control elements.

Common chemistry and physics topics include pressure, heat transfer, viscosity, atoms and elements, compounds, hydrocarbons, temperature, heat and energy, specific gravity, bonding, solutions, distillation, fluid flow, density, reactions, molecules, compounds, mixtures, and matter.

Introduction to Process Technology • Chapter 2

Chapter 2 — Review Questions

1. What are the primary responsibilities of a process operator?

2. Identify the future trends that have been predicted for the process industry.

3. Describe the basic principles of process safety management.

4. List the key elements of HAZCOM.

5. List the basic concepts of modern quality control.

6. List the process equipment found in the chemical processing industry.

7. List the basic systems found in the chemical processing industry.

8. What are the steps involved in process operations?

9. Describe three petrochemical processes.

10. Describe three refinery processes.

11. List the primary responsibilities of a process technician during start-up and shutdown.

12. Explain modern process control.

13. Match the principles of chemistry and physics on the right with the equipment on the left.
    - pumps — D
    - compressors — C
    - steam turbines — F
    - heat exchangers — E
    - cooling towers — A
    - reactors — G
    - distillation — B
    - boilers — h
    - furnaces — ___

    a. evaporation
    b. boiling point
    c. air pressure
    d. fluid flow
    e. heat transfer
    f. kinetic energy
    g. chemical bonding
    h. boiling water

14. List five elements of a control loop.

15. List the tools of modern troubleshooting.

16. What are the five basic skills of equipment training?

17. According to OSHA and the EPA, what is the key to preventing catastrophic emergencies inside of the chemical processing industry?

Chapter 3
Safety, Health, and Environment

OBJECTIVES

After studying this chapter, the student will be able to:
- *Describe the general safety rules used by the plants.*
- *Describe the process safety management "PSM" standard.*
- *Describe the hazard communication standard.*
- *Describe physical and health hazards.*
- *Describe toxicology and the terms associated with it.*
- *Describe air purifying and air supplying respirators.*
- *Identify personal protective equipment found in a plant.*
- *Describe PPE and the four levels of personal protective equipment.*
- *Describe the principles of hearing protection.*
- *Describe typical plant permit systems.*
- *Analyze confined space entry procedure.*
- *Describe a lock-out, tag-out permit.*
- *Describe the principles of fire prevention, protection and control.*
- *Evaluate the different types of fire extinguishers.*
- *Describe the sections of DOT.*
- *Describe HAZWOPER.*

Safety, Health, and Environment • Chapter 3

KEY TERMS

Air-purifying respirator—mechanically filters or absorbs airborne contaminants.

Air-supplying respirator—provides the user with a contaminant-free air source.

Department of Transportation (DOT)—governmental agency empowered to regulate the transportation of goods on our public roads and highways.

Emergency response—a written plan that documents how specific individuals respond during an emergency situation.

First-aid incident—A first-aid incident is any incident that requires any level of medical attention. This includes minor scrapes and bruises.

First responder—The first responder level is the first two levels of emergency response as described by HAZWOPER 29 CFR 1910.120. The first responder awareness and operations levels have a series of structured responsibilities. The awareness level teaches a technician how to recognize a hazardous chemical release, the hazards associated with the release, and how to initiate the emergency response procedure. The operations level teaches a technician how to safely respond to a release and prevent its spread.

HAZCOM—Hazard communication standard known as the "workers' right to know."

HAZWOPER—hazardous waste operations and emergency response.

Housekeeping—is closely associated with safety in the chemical processing industry. Process technicians are required to keep their immediate areas clean.

Lock-out, tag-out—a term used to describe a procedure for locking out and tagging equipment that falls under the control of hazardous energy, 29 CFR1910.147.

Permit system—a regulated system that uses a variety of permits for various applications. The more common applications are cold work, hot work, confined space entry, opening/blinding, permit to enter, lock-out, tag-out.

Personal Protective Equipment (PPE)—Equipment used to protect a technician from hazards found in a plant. OSHA and EPA have identified four levels of PPE that could be required during an emergency situation. Level A provides the most protection while level D requires the least.

Physical hazard—a term applied to a chemical that statistically falls into one of the following categories: combustible liquid, compressed gas, explosive or flammable, organic peroxide, oxidizer, pyrophoric, unstable or water reactive.

Process safety management standard (PSM)—designed to prevent the catastrophic release of toxic, hazardous, or flammable materials that could lead to a fire, explosion, or asphyxiation.

Respiratory protection—a standard designed to protect employees from airborne contaminants.

3.2 Basic Safety

The philosophy behind most modern safety programs involves the prevention of accidents. Successful accident prevention depends upon three basic elements: safe working environments, safe working practices, and effective leadership. Safety programs for process technicians usually include elements of the following topics:

- HAZCOM—workers' right to know about the chemicals they use
- HAZWOPER—hazardous waste operations and emergency response
- Respiratory Protection
- The Permit System
- Process Safety Management
- Personal Protective Equipment
- Hearing Conservation
- Fire Prevention and Protection
- Department of Transportation (DOT)
- Environmental Standards
- Basic Principles of Safety & Contractor Safety
- Lock-Out, Tag-Out and Confined Space Awareness

General safety rules are designed to protect human life, the environment, and physical equipment or facilities. Before entering a refinery or chemical plant a simple overview of the general plant safety rules is conducted. These include:

1. Do not go to a fire, explosion scene, accident, or vapor release unless you have specific duties or responsibilities.
2. Obey all traffic rules.
3. Do not park in designated fire lanes.
4. Report injuries immediately.
5. Stay clear of suspended loads.
6. Smoking and matches are not permitted in most sections of a plant.
7. Drink from designated water fountains and potable water outlets.
8. Use the right tool for the right job.
9. Report to the designated equipment owner before entering an operating area. Stay in your assigned area.
10. Illegal drugs and alcohol are not permitted in the plant.
11. Firearms and cameras are not allowed in the plant.
12. Take steps to remove hazardous conditions.

Safety, Health, and Environment • Chapter 3

13. Review and follow all safety rules and procedures including:
    - Personal protective equipment
    - Hazard communication
    - Respiratory protection
    - Permit system
    - Hazardous waste operations and emergency response
    - Housekeeping
    - Fire prevention
14. Know and understand the following alarms and rules associated with:
    - Vapor release
    - Fire or explosion
    - Evacuation
    - All clear

3.3 OSHA

In 1970 a landmark piece of legislation was passed that required the chemical processing industry to make safety and health on the job a matter of federal law. The Occupational Safety and Health Act (OSHA) brought in sweeping changes that affected 4 million American businesses and more importantly 57 million employees and their families. In 1969 there were 2.5 million disabling injuries and 14,000 deaths that were directly linked to safety and health violations.

The purpose of OSHA is to (1) remove known hazards from the workplace that could lead to serious injury or death and (2) ensure safe and healthful working conditions for American workers. The coverage of the legislation is extensive in scope. The Occupational Safety and Health Act applies to four broad categories: agriculture, construction, general industry, and maritime. There are three primary agencies responsible for the administration of the Occupational Safety and Health Act (See Figure 3.3–1):
1. NIOSH—National Institute for Occupational Safety and Health
2. OSHA—Occupational Safety and Health Administration
3. OSHRC—Occupational Safety and Health Review Commission

OSHA
OCCUPATIONAL SAFETY & HEALTH ACT

NIOSH
National Institute for Occupational Safety & Health
- Safety & Health Research
- Recommends New Standards

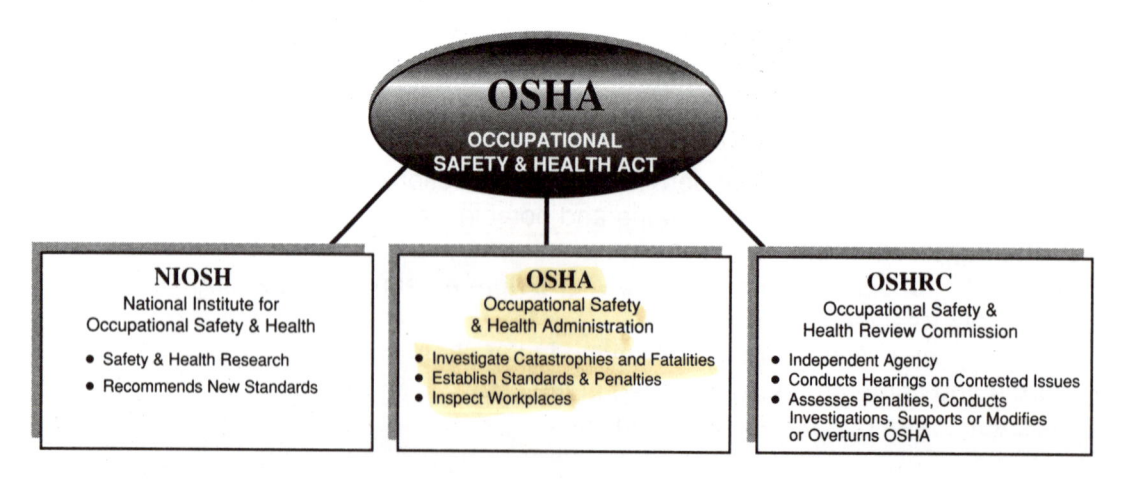

OSHA
Occupational Safety & Health Administration
- Investigate Catastrophies and Fatalities
- Establish Standards & Penalties
- Inspect Workplaces

OSHRC
Occupational Safety & Health Review Commission
- Independent Agency
- Conducts Hearings on Contested Issues
- Assesses Penalties, Conducts Investigations, Supports or Modifies or Overturns OSHA

Figure 3.3-1 *OSHA*

Chapter 3 • *Safety, Health, and Environment*

3.4 The PSM Standard

After the ARCO and Phillips plant explosions in 1989 and 1990, OSHA and the EPA went to work on a new standard that would limit the possibility of this happening again. After years of research and investigation into the causes of industrial explosions, fires, and vapor releases the government issued the Process Safety Management standard. Figure 3.4–1 illustrates the key elements of the standard.

EMPLOYEE PARTICIPATION
- Written Program
- How Employees will Access Hazard Identification System
  - identify hazards
  - gather information
  - communication system

OPERATIONS PROCEDURES
- Operations and Maintenance
- Reflect current work practices
- Process Properties
- Hazards
- Start-up, Shutdown
- Change of Chemicals

PROCESS SAFETY
- Process Flow Diagram
- Equipment, Process Description, Limitations
- Consequences of Deviation
- Safety and Relief Devices
- Electrical Classifications
- Characteristics of Chemicals
- Process Chemistry
- Mixing Chemicals

EMPLOYEE TRAINING
- Process Overview
- Training Records and Method
- Attendance & Competency
- Training materials reflect current work practices
- Control access to unit
- Refresher training
- Contractors must inform and train their employees and document that training

PROCESS HAZARD ANALYSIS
- SOP, Safety, Training up front
- Identify unit hazards
- Identify causes & consequences
  - fires
  - vapor releases
  - explosions

HOT WORK PERMIT
- Protect from fires and explosions
- Specific procedures for hot work
- Defined as welding, cutting, spark producing

INCIDENT INVESTIGATION
For catastrophic events:
- Assemble team within 48 hours
- Address all findings
- Correct action items

EMERGENCY PLANNING
- Emergency response plan
  - designated meeting points
  - key contacts
- Emergency response
  - roles and responsibilities
- Written action plans

Figure 3.4-1 *Process Safety Management*

3.5 The Hazard Communication Program

Government Mandate: Hazard Communication 29 CFR 1910.1200
A fundamental principle of the chemical hazard communication program is that informed people are less likely to be injured by chemicals and chemical processes than uninformed people. According to the standard all of the chemical inventories and processes within a chemical plant or refinery must be evaluated for potential hazards and risks. Where a risk is found essential information and training is required for all people affected. A chemical hazard communication program is composed of two essential parts: the written program (which addresses chemical manufacturers) and employee training, Figure 3.5–1.

Requirements of the Standard (Documentation)
Since the chemical processing industry manufactures chemicals and employs technicians, the CPI is responsible for both sections of the OSHA standard that addresses chemical manufacturers employer requirements and user responsibilities. Chemical manufacturers are required by the HAZCOM standard to:
- Analyze and assess the hazards associated with the chemical,
  and develop written procedures for evaluating chemicals.
- Document the hazard, and develop material safety data sheets and warning labels.
- Disseminate the information to affected individuals.
- Label, tag, and attach warning documentation to chemicals leaving the workplace.

Employers are responsible for the development of a written hazard communication program, a hazardous chemical inventory list, and associated material safety data sheets. This written program should be designed so that it is given to the new employee upon initial assignment. The materials should be site specific, readily accessible by plant personnel, and include an evergreen feature that will keep it up to date. Employers are also required to provide training to employees about the hazards of the chemicals they will be working with, how to read an MSDS, how to select and use personal protective equipment and how to read and use one of the three standard labeling systems: DOT—Department of Transportation, HMIS—Hazardous Materials Identification System, and NFPA—National Fire Protection Association.

Delivery of the Standard to Employees
The HAZCOM standard requires employers to provide information or training to employees about their plant's hazard communication program. Fundamental information that must be provided to a process technician includes: the key elements of the HAZCOM standard; their plant's written hazard communication program; a detailed hazardous chemical inventory list; and associated material safety data sheets (MSDS) along with warning labels, tags, and signs. Information should be included on how to access the HAZCOM system, chemical inventory list, and MSDS twenty-four hours a day, seven days a week. Employers are required by law to provide open access to HAZCOM materials. This is why the HAZCOM standard is frequently referred to as the "Workers' Right to Know Act."

The chemical processing industry initiates the delivery of HAZCOM training upon initial assignment to the plant. Training focuses on the physical and health hazards associated with exposure to chemicals. Additional information is provided on the toxicology, physical properties, and hazards associated with handling, storing, and transporting chemicals. New technicians are required

Chapter 3 • *Safety, Health, and Environment*

HAZCOM
29 CFR 1910.1200
OSHA

HAZCOM PROGRAM
must include:
- Container labeling and warnings
- MSDS
- Employee training

WRITTEN PROGRAM
(Documentation)

Chemical Manufacturers
- Analyze Chemical Hazards
- Develop Written Procedures for Evaluating Chemicals
- Document Hazards
- Develop MSDS & Warning Labels
- Disseminate Information
- Label, Tag, Attach Documentation to Chemicals Leaving Workplace

Chemical Lists
- Plant Chemical Inventory List
- Provided to all Employees
- Toxicology

Material Safety Data Sheets
- Chemical Hazards
- Physical Hazards
- Product Identification
- PPE
- Storage & Handling
- Reactivity

Target Critical Operations
- Hazardous Chemicals
- Hazardous Operations

EMPLOYEE TRAINING

Physical Hazards
- Combustible
- Flammable
- Explosive
- Compressed
- Oxidizer
- Pyrophoric
- Unstable
- Water Reactive

Chemical Hazards
- Carcinogen
- Mutagen
- Teratogen
- Asphyxiation
- Corrosive
- Toxic
- Neurotoxic
- Target Organ Effects

Personal Protective Equipment
- Hard Hat
- Safety Glasses
- FRC
- Monogoggles
- Respirators
- Gloves
- Safety Shoes
- Radio

Release Detection
- Methods used to detect the release of Hazardous Chemicals
- Human Senses
- Detectors

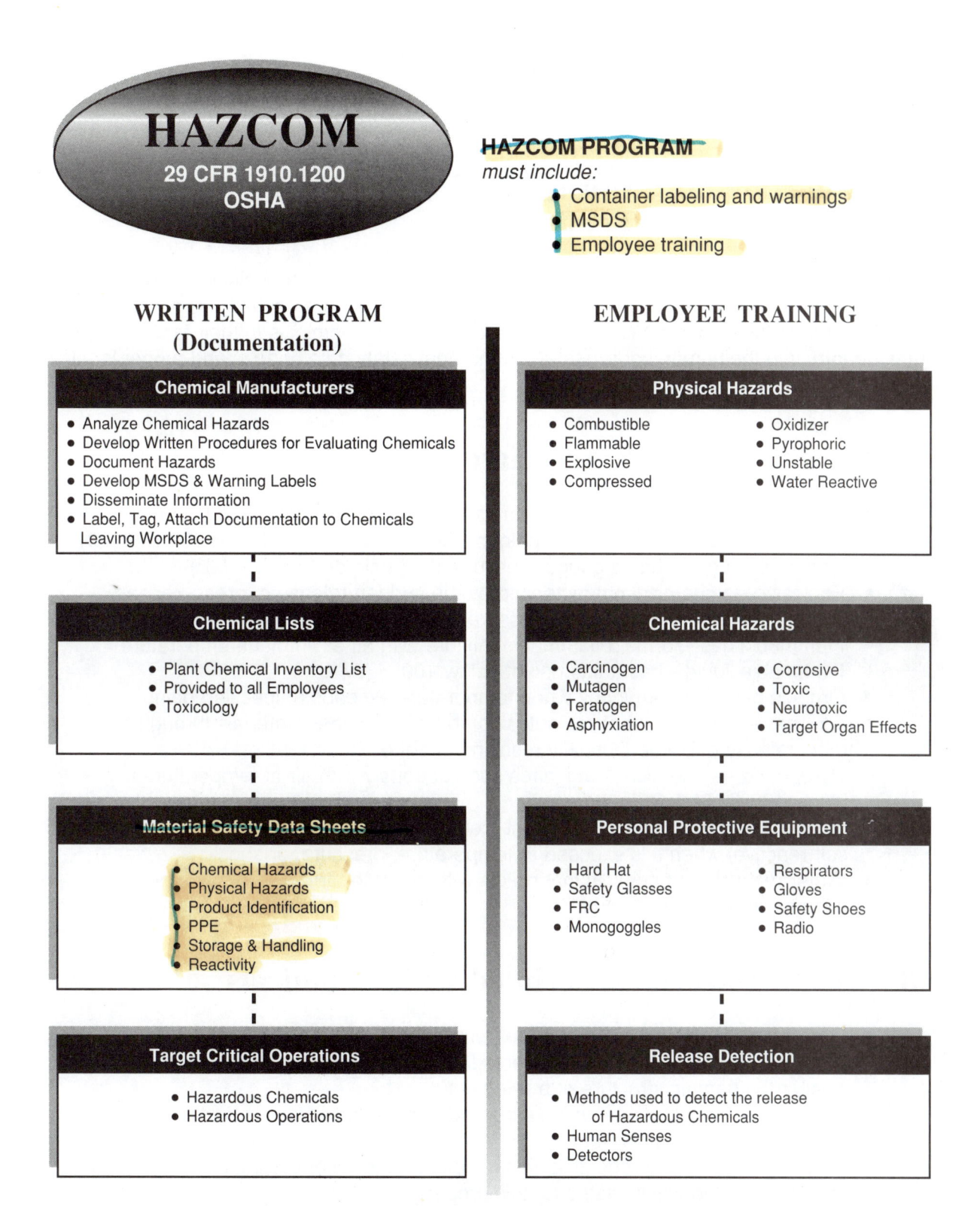

Figure 3.5-1   *HAZCOM*

to review company procedures used to protect employees from hazardous chemicals and specific operations are identified that may expose an employee to a chemical. The training section also includes the selection and use of personal protective equipment (PPE) and the methods and observations utilized to detect the release of hazardous chemicals.

3.6 Safe Handling, Storage, and Moving of Hazardous Chemicals

Transporting, storing, and handling chemicals require that process technicians understand the systems, equipment, and technology they are working with, the physical hazards associated with chemicals in their facility, the health hazards associated with chemicals in their facility, chemical routes of entry into the human body, use of the material safety data sheets, and proper labeling, signs, and tags usage.

3.7 Physical Hazards Associated with Chemicals

A physical hazard is defined as a chemical that falls into one of the following categories:
- Combustible liquid—has a flash point between 100°F (38.8°C) and 200°F (93°C)
- Compressed gas—has a gauge pressure of 40 psig at 70°F (21.1C°)
- Explosive—a chemical characterized by the sudden release of pressure, gas, and heat when it is exposed to pressure, high temperature, or sudden shock
- Flammable gas—forms a flammable mixture with air at ambient temperature
- Flammable liquid—has a flash point below 100°F (37°C)
- Organic peroxide—explodes when temperature exceeds a specified point
- Oxidizer—a chemical that promotes combustion in other materials through the rapid release of oxygen, usually resulting in a fire
- Pyrophoric—a chemical that ignites spontaneously with air at temperatures below 130°F (54.4°C)
- Unstable—a chemical that will react (condense, decompose, polymerize, or become self-reactive) when it is exposed to temperature, pressure, or shock
- Water reactive—a chemical that reacts with water to form a flammable or hazardous gas

3.8 Health Hazards Associated with Chemicals

- Carcinogens—known cancer-causing substances
- Mutagen—a chemical that is suspected to have the properties required to change or alter the genetic structure of a living cell
- Teratogen—a substance that is suspected to have an adverse effect on the development of a human fetus
- Reproductive toxin—a chemical that inhibits the ability of a person to have children; chemicals are routinely tested for this property
- Asphyxiation—occurs when oxygen is removed or displaced by a chemical or when a chemical blocks or impedes the ability of a person to use oxygen

Chapter 3 • *Safety, Health, and Environment*

- Anesthetic—dulls the senses (example: alcohol)
- Neurotoxic—slows brain down (example: lead and mercury)
- Allergic response—a negative reaction to a chemical that triggers a physical response of discomfort, injury, or death
- Irritants—chemicals that cause temporary discomfort when they come into contact with human tissue
- Sensitizer—a chemical that effects the nerves (example: phenol is absorbed through the skin and will sensitize the affected area)
- Corrosive—a chemical that causes severe damage to human tissue (example: sulfuric acid)
- Toxic—a term applied to a chemical that has been determined to have an adverse health impact
- Highly toxic—a term applied to a chemical that requires only a small amount to be lethal
- Target organ effects—a term applied to a chemical that contacts the body at a remote location and is transferred to another area of the body where it has an adverse effect on a specific organ

Hazardous chemicals can enter the human body through:
- Inhalation
- Absorption (skin contact)
- Ingestion
- Injection

Physical hazards and health hazards in a chemical plant or refinery must be quickly recognizable to process technicians. Recognizing a hazard and knowing how to respond are key elements of a technician's training. Figure 3.8–1 illustrates this.

Figure 3.8-1 *Hazard Recognition*

3.9 The Material Safety Data Sheet (MSDS)

It has been estimated that one out of every four workers in the United States handles a chemical. The development of the material safety data sheet (MSDS) is the responsibility of the chemical manufacturer. A typical material safety data sheet has nine or ten sections, as follows:

1. Product Identification & Emergency Information
2. Hazardous Ingredients
3. Health Information & Protection or Hazards Identification
4. Fire and Explosion Hazard
5. Physical Data & Chemical Properties
6. Spill Control Procedure
7. Regulatory Information
8. Reactivity Data
9. Storage and Handling
10. Personal Protective Equipment

3.10 Toxicology

Toxicology is the science that studies the noxious or harmful effects of chemicals on living substances. The fundamentals of toxicology include a relationship between dose and response. Dose is the amount of chemical entering or being administered to a subject. Response is the toxic effect the dose has upon the subject.

3.11 Respiratory Protection Programs

The Occupational Safety and Health Administration requires employers who use and issue respirators to develop a written respiratory protection program. Employees must receive proper training in respiratory protection. Process technicians use two basic types of respirators: (1) air purifying and (2) air supplying.

Air Purifying Respirators
Air purifying respirators are either half-face or full-face. Half-face air purifying respirators are designed to cover the mouth and nose while full-face respirators form a positive seal around the eyes, nose and mouth. These respirators are designed to remove specific contaminants or organic vapors from the air. These concentrations may range from 5 to 50 times the normal exposure limit allowed by law.

Air Supplying Respirators
Air supplying respirators are either SCBA (self-contained breathing apparatus) or hose line respirators. These respirators are designed to be used in oxygen deficient atmospheres.

3.12 Personal Protective Equipment (PPE)

Human beings have over 19 square feet of surface area and breathe over 3,000 gallons of air per day. Since chemical exposure comes through inhalation, ingestion, injection, and skin contact, protective measures need to be in place. Personal protective equipment provides an effective means for protecting technicians from hazardous situations. Engineering and environmental controls provide another layer of protection. The primary purpose of PPE is to prevent exposure to hazards when engineering or environmental controls cannot be used.

Typical outerwear worn by process technicians include:
- Safety hats
- Safety glasses
- Fire retardant clothing
- Safety shoes
- Hearing protection
- Gloves
- Face shield
- Chemical monogoggles
- Slicker suits
- Radio
- Respirators
- Chemical suits
- Totally encapsulating chemical protective suits

3.13 Emergency Response

The chemical manufacturing industry defines emergency response as a loss of containment for a chemical or the potential for loss of containment that results in an emergency situation requiring an immediate response. Examples of emergency response situations include fires, explosions, vapor releases, and reportable quantity chemical spills.

The levels of response have been determined by the chemical processing industry to be:
- First responder awareness level—individuals are trained to respond to a hazardous substance release, initiate an emergency response, evacuate the area, and notify proper authorities
- First responder operations level—individuals trained to respond with an aggressive posture during a chemical release by going to the point of the release and attempting to contain or stop it

Emergency Response—Four Levels of PPE

Emergency response has four levels of personal protective equipment according to the Environmental Protection Agency and the Occupational Safety and Health Administration.

1. Level A requires the highest level of PPE protection by requiring a technician to don a totally encapsulating chemical protective suit.

2. Level B deals with chemical exposures that are not considered to be extremely toxic unless they are absorbed through the skin. In this case, a non-air tight chemical protective suit may be worn. Typically, the openings on a non-air tight chemical suit are taped to limit exposure.
3. Level C is used when the hazard is determined not to adversely affect exposed skin.
4. Level D provides the minimal amount of protection to a process technician. Level D protection is determined by individual companies since the standard personal protective equipment is the work uniform.

3.14 Plant Permit System

The types of permits used in the chemical processing industry include:
- Cold Work Permit—routine maintenance and mechanical work
- Hot Work Permit—any maintenance procedure that produces a spark, excessive heat, or requires welding or burning
- Opening/Blinding Permit—removing blinds, installing blinds or opening vessels, lines, and equipment
- Permit to Enter—is designed to protect employees from oxygen-deficient atmospheres, hazardous conditions, power driven equipment, & toxic and flammable materials
- Unplugging Permit—barricades area, clears lines for unplugging, informs personnel, issues opening blinding permit, issues unplugging permit
- Energy Isolation Procedure—isolates potentially hazardous forms of energy: electricity, pressurized gases and liquids, gravity and spring tension
- Lock-out, Tag-out Procedure—a standard designed to shut down a piece of equipment at the local start/stop switch, turn the main breaker off, attach lock-out adapter and process padlock, try to start the equipment, and tag-out and record in lock-out logbook

3.15 Classification of Fires and Fire Extinguishers

The fire classification system is designed to simplify the selection of fire-fighting techniques and equipment.
- Class A fires involve the burning of combustible materials such as wood, paper, plastic, cloth fibers, and rubber.
- Class B fires involve combustible and flammable gases and liquids and grease.
- Class C fires are categorized as electrical fires. This involves energized equipment and class A, B, and D materials that are located near the fire.
- Class D fires cover combustible metals such as aluminum, magnesium, potassium, sodium, titanium, and zirconium.

The five most common fire extinguishers found in the chemical processing industry and their range of effectiveness are as follows:
1. CO_2 extinguishers are effective on class B and C fires because they cool and displace oxygen.

Chapter 3 • *Safety, Health, and Environment*

2. Dry chemical fire extinguishers are effective on class A, B, and C fires because they displace oxygen.
3. Foam fire extinguishers are used to control flammable liquid fires. The foam forms an effective barrier between the flammable liquid and the needed oxygen for combustion. Foam extinguishers are effective on class A and B fires.
4. Halon fire extinguishers are designed for use on class A, B, and C fires.
5. Water fire extinguishers are designed for use on class A fires only.

Figure 3.15–1 shows some common fire extinguishers.

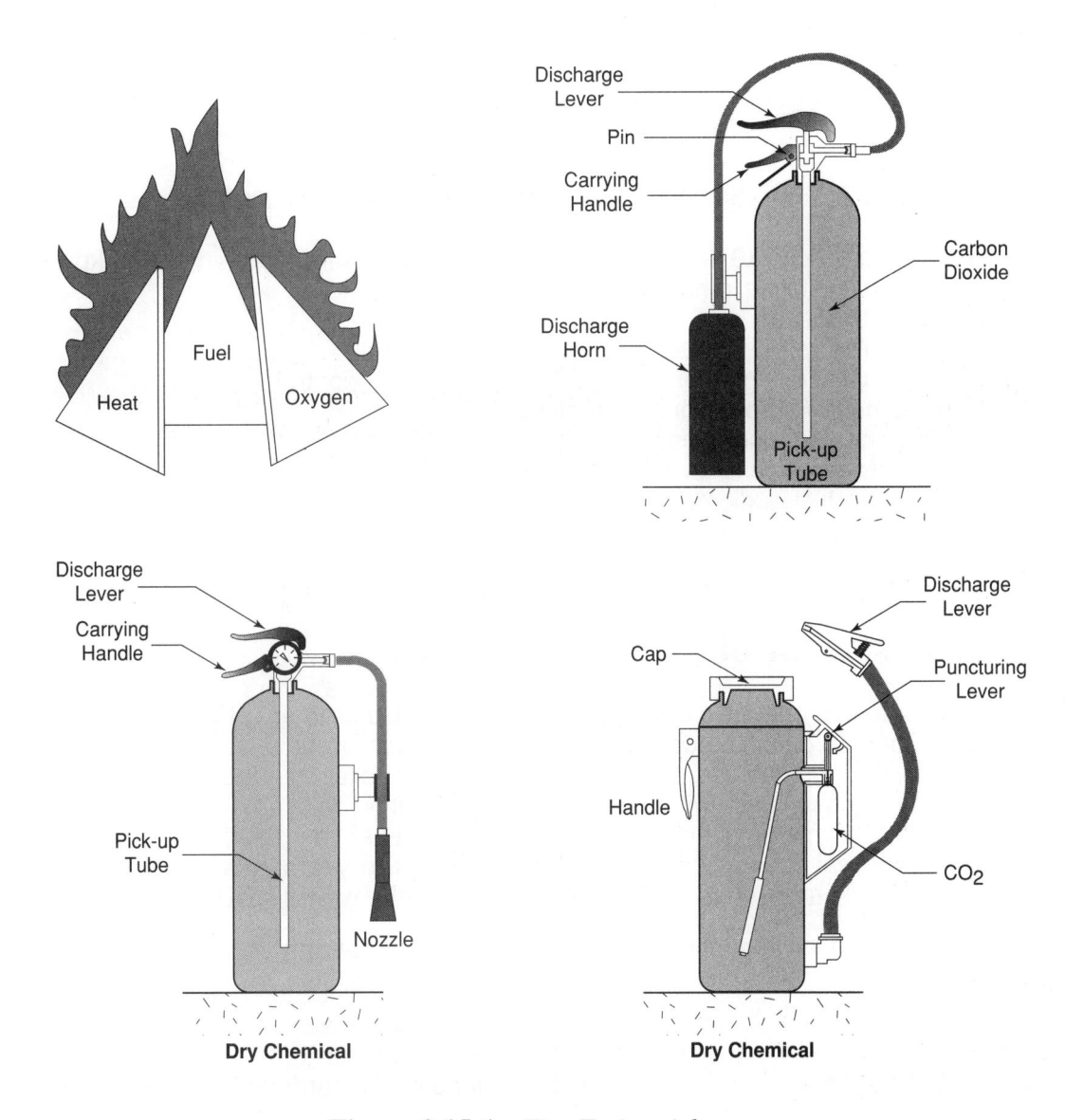

Figure 3.15-1 *Fire Extinguishers*

3.16 HAZWOPER

The term HAZWOPER is used to describe OSHA's Hazardous Waste Operations and Emergency Response standard. HAZWOPER is broken down into the following areas:
- Emergency Response—first responder awareness level, first responder operations level
- Hazardous Waste Operations—incident command system, scene safety and control, spill control and containment, decontamination procedures, emergency termination or all clear
- Hazard Protection, Prevention, and Control
    terms and definitions
    PPE levels
    identifying hazardous materials
    hazards initiating an emergency response
    avoiding hazards
    entry of hazardous materials into the body
    unit monitors and field survey instruments

3.17 Hearing Conservation and Industrial Noise

When OSHA was enacted in 1970, federal regulations for controlling noise in the workplace were implemented. This new standard has two major components:
1. Maximum noise exposure
2. Actions that employers must take if the limits are exceeded:
   - Reduce noise using engineering and administrative controls
   - Provide hearing protection for employees
   - Implement a hearing conservation program (HCP)
       a. Monitor sound levels
       b. Conduct audiometric tests
       c. Provide hearing protection
       d. Provide training

3.18 Department of Transportation

Shipments of hazardous materials are regulated by the US Department of Transportation (DOT), Figure 3.18–1. These regulations contain specific information on how hazardous materials are identified, placarded, documented, labeled, marked, and packaged. Hazardous material shipments that are not in compliance with federal regulations will be delayed and can result in severe penalties. In civil cases, marking the wrong name on a container can result in fines up to $25,000 per violation. In criminal cases, fines up to $500,000 and five years in jail can be imposed for intentionally shipping a hazardous chemical and attaching the wrong MSDS.

Materials are classified for transportation using nine different categories:

1. Explosive
2. Gases
3. Flammable Liquids
4. Self Reactive
5. Oxides and Peroxides
6. Poisonous and Infectious Materials
7. Radioactive Materials
8. Corrosive Materials
9. Miscellaneous Hazardous Materials

Chapter 3 • *Safety, Health, and Environment*

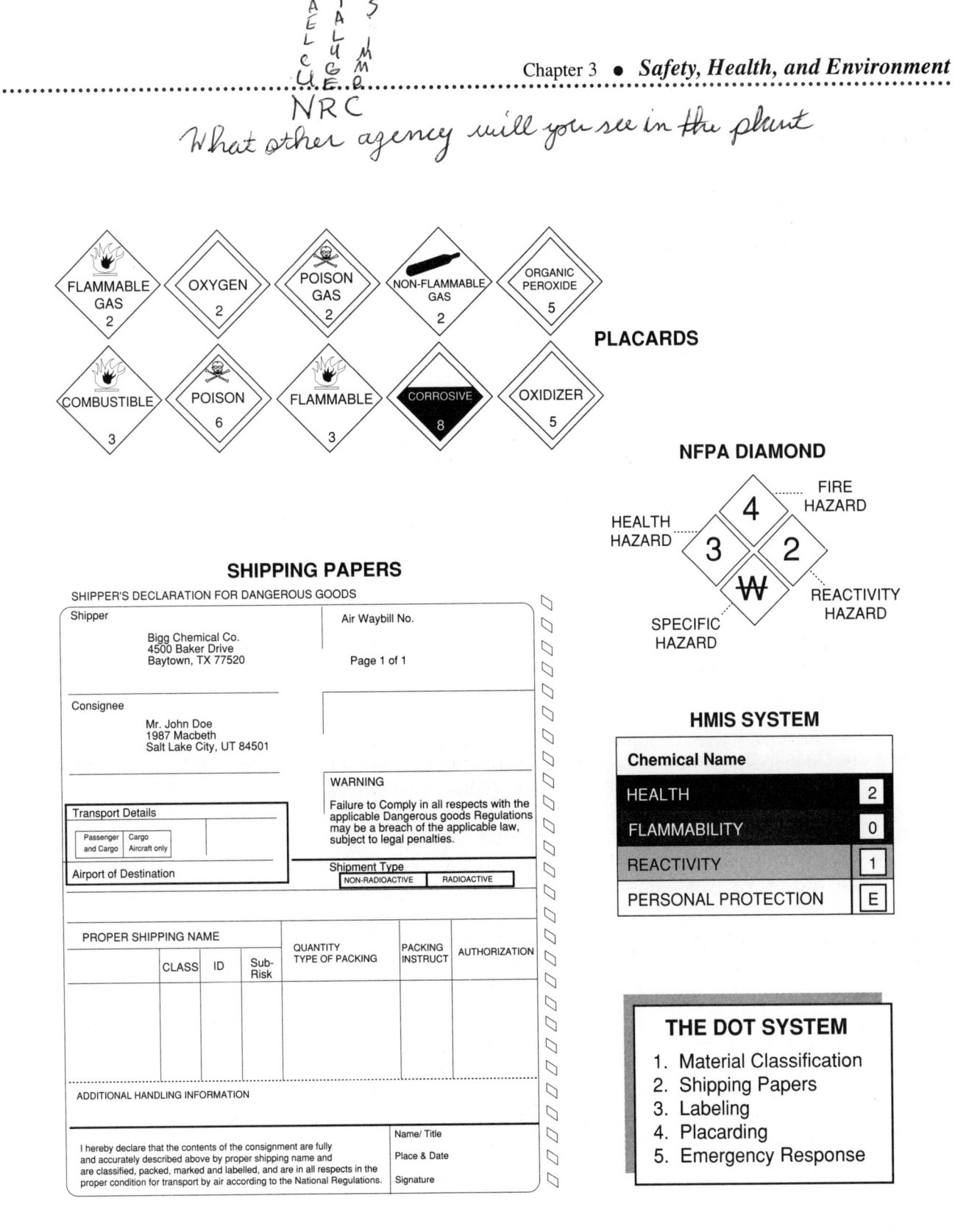

PLACARDS

NFPA DIAMOND

FIRE HAZARD
HEALTH HAZARD
REACTIVITY HAZARD
SPECIFIC HAZARD

SHIPPING PAPERS

HMIS SYSTEM

Chemical Name	
HEALTH	2
FLAMMABILITY	0
REACTIVITY	1
PERSONAL PROTECTION	E

THE DOT SYSTEM

1. Material Classification
2. Shipping Papers
3. Labeling
4. Placarding
5. Emergency Response

Figure 3.18-1 *DOT Labels, Signs, and Placards*

57

Summary

OSHA and the EPA developed the process safety management standard to prevent the catastrophic release of toxic, hazardous, or flammable materials that could lead to a fire, explosion, or asphyxiation. Several critical elements of the PSM standard include employee training, operations procedures, process safety, employee participation, and hot work.

A fundamental principle of the chemical hazard communication program (HAZCOM) is that informed people are less likely to be injured by chemicals and chemical processes than uninformed people. A chemical hazard communication program is composed of both information and training.

Chemical exposure comes from inhalation, ingestion, injection and skin contact. Personal protective equipment (PPE) provides an effective means for protecting technicians from hazardous situations. Engineering and environmental controls provide another layer of protection. The primary purpose of PPE is to prevent exposure to hazards when engineering or environmental controls cannot be used. Process technicians use two basic types of respirators: air purifying and air supplying. Hearing conservation can be broken down into two major components: maximum noise exposure and actions that employers must take if the limits are exceeded.

The types of permits used in the chemical processing industry include cold work permit, hot work permit, opening/blinding permit, permit to enter, unplugging permit, energy isolation procedure, and lock-out, tag-out procedure.

Fires are classified as Class A, B, C, and D. The most common fire extinguishers found in industry are CO_2, dry chemical, foam, halon, and water fire extinguishers.

The term HAZWOPER is used to describe OSHA's Hazardous Waste Operations and Emergency Response standard. HAZWOPER is broken down into the areas of emergency response, hazardous waste operations, and hazard protection, prevention and control.

Shipments of hazardous materials are regulated by the U.S. Department of Transportation (DOT). These regulations contain specific information on how hazardous materials are identified, placarded, documented, labeled, marked, and packaged.

Chapter 3

Review Questions

1. Describe the important features of the HAZCOM program.

2. Describe the important features of HAZWOPER.

3. What personal protective equipment does a process technician typically wear?

4. What is the respiratory protection standard?

5. What three primary agencies are responsible for the administration of the Occupational Safety and Health Act?

6. What is the Occupational Safety and Health Administration?

7. Describe the major features of PSM.

8. What is emergency response?

9. Explain the importance of the process safety management standard.

10. What is toxicology, and how are dose and response important?

11. Describe the role of DOT in ensuring the safety of hazardous materials.

12. What is a physical hazard?

13. Identify the physical properties and hazards associated with handling, storing, and transporting chemicals.

14. What is a fundamental principle of the chemical hazard communication program?

15. What are the two basic types of air purifying respirators?

16. What are the basic types of air respirators?

17. Describe the key elements of the permit system.

18. What do you think are the ten most important general safety rules for a chemical plant?

19. What are the critical elements of hearing conservation, including employer's responsibilities?

20. What do you think is the most important safety rule?

Chapter 4

Basic Process Principles

OBJECTIVES

After studying this chapter, the student will be able to:

- *Describe key terms and definitions used in basic process principles.*

- *Describe and apply the basic principles of pressure.*

- *Analyze the scientific principles of heat, heat transfer, and temperature.*

- *Examine the principles of fluid flow in process equipment.*

- *Solve basic mathematical problems encountered in industry.*

Basic Process Principles • Chapter 4

KEY TERMS

Addition—a term applied to a mathematical operation for combining numbers.

Algebra—a branch of mathematics that uses letters to represent numbers and signs to represent operations. It is a kind of universal arithmetic or simply mathematics using letters.

Bernoulli's principle—stated that in a closed process with a constant flow rate, changes in fluid velocity (kinetic energy) decrease or increase pressure, kinetic energy and pressure energy changes correspond to pipe size changes, pipe diameter changes cause velocity changes, and pressure energy, kinetic energy, or "fluid velocity," and pipe diameter changes are related.

Boyle's law—Robert Boyle was an Irish scientist who developed the law that describes how the volume of a gas responds to pressure changes. Its basic principles are that pressure decreases volume and moves gas molecules closer together, the higher the pressure the smaller the volume, and gas volume decreases by one half when pressure doubles.

Conversion tables—display equivalent units of measure.

Dalton's law—$P_{total} = P_1 + P_2 + P_3$, states that the total pressure of a gas mixture is the sum of the pressures of the individual gases.

Decimal point—the period between whole numbers and fractional numbers.

Denominator—the bottom number in a fraction.

Dimensional analysis—conversion within one system of units or to another system of units. Example: English to International System (SI).

Division—a term applied to a mathematical operation for determining how many times one number is contained in another number.

Divisor—the number doing the dividing.

Fluid flow—characterized by fluid particle movements, such as laminar and turbulent.

Fluid pressure—the pressure exerted by a confined fluid. Fluid pressure is exerted equally and perpendicularly to all surfaces confining it.

Fraction— a part of a whole amount.

Grouping symbols—used to separate functions in an equation.

Chapter 4 • *Basic Process Principles*

Heat—a form of energy caused by increased molecular activity.

Heat transfer—heat is transmitted through conduction (heat energy is transferred through a solid object; e.g., a heat exchanger), convection (requires fluid currents to transfer heat from a heat source; e.g., the convection section of furnace or economizer section of boiler), and radiation (the transfer of energy through space by the means of electromagnetic waves; e.g., the sun).

Lowest common denominator (LCD)—the smallest whole number that can be used to divide two or more denominators.

Mixed number—a whole number and a fraction.

Multiplication—the process of adding a number to itself a specified number of times.

Numerator—the top number in a fraction.

Pascal's law—Its basic principles are that pressure in a fluid is transmitted equally in all directions, molecules in liquids move freely, and molecules are close together in a liquid.

Percent—a fractional amount expressed in terms of "parts per one hundred."

Pressure—force or weight per unit area (Force ÷ Area = Pressure). Pressure is measured in pounds per square inch.

Subtraction—a term applied to a mathematical operation for deducting one number from another.

Temperature—the hotness or coldness of a substance.

4.2   Basic Principles of Pressure

Pressure is defined as force or weight per unit area. (Force ÷ Area = Pressure). The term *pressure* is typically applied to gases or liquids. Pressure is measured in pounds per square inch. Atmospheric pressure is produced by the weight of the atmosphere as it presses down on an object resting on the surface of the Earth. Pressure is directly proportional to height: the higher the atmosphere, gas, or liquid, the greater the pressure. At sea level, atmospheric pressure equals 14.7 psi.

Boiling Point and Vapor Pressure
The boiling point of a substance is the temperature at which the vapor pressure exceeds atmospheric pressure, bubbles become visible in the liquid, and vaporization begins.

Molecular motion in water vapor produces pressure and increases as heat is added to the liquid.

Basic Process Principles • Chapter 4

The vapor pressure of a substance can be linked directly to the strength of the molecular bonds of a substance. The stronger the bonds or molecular attraction, the lower the vapor pressure. If a substance has a low vapor pressure, it will have a high boiling point. For example, gold changes from a solid to a liquid at 1,947 degrees F (1,064 degrees C) and boils when the temperature reaches 5,084 degrees F (2,807 degrees C). Water changes from a solid to a liquid at 32 degrees F (0 degree C) and boils when the temperature reaches 212 degrees F (100 degrees C).

Liquids do not need to reach their boiling points in order to begin the process of evaporation. For example, a pan of water placed outside on a hot summer day (98 degrees F [36.66 degrees C]) will evaporate over time. The sun increases the molecular activity of the water vapor, and some of the molecules escape into the atmosphere. Wind currents enhance the process of evaporation by sweeping away water molecules that are replaced by other water molecules.

Pressure Impact on Boiling
Pressure directly affects the boiling point of a substance. As the pressure increases,
- the boiling point increases.
- the escape of molecules from the surface of the liquid is reduced proportionally.
- the gas or vapor molecules are forced closer together.
- the vapor phase above a liquid could be forced back into solution.

This is an important fact for a process technician to understand. A change in pressure shifts the boiling points of raw materials and products. Pressure problems are common in industrial manufacturing environments and must be controlled.

Vacuum
Atmospheric pressure is 14.7 psi, so any pressure below this is referred to as a vacuum. Vacuum affects the boiling point of a substance in the opposite way that positive pressure does.

Vacuum Systems
- lower the boiling point of a substance.
- enhance the molecular escape of a liquid.
- reduce energy costs.
- reduce molecular damage due to overheating.
- reduce equipment damage.

Blaise Pascal's Law
Blaise Pascal was a French scientist who discovered that pressure in a fluid is transmitted equally in all directions. Pascal successfully described the effects of pressure in a liquid and established the scientific foundation for hydraulics. Key facts for process technicians to remember are that pressure in a fluid is transmitted equally in all directions, molecules in liquids move freely, and molecules are close together in a liquid.

Robert Boyle's Law
Robert Boyle was an Irish scientist who developed the law that describes how the volume of a gas responds to pressure changes. Key facts for process technicians to remember are that pres-

Chapter 4 • *Basic Process Principles*

sure decreases volume and moves gas molecules closer together; the higher the pressure the smaller the volume; and gas volume decreases by one half when pressure doubles.

Determining Pressures Produced by Liquids
The pressure a liquid exerts on a container is determined by the height and the density of the fluid. The pressure exerted by a 20-ft column of mercury would be more than a 20-ft column of water. The specific gravity of mercury (Hg) is much higher than water.

Pressure Problems
Pressure problems can be correctly calculated by using the following formula:

Pressure = Force ÷ Area

EXAMPLE 1: stone   Calculate the pressure produced by a 1,000 lb. stone block.

20-in. length × 20-in. width

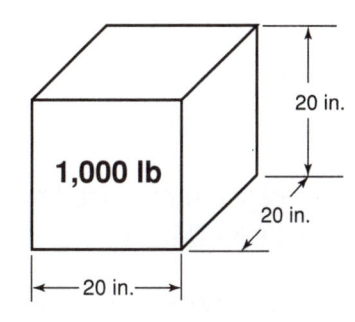

Figure 4.2-1 *Stone Block*

Solution:
The area occupied by the stone = 400 sq in.
20-in. length × 20-in. wide = 400 sq in.
Pressure = 1,000 ÷ 400 = 2.5 psi
The psi at the base of the stone: = 2.5 psi

EXAMPLE 2: water Calculate the pressure produced by one cubic foot of water (62.4 lb) in a 1-ft length × 1-ft width × 1-ft height vessel.

Solution:
1-ft length × 1-ft wide = 1 sq ft. or 144 sq in.
1-ft^3 H$_2$O = 62.4 lb H$_2$O
Pressure = 62.4 ÷ 144 = 0.433 psi

Note: Each additional foot of water will add 0.433 psi. A common formula used to figure pressure is   **Height × 0.433 × specific gravity = Pressure.**

Basic Process Principles • Chapter 4

EXAMPLE 3: gasoline   Calculate the pressure produced by one cubic foot of gasoline (0.75 specific gravity [sg]) in a 1-ft long × 1-ft wide × 1-ft high vessel.

Solution: 1-ft length × 1-ft wide × 1-ft height = 1 sq ft. or 144 sq in.
= 62.4 lb H_2O × 0.75 specific gravity
Pressure = 47 ÷ 144 = 0.327 psi

Note: Each additional foot of gasoline will add 0.327 psi.

EXAMPLE 4: water   Calculate the pressure produced by water (62.4 lb) in a 6-ft high vessel.

Solution: 1 sq ft or 144 sq in. = 62.4 lb H_2O
62.4 lb × 6 ft ÷ 144 sq in. = 2.6 psi

Now try: Height × 0.433 × Specific Gravity = pressure
6 foot × 0.433 × 1 = 2.6

EXAMPLE 5: water   Calculate the pressure produced by water (62.4 lb) in a 200-ft high vessel.

Solution: 200 ft × 0.433 × 1 = 86.6 psi

EXAMPLE 6: gas   Calculate the pressure exerted on the bottom of a 20-ft distillation column filled with gasoline. Add 100 psi to the column, giving a top gauge reading of 100 psi. What will be the bottom gauge reading in psi?

Solution: To calculate the bottom pressure of the distillation column, two variables must be considered:
the pressure of the gasoline
20 ft × 0.433 × 0.75 = 6.5 psi
plus the pressure added to the column: 100 psi.

The answer is 6.5 psi + 100 psi = 106.5 psi.

EXAMPLE 7:   Calculate the pressure exerted on a 20 ft column filled with 10 ft of gasoline. The vapor pressure of gasoline at 100° F is 12 psi.

Chapter 4 • *Basic Process Principles*

Solution: 10 ft × 0.433 × 0.75 = 3.25 psi
3.25 + 12 psi = 15.25 psi

The answer is 15.25 psi.

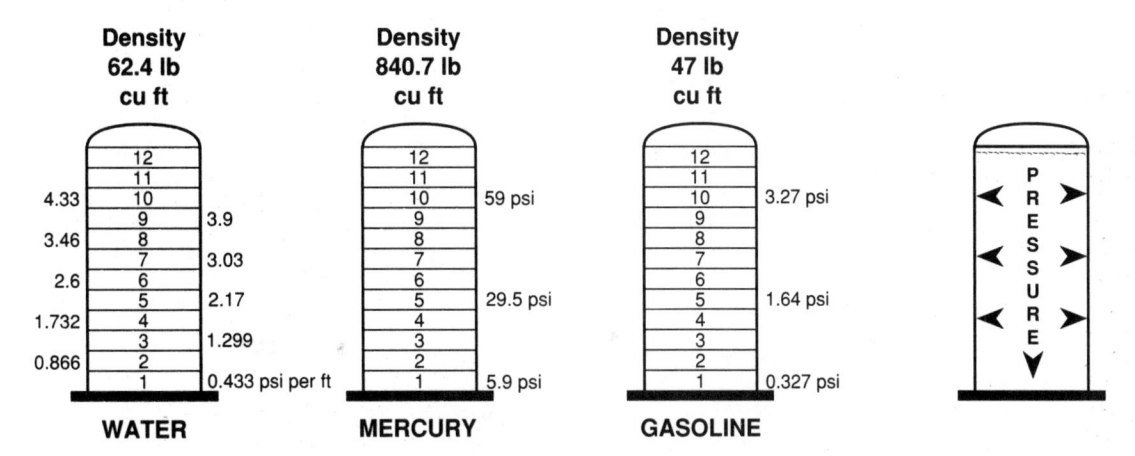

Figure 4.2-2 *Liquid Pressure Principles 1, 2, 3*

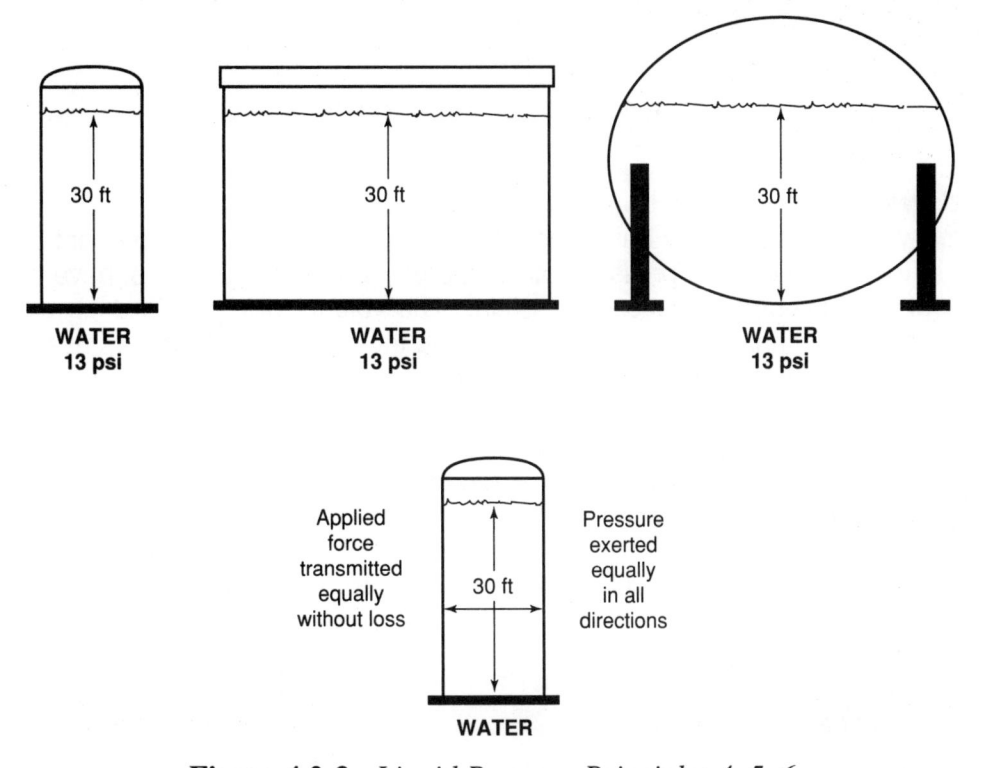

Figure 4.2-3 *Liquid Pressure Principles 4, 5, 6*

67

Basic Process Principles • Chapter 4

Principles of Liquid Pressure
The principles of liquid pressure are: (See Figures 4.2–2 and 4.2–3)
1. Liquid pressure is directly proportional to its density.
2. Liquid pressure is proportional to the height of the liquid.
3. Liquid pressure is exerted in a perpendicular direction on the walls of a vessel.
4. Liquid pressure is exerted equally in all directions.
5. Liquid pressure at the base of a tank is not affected by the size or shape of the tank.
6. Liquid pressure transmits applied force equally, without loss, inside an enclosed container.

Absolute, Vacuum, and Gauge Pressure
Three different types of pressure gauges can be found in industrial environments: absolute (psia), gauge (psig), and vacuum (psiv). Absolute pressure is equal to gauge pressure plus local atmospheric pressure (14.7 psi). Gauge pressure is equal to the absolute pressure minus the local atmospheric pressure (14.7 psi). Vacuum is measured typically in inches of mercury (in. Hg). Any pressure below atmospheric pressure (14.7 psi) is referred to as vacuum.

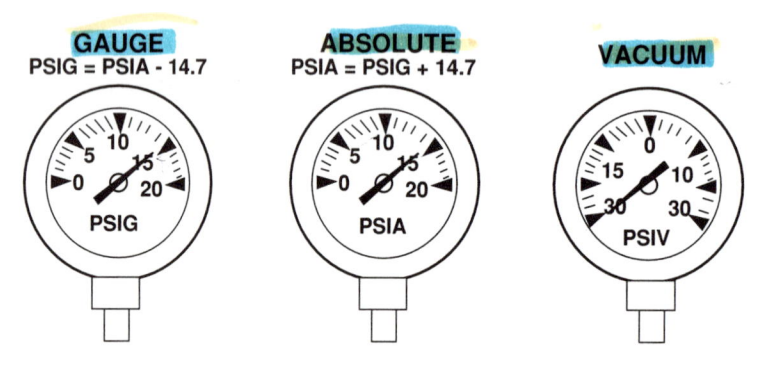

Figure 4.2-4 *PSIA–PSIG–PSIV*

Gases and Pressure
Liquids typically are considered to be noncompressible even though a 10 percent decrease in volume can be observed when a pressure of 65,000 psi is applied. Gases behave much differently than liquids. Gases are very compressible. The volume of a gas is determined by the shape of the vessel containing it, the temperature, and the pressure. Operators use these three factors in the control and storage of gases.

Gas Laws
Dalton's law: $P_{total} = P_1 + P_2 + P_3$ states that the total pressure of a gas mixture is the sum of the pressures of the individual gases.

The Ideal Gas Law ($PV = nRT$) calculates the pressure, temperature, volume, or moles of any gas.
- P = pressure of the gas
- V = volume
- n = moles of gas

T = temperature in Kelvins (K)
R = ideal gas constant (0.08206 L × atm/mol × K)

The combined gas law calculates changes in a gaseous substance from one condition to another.

$$\frac{P_1 V_1}{T_1} = \frac{P_2 V_2}{T_2}$$

4.3 Heat, Heat Transfer, and Temperature

Heat is a form of energy caused by increased molecular activity. A basic principle of heat states that it cannot be created or destroyed, only transferred from one substance to another. Heat energy moves from areas of hot to cold, transferring energy as it goes. This process will continue until the heat energy has been equally distributed. A stone thrown into a still pool of water sends ripples out in all directions. Heat energy moves in a similar pattern.

Heat is measured in energy units called British Thermal Units (Btus). A Btu is the amount of heat needed to raise one pound of water one degree F. Another common unit used in industrial manufacturing is the calorie. One calorie is roughly equal to the heat energy required to raise the temperature of one gram of water one degree Celsius.

The effects of absorbed heat are:
- increase in volume
- increase in temperature
- change of state (solid, liquid, gas)
- chemical change (matches)
- electrical transfer (thermocouple)

Heat comes in a variety of forms:
- **sensible heat**—heat that can be sensed or measured; increase or decrease in temperature.
- **latent heat**—hidden heat that does not cause a temperature change.
- **latent heat of fusion**—heat required to melt a substance; heat removed to freeze a substance.
- **latent heat of vaporization**—heat required to change a liquid to gas.
- **latent heat of condensation**—heat removed to condense a gas.
- **specific heat**—the required Btus needed to raise one pound of a specific substance one degree F.

Heat is transmitted through:
- **conduction**—when heat energy is transferred through a solid object; e.g., a boiler.
- **convection**—requires fluid currents to transfer heat from a heat source; e.g., upper convection section of a furnace.
- **radiation**—the transfer of energy through space by the means of electromagnetic waves; e.g., the sun.

Basic Process Principles ● Chapter 4

Temperature

By measuring the hotness or coldness of a substance, we determine temperature. Process operators use a variety of temperature systems. The four most common systems are described here:

	Water Boils	*Water Freezes*	**Conversion Formulas**
Kelvin (K)	373 K	273 K	K = °C + 273
Celsius (°C)	100° C	0° C	°C = °(F −32) ÷ 1.8
Fahrenheit (°F)	212° F	32° F	°F = 1.8° C + 32
Rankine (°R)	672° R	492° R	°R = °F + 460

Key Points To Remember

- Heat is a form of energy caused by increased molecular activity that cannot be created or destroyed, only transferred from one substance to another.
- The hotness or coldness of a substance determines the temperature.
- Heat is measured in Btus, and temperature is measured by K, C, F, and R.
- Temperature and heat are not the same.

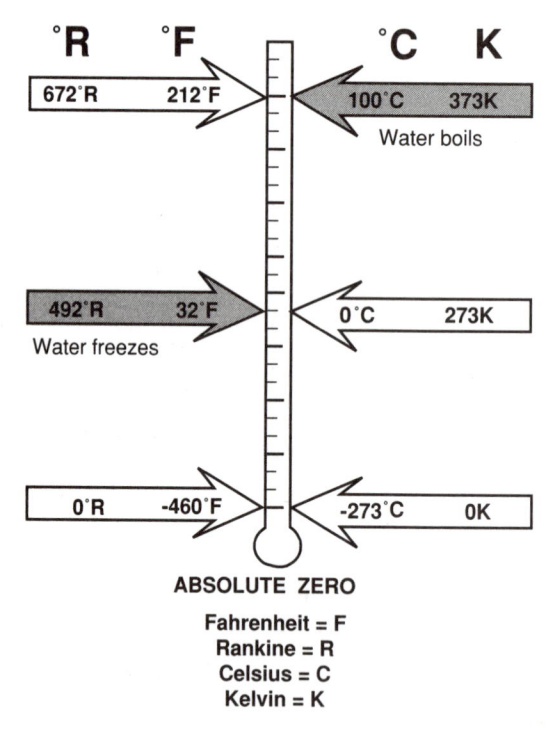

Figure 4.3-1 *Temperature Scales*

4.4 Fluid Flow

Modern industrial process plants are connected by a complex network of pipes, valves, pumps, and tanks. Centrifugal and positive displacement pumps are used to transfer fluids from place

Chapter 4 • **Basic Process Principles**

to place inside and outside the plant. The combination of pumps and pipes closely resembles the way the human heart pumps fluid into arteries and veins.

Fluids assume the shape of the container they occupy. A fluid can be classified as a liquid or a gas. When a liquid is in motion, it remains in motion until it reaches its own level or is stopped. Fluid flow is a critical process used in the day-to-day operation of all plants.

Bernoulli's Principle

The Swiss scientist, Daniel Bernoulli, developed a key scientific principle for fluid flow. Bernoulli's principle stated that in a closed process with a constant flow, rate changes in fluid velocity (kinetic energy) decrease or increase pressure, kinetic energy and pressure energy changes correspond to pipe size changes, pipe diameter changes cause velocity changes, and pressure energy, kinetic energy or "fluid velocity," and pipe diameter changes are related.

Density

Industry uses four different ways to express a fluid's heaviness: density, specific gravity, baume, and API.

The density of a fluid is defined as the mass of a substance per unit volume. Density measurements are used to determine heaviness. For example, one gallon of water weighs 8.33 lb, one gallon of crude oil weighs 7.20 lb, and one gallon of gasoline weighs 6.15 lb.

Viscosity

Another common term used by industry to describe the flow characteristics of a substance is *viscosity*. Viscosity is defined as a fluid's resistance to flow.

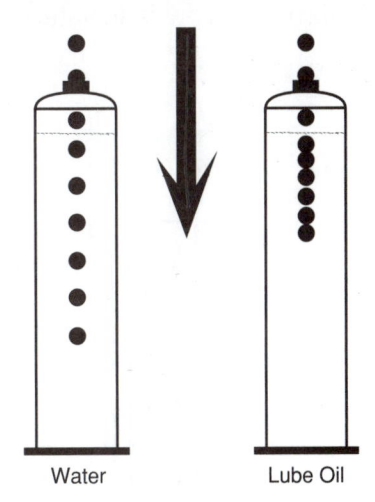

Figure 4.4-1 *Viscosity*

Specific Gravity

Specific gravity (sg) is defined as the ratio of a fluid's density (liquid or gas) to the density of water or air. It is a common mistake for operators to confuse specific gravity with density. This is easy to understand because specific gravity is a method for determining the heaviness of a fluid.

71

Basic Process Principles • Chapter 4

Density is the heaviness of a substance while specific gravity compares this heaviness to a standard and then calculates a new ratio. Most hydrocarbons have a specific gravity below 1.0.

Key points to remember:
- The specific gravity of water is 8.33 lb/gal. ÷ 8.33 = 1.0.
- The specific gravity of gasoline is 6.15 lb/gal. ÷ 8.33 = 0.738.
- The density of one gallon of water is 8.33 lb/gal.
- The density of air is 0.08 lb/cu ft.
- Density is calculated by weighing unit volumes of a fluid at 60 degrees F (15.55 degrees C).

Baume Gravity
Baume gravity is the standard used by industrial manufacturers to measure nonhydrocarbon heaviness.

API Gravity
The American Petroleum Institute (API) applies gravity standards to measure the heaviness of a hydrocarbon. A specially designed hydrometer, marked in units API, is used to determine the heaviness or density of a hydrocarbon. High API readings indicate low fluid gravity.

Turbulent and Laminar Flow
Two major classifications of fluid flow are laminar and turbulent. Laminar flow, or streamline flow, moves through a system in thin cylindrical sheets of liquid flowing inside one another. This type of flow has little, if any, turbulence in it. Laminar flow usually exists at lower flow rates. As flows increase, the laminar flow pattern breaks into turbulent flow. Turbulent flow is the random movement or mixing of fluids. Once the turbulent flow is initiated, molecular activity speeds up until the fluid is uniformly turbulent.

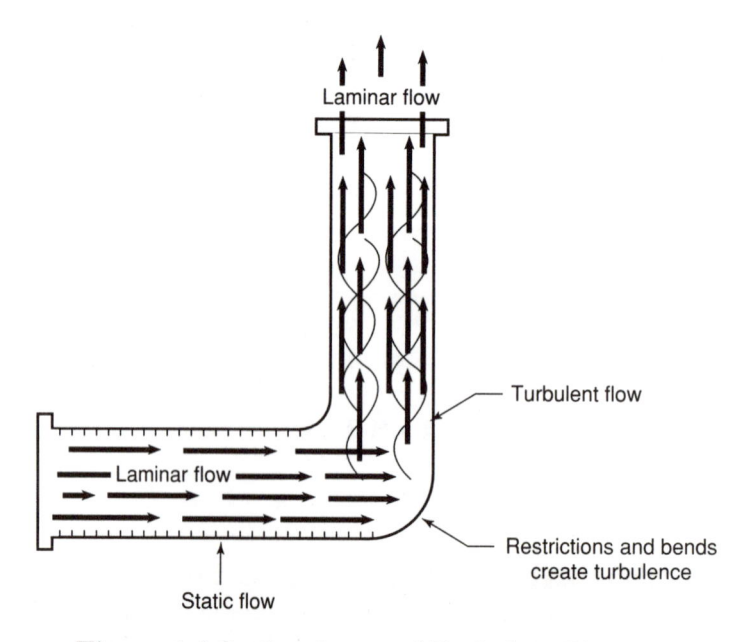

Figure 4.4-2 *Laminar and Turbulent Flow*

Chapter 4 • *Basic Process Principles*

Turbulent flow allows molecules of fluid to mix more readily and absorb heat. Laminar flow promotes the development of static film, which acts as an insulator. Turbulent flow decreases the thickness of static film.

Forms of Liquid Energy
Liquid energy can be in the form of kinetic energy (fluid motion), pressure and potential energy (stored energy, liquid head, internal pressure), or heat energy (fluid friction).

Fluid Energy Conversions
- Steam turbine—steam pressure energy is converted to kinetic energy; kinetic energy is converted to rotational or mechanical energy.
- Boiler—heat energy is transferred to water; water boils, creating steam energy. Steam energy creates pressure energy. Steam and pressure energy are used in distillation, heat exchangers, reactors, laminating, extrusion, and steam turbines.
- Furnace—heat energy is transferred to the charge.
- Distillation Tower—heat energy is transferred to feed, which separates the individual components by boiling point. Condensation and vaporization occur along the temperature gradient of the tower.
- Energy is converted into kinetic energy. As fluid slows, it is converted into pressure energy.

Measuring Flow Rate
Flow rate (gpm—gallons per minute) equals volume per unit of time. Velocity (fps—feet per second, fpm—feet per minute, fph—feet per hour) equals distance per unit of time.

Flow of Solids
A variety of gases are used to transfer solids from one location to another: nitrogen, air, chlorine, and hydrogen. When properly fluidized, solids respond like fluids. Solid transfer requires small, granular, porous solids that respond positively to aeration. Several examples of this procedure are modern plastics manufacturing (granules, powder, flakes), catalytic cracking units, and vacuum systems.

4.5 Basic Math for Process Technicians

Basic mathematics is encountered typically on the pre-employment tests administered by most plants. Simple mathematics functions appear to be the primary disqualifier for potential applicants. The widespread use of calculators and the years since eighth-grade mathematics require most people to review quickly these rusty skills or risk being eliminated from being asked to interview. Process technicians use a variety of mathematical functions to perform their normal job responsibilities. Some of these functions include:

Phase 1: Pre-Employment Skills Required

Skills required.
- addition
- subtraction
- multiplication

73

Basic Process Principles • Chapter 4

- division
- fractions (addition, subtraction, multiplication, and division)
- decimals (addition, subtraction, multiplication, and division)
- percentage
- averaging
- mechanical aptitude
- equations (algebraic expression)
- canceling
- ratio
- proportion, direct and inverse
- constants and variables
- factors and factoring
- exponents
- grouping

Phase 2: On the Job

Process technicians frequently use the following scientific functions to perform their jobs.
- area
- volume
- volumetric flow rate
- X-Y graphs
- bar graphs
- pie graphs
- strip charts
- trends
- word problems
- pressure in fluids: Force = Area × Height × Density
  Pressure = Force ÷ Area; Pressure = Height × Density
- Specific Weight of Liquid = Weight of Liquid ÷ Weight of Water
- Work, force, and distance: W = Force × Density,
  Mechanical Advantage = Resistance + Effort
- Levers
- Boyle's law: $P_1 V_1 = P_2 V_2$
- motion of bodies: $v = s \div t$, $s = vt$
- heat transfer

Phase 1: Pre-employment

Mathematics is an important part of operating a process unit. Flow rates need to be calculated, filling ratios checked, conversion tables used, additive recipes blended, and special equations applied to industrial processes.

1.   1,545
      + 2,000
          3,545

Chapter 4 • Basic Process Principles

2.   1,245
    −   456
         789

3. 8,768 ÷ 234 = 37.47

4. Calculate the mean average of the following numbers:

    125,678
      2,345
        234
      1,429

STEP 1
  125,678
    2,345
      234
+ 1,429
  129,686

STEP 2
129686 ÷ 4 = 32,421.5

5. 467,897 × 34 = 15,908,498

6. 0.4568 × 9,457 = 4,319.96

7. Convert $\frac{39}{19}$ to a mixed number.

    Divide 39 by 19, and the answer is $2\frac{1}{19}$.

8. Convert $1\ ^4/_8$ to a fraction.

STEP 1
1 × 8 + 4 = the numerator
      12 = the numerator

STEP 2
Put 12 over eight. The answer is $\frac{12}{8}$.

9. $\frac{4}{7} + \frac{9}{8}$

Basic Process Principles • Chapter 4

STEP 1
When adding or subtracting fractions, first find the lowest common denominator, (LCD).

Find a number that both 7 and 8 can divide into:
$7 \times 8 = 56$

STEP 2
Write equivalent fractions with a common denominator.
$$\frac{4}{7} = \frac{32}{56}$$

$$\frac{9}{8} = \frac{63}{56}$$

STEP 3
Add the numerators.
$$\frac{32}{56} + \frac{63}{56} = \frac{95}{56}$$

STEP 4
Convert to a mixed number.
$$\frac{95}{56} = 1\frac{39}{56}$$

10. $\dfrac{9}{2} \div \dfrac{9}{4}$

STEP 1
When dividing fractions, invert the divisor.
$\dfrac{9}{4}$ inverted is $\dfrac{4}{9}$.

STEP 2
Multiply.
$$\frac{9}{2} \times \frac{4}{9} = \frac{36}{18} = 2$$

STEP 3
Convert to a whole number.
$$\frac{36}{18} = 2$$

11. $\dfrac{18}{12} \times \dfrac{34}{2} = \dfrac{612}{24} = \dfrac{51}{2} = 25\dfrac{1}{2}$

12. $\dfrac{4}{3} - \dfrac{3}{9}$

STEP 1
Write equivalent fractions with a common denominator.

STEP 2
Subtract the numerators and convert.

76

Chapter 4 • *Basic Process Principles*

$$\frac{4}{3} = \frac{12}{9} \qquad\qquad \frac{12}{9} - \frac{3}{9} = \frac{9}{9}$$

$$\frac{3}{9} = \frac{3}{9} \qquad\qquad \frac{9}{9} = 1$$

13. $123.678 + 0.0043 = 123.68$

14. $454.67 \div 12.34 = 36.85$

15. A tank has a 1,400-lb mixture of water and salt in it. Of the mixture, 18 percent is salt. How many pounds of salt are in it?

    $1,400 \times 0.18 = 252$ lb of salt

16. Product Tank 1403 has a total capacity of 400,000 gal. At 1:00 AM Tk-1403 has 60,000 gal in it. Your product pump is filling the tank at 2.2 gal /minute. How many hours do you have before the tank runs over?

STEP 1	**STEP 2**	**STEP 3**
400,000	2.2	$340,000 \div 132 = 2,575.75$ h

17. $(10^2)^2 = (10 \times 10 = 100)^2$, $\quad 100 \times 100 = 10,000$

18. Convert 0.45 to a percentage. The answer is 45%.

19. Convert 115% to a decimal. The answer is 1.15.

ALGEBRA

Algebra is used to solve many simple problems encountered by process technicians. Basic mathematics is useful but inadequate when dealing with all process problems. Algebra uses letters and symbols to represent variables that are known and unknown. This form of mathematics allows unknown variables to be identified by following well defined principles.

PRINCIPLE 1

An algebraic equation is structured like a balance scale. The products on the left equal the products on the right.
For example: $\quad 6x = 30$ or $6 (?) = 30$
   Solution: $\quad x = 5$

PRINCIPLE 2

When solving for unknowns, the opposite function must be used. Addition and subtraction are opposite, and multiplication and division are opposites.
Solve for *x*:

$$x + 5 = 2$$
$$x + 5 - 5 = 2 - 5 \text{ (The opposite of addition is subtraction)}$$
$$x = -3$$

Basic Process Principles • Chapter 4

20. Solve for x:
$$4x = 20$$
$$\frac{4x}{4} = \frac{20}{4}$$
$$x = 5$$

21. Solve for x:
$$x - 10 = 14 - 3$$
$$x - 10 + 10 = 11 + 10$$
$$x = 21$$

22. Solve for x:
$$2x = 8 + x$$
$$2x - x = 8 + x - x$$
$$x = 8$$

23. Solve for x:
$$x - 2 = 8$$
$$x - 2 + 2 = 8 + 2$$
$$x = 10$$

24. Solve for x:
$$x + 14 = 16$$
$$x + 14 - 14 = 16 - 14$$
$$x = 2$$

25. Solve for x:
$$3 + 6(2 + x) = 45$$
$$3 + 12 + 6x = 45$$
$$15 + 6x = 45$$
$$15 - 15 + 6x = 45 - 15$$
$$6x = 30$$
$$\frac{6x}{6} = \frac{30}{6}$$
$$x = 5$$

Phase 3: On the Job

26. A rectangular tank is 30 ft long, 16 ft tall, and 6 ft wide. What is the volume?
$V = LWH$
$30 \times 16 \times 6 = 2,880$

27. A vertical tank is 30 ft tall with a diameter of 10 ft. Product level is 15 ft. What is the volume of the product?
$V = \pi r^2 h$

$V = 3.1416 \times 5^2 \times 15 \text{ ft}$
$V = 1,178.1$

28. The product level in drum 1201 was 950 cubic feet at 4 AM. At 8 AM, D-1201 has 1950 cu ft of product. No fluid was removed from the drum.

    Calculate the flow rate into D-1201.

    $$V_{in} = \frac{V_f - V_i}{t} \qquad \frac{1950 - 950}{4} = 250$$

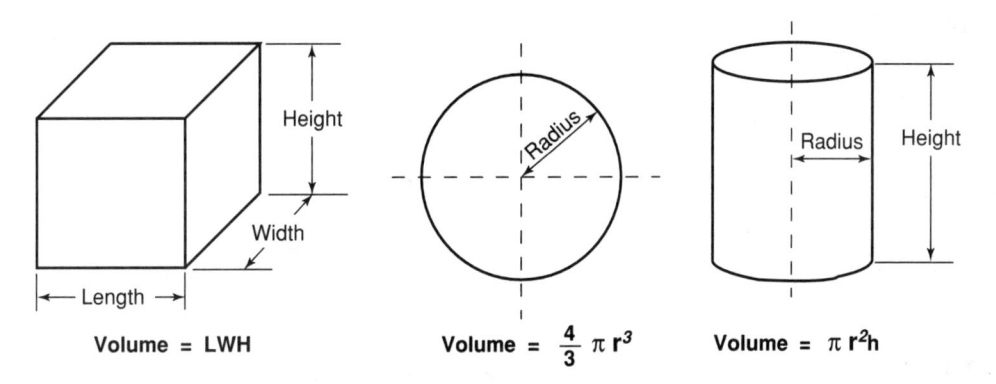

Volume = LWH

Volume = $\frac{4}{3} \pi r^3$

Volume = $\pi r^2 h$

Figure 4.5-1 *Volume Formulas*

Summary

Pressure is defined as force or weight per unit area. (Force ÷ Area = Pressure). Pressure is typically applied to gases or liquids and is measured in pounds per square inch. Produced by the weight of the atmosphere as it presses down on an object resting on the surface of the earth, pressure is directly proportional to height; the higher the atmosphere, gas or liquid, the greater the pressure. Pressure equals 14.7 pounds per square inch at sea level (atmospheric pressure).

The boiling point of a substance is the temperature at which the vapor pressure exceeds atmospheric pressure, bubbles become visible in the liquid, and vaporization begins.

Vapor pressure is the weight of a liquid's vapor. The vapor pressure of a substance can be directly linked to the strength of the molecular bonds of a substance. The stronger the bonds or molecular attraction, the lower the vapor pressure. If a substance has a low vapor pressure, it will have a high boiling point.

As the pressure increases, the boiling point increases and the escape of molecules from the surface of the liquid is reduced proportionally. The gas or vapor molecules are forced closer together. The vapor phase above a liquid could be forced back into solution.

Basic Process Principles • Chapter 4

Atmospheric pressure is 14.7 pounds per square inch, so any pressure below this is referred to as a vacuum. Vacuum affects the boiling point of a substance in the opposite way that positive pressure does. It lowers the boiling point of a substance, enhances molecular escape of liquid, and reduces energy costs, molecular damage due to overheating, and equipment damage.

Robert Boyle was an Irish scientist who developed the law that describes how the volume of a gas responds to pressure changes. The basic principles of Boyle's law are that pressure decreases volume and moves gas molecules closer together, the higher the pressure the smaller the volume, and gas volume decreases by one-half when pressure doubles.

Pascal's law states that pressure in a fluid is transmitted equally in all directions. Its basic principles are that pressure in a fluid is transmitted equally in all directions, molecules in liquids move freely, and molecules are close together in a liquid. The pressure a liquid exerts on a container is determined by the height and the weight of the fluid.
(Height $\times 0.433 \times$ specific gravity = pressure)

The principles of liquid pressure are:
- Liquid pressure is directly proportional to its density.
- Liquid pressure is proportional to the height of the liquid.
- Liquid pressure is exerted in a perpendicular direction on the walls of a vessel.
- Liquid pressure is exerted equally in all directions.
- Liquid pressure at the base of a tank is not affected by the size or shape of the tank.
- Liquid pressure transmits applied force equally, without loss, inside an enclosed container.

Three different types of pressure gauges can be found in industrial environments: *absolute* (psia), *gauge* (psig), and *vacuum* (psiv). Absolute pressure is equal to gauge pressure plus local atmospheric 14.7. Gauge pressure is equal to the absolute pressure minus the local atmospheric pressure 14.7. Vacuum is measured typically in inches of mercury. Any pressure below atmospheric pressure 14.7 is referred to as vacuum.

Liquids are considered to be noncompressible. Gases are described as very compressible. Dalton's law ($P_{total} = P_1 + P_2 + P_3$) . . . states that the total pressure of a gas mixture is the sum of the pressures of the individual gases.

Heat is a form of energy caused by increased molecular activity. A basic principle of heat states that it cannot be created or destroyed, only transferred from one substance to another. Heat energy moves from areas of hot to cold, transferring energy as it goes.

Heat is measured in energy units called British thermal units (Btus) A Btu is the amount of heat needed to raise one pound of water one degree.

The effects of absorbed heat are:
- increase in volume
- increase in temperature
- change of state (solid, liquid, or gas)
- chemical change (matches)
- electrical transfer (thermocouple)

Heat comes in a variety of forms. Sensible heat can be sensed or measured. Temperature can be increased or decreased. Latent heat is hidden heat. It does not cause a temperature change. Latent heat of fusion is required to melt a substance. Heat is removed to freeze a substance. Latent heat of vaporization is required to change a liquid to gas. Latent heat of condensation is required to condense a gas. Specific heat is the required BTUs needed to raise one pound of a specific substance one degree F.

Heat is transmitted through conduction (when heat energy is transferred through a solid object, e.g., a boiler), convection (requires fluid currents to transfer heat from a heat source, e.g., the upper convection section of a furnace), and radiation (the transfer of energy through space by the means of electromagnetic waves, e.g., the sun).

By measuring the hotness or coldness of a substance, we determine temperature. Process operators use a variety of temperature systems. The four most common are Kelvin (K), Celsius (C), Fahrenheit (F), and Rankine (R).

Temperature conversion formulas are available to be used by process technicians.

A fluid can be classified as a liquid or a gas. When a liquid is in motion, it will remain in motion until it reaches its own level or is stopped.

Bernoulli's principle states that in a closed process with a constant flow rate, changes in fluid velocity (kinetic energy) decrease or increase pressure. Kinetic energy and pressure energy changes correspond to pipe size changes. Pipe diameter changes cause velocity changes. Pressure energy, kinetic energy or "fluid velocity," and pipe diameter changes are related.

Process technicians use four different ways to express a fluid's heaviness: density, specific gravity, baume, and API.

The density of a fluid is defined as the mass of a substance per unit volume. Density measurements are used to determine heaviness.

Another common term used by process technicians to describe the flow characteristics of a substance is viscosity. Viscosity is defined as a fluid's resistance to flow.

Specific gravity is defined as the ratio of a fluid density (liquid or gas) to the density of water or air. It is a common mistake for operators to confuse specific gravity with density. This is easy to understand because specific gravity is a method for determining the heaviness of a fluid. Density is the heaviness of a substance whereas specific gravity compares this heaviness to a standard and then calculates a new ratio. Most hydrocarbons have a specific gravity below 1.0.

Baume gravity is the standard used by industrial manufacturers to measure nonhydrocarbon heaviness.

The American Petroleum Institute applies API gravity standards to measure the heaviness of a hydrocarbon. A specially designed hydrometer, marked in units API, is used to determine the heaviness or density of a hydrocarbon. High API readings indicate low fluid gravity. Two major classifications of fluid flow are laminar and turbulent. Laminar flow, or streamline flow, moves through a system in thin cylindrical sheets of liquid flowing inside one another. Turbulent flow is the random movement or mixing of fluids. Turbulent flow allows molecules of fluid to mix more readily and absorb heat. Laminar flow promotes the development of static film, which acts as an insulator. Turbulent flow decreases the thickness of static film.

Industrial forms of liquid energy include kinetic energy (fluid motion), pressure and potential energy (stored energy, liquid head, internal pressure), and heat energy (fluid friction).

Process technicians use a variety of mathematical functions to perform their normal job responsibilities.

Chapter 4

Review Questions

1. Bernoulli's principle stated that in a closed process with a constant flow rate
    a. changes in fluid velocity (kinetic energy) decrease or increase pressure.
    b. kinetic energy and pressure energy changes correspond to pipe size changes.
    c. pipe diameter changes cause velocity changes.
    (d.) all of the above.

2. As the pressure increases inside a confined space
    a. the boiling point increases.
    b. the escape of molecules from the surface of the liquid is increased proportionally.
    c. the gas or vapor molecules are forced closer together.
    (d.) a and c.

3. Solve for y: $62 = 13y - 3$

4. Solve for x: $2x = 9$

$$62 = 13y - 3$$
$$+3 \qquad +3$$
$$\frac{65}{13} = \frac{13y}{13} \qquad y = \frac{65}{13}$$

$$\frac{2x}{2} = \frac{9}{2} \qquad x = \frac{9}{2}$$

5. Pressure is directly proportional to
    (a.) height.
    b. sound.
    c. specific gravity.
    d. mathematics.

6. Atmospheric pressure is
    a. 14.3 psi.
    b. 14.5 psi.
    (c.) 14.7 psi.
    d. 15.7 psi.

7. True or (False?) Heat and temperature are basically the same thing.

8. An example of fluid flow is
    (a.) turbulent.
    b. gravity.
    c. kinetic.
    d. potential.

Basic Process Principles • Chapter 4

9. Boyle's law describes how
    a. the volume of a gas responds to pressure changes.
    b. pressure in a fluid is transmitted equally in all directions.
    c. the volume of a liquid responds to pressure changes.
    d. kinetic energy and pressure energy changes correspond to pipe size changes.

10. True or False? Liquids do not need to reach their boiling points to begin the process of evaporation.

11. Calculate the pressure produced by a 2000-lb stone block,
    12-in. length × 12-in. width × 12-in. height
    Pressure = Force (weight) ÷ Area

12. Calculate the pressure exerted on a 26-ft column filled with 13 ft of gasoline. The vapor pressure of gasoline at 100° F is 12 psi.

```
   12
  ×12
  ---
   24
   24
  ---
  264
```

84

Chapter 5

Equipment One

OBJECTIVES

After studying this chapter, the student will be able to:

- *Describe the basic hand tools used in industry.*
- *Identify and describe the valves used in industry.*
- *Describe the various types of storage and piping used in the chemical processing industry.*
- *Identify the operation and primary components of a centrifugal pump.*
- *Explain the operation and types of positive displacement pumps.*
- *Describe dynamic and positive displacement compressors.*
- *Describe how a steam turbine works.*
- *Describe the purpose of seals, bearings, and lubrication.*

KEY TERMS

Basic hand tools—a term used to describe the typical tools that process technicians use to perform their job activities.

Compressors—come in two basic designs, (1) positive displacement (rotary and reciprocating) and (2) dynamic (axial and centrifugal). A compressor is designed to accelerate or compress gases.

Filters—remove solids from fluids.

Process Instruments—control processes and provide information about pressure, temperature, levels, flow and analytical variables.

Pumps—used to move liquids from one place to another. Pumps come in two basic designs, (1) positive displacement (rotary and reciprocating) and (2) centrifugal.

Steam turbines—used as drivers to turn pumps, compressors, and electric generators. A steam turbine is an energy conversion device that converts steam energy (kinetic energy) to useful mechanical energy.

Steam traps—remove condensate from steam systems.

Strainer—a device used to remove solids from a process before they can enter a pump and damage it.

Tanks and pipes—store and contain fluids. Tank designs include spherical, open top, floating roofs, drums, and closed tanks.

Chapter 5 • *Equipment One*

5.2 Basic Hand Tools

Basic hand tools are the typical tools that process technicians use to perform their job activities, Figure 5.2–1. Union plants may have limitations on the type of work a process technician may perform. In these types of plants the process technician may not be allowed to cross crafts and use hand tools except on a limited basis. In non-union plants hand tool usage is limited to a minor role since skilled craftspersons are available for complex jobs. Process technicians are required to perform routine maintenance on their unit since most mechanical craftspersons work the day shift and leave the evening and night shift open for call outs. When a call out is required the company typically pays time and a half. Besides the money issue it takes time for the maintenance staff to return to the work site. Because of these conditions companies require routine maintenance on the off shift to be handled by the process technician. In some cases a little minor maintenance can prevent major equipment damage.

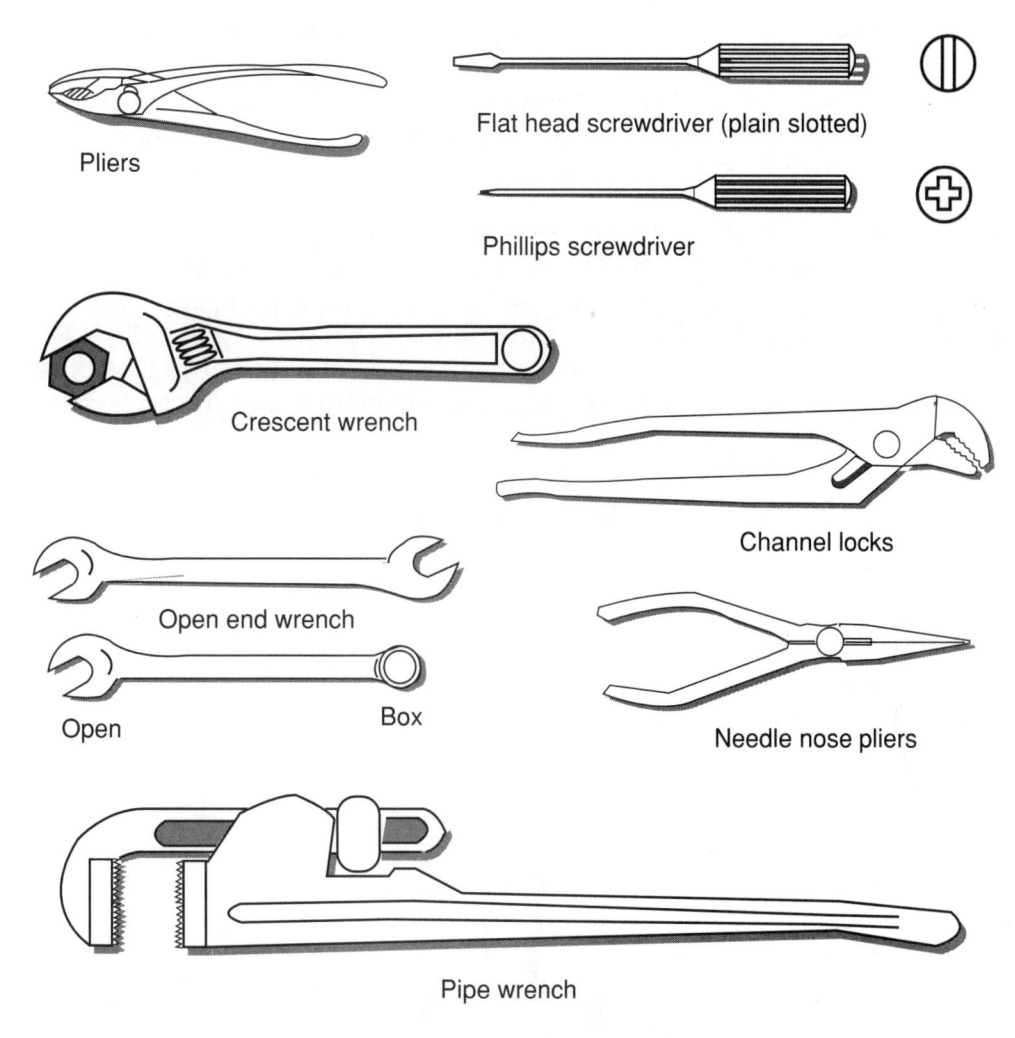

Figure 5.2-1 *Basic hand tools*

Here is a list of some basic hand tools:

Pliers	Wire cutters	Needle nose pliers
Channel locks	Vice grips	Phillips screwdriver
Flat head screwdriver	Pipe wrench	Crescent wrench
Ratchet and socket sets	Hammer	Utility knife
Chisels	File	Wire brush
Hack saw	Level	
Allen wrenches	Wrenches—metric and English, open, box, combination	

5.3 Valves

Gate Valves

[handwritten: generally on-off service]

A gate valve places a movable metal gate in the path of a process flow in a pipe line. Gate valves come in two designs: (1) rising stem and (2) non-rising stem. Located at the top of a closed gate valve is the hand wheel. The hand wheel is attached to a threaded stem. As the hand wheel is turned counterclockwise the stem in the center of the hand wheel begins to rise. This lifts the gate out of the valve body and allows product to flow. Another type of rising stem valve is threaded at the bottom of the stem. In this type of valve the hand wheel is firmly attached to the stem and rises with it as the valve is opened.

A non-rising stem gate valve has a collar that keeps the stem from moving up or down. The hand wheel is firmly attached to the stem of a non-rising gate. Turning the hand wheel screws the stem into or out of the gate. The basic components of a gate valve are illustrated in Figure 5.3–1.

Globe Valves

[handwritten: used for throttling   most often automated]

A globe valve places a movable metal disc in the path of a process flow. This type of valve is the most common used for throttling service. The disc is designed to fit snugly into the seat and stop

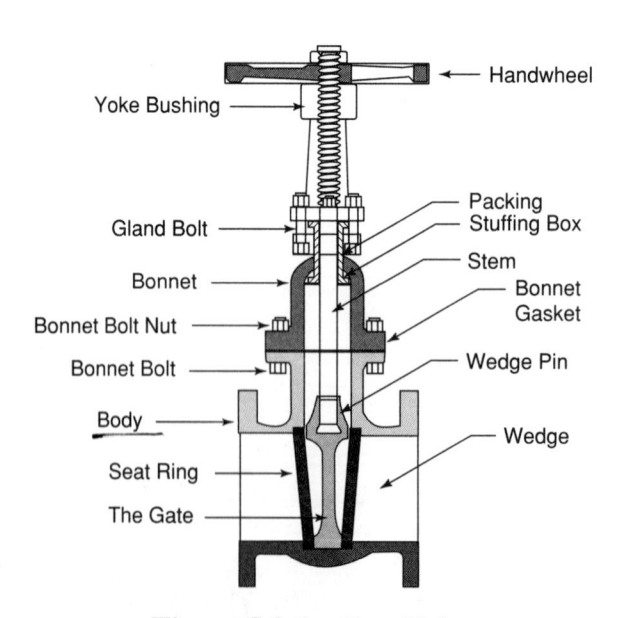

Figure 5.3-1 *Gate Valve*

Figure 5.3-2 *Globe Valve*

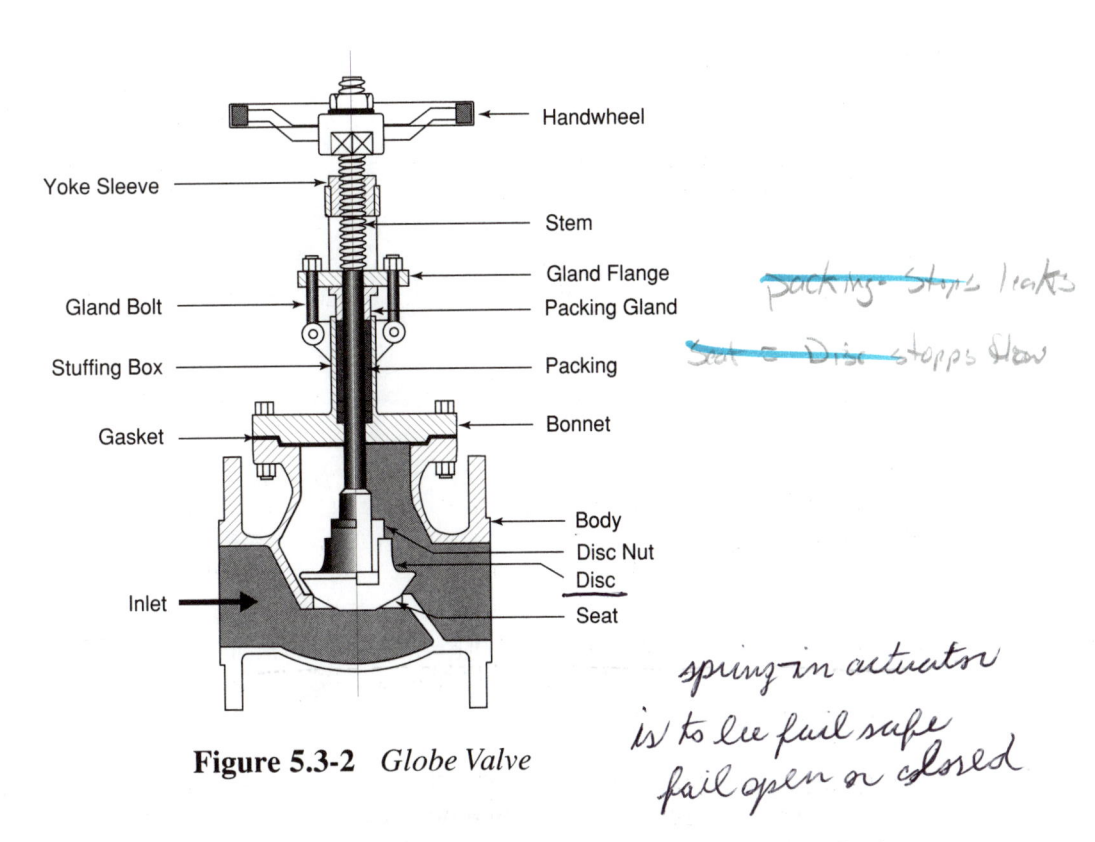

flow. Process fluid enters the globe valve and is directed through a 90 degree turn to the bottom of the seat and disc. As the fluid passes by the disc it is evenly dispersed.

Globe valves are designed to be installed in high use areas. If a globe valve is installed in a low use area it tends to plug up even though it has a self-cleaning type design. Globe valves can be found in the following designs: typical globe valve with ball, plug, or composition element; needle valve; and angle valve. Globe valves and gate valves have very similar components as illustrated in Figure 5.3–2.

Ball Valves
Ball valves, Figure 5.3–3, take their name from the ball shaped, movable element in the center of the valve. Unlike the gate and globe valve, a ball valve does not lift the flow control device out of the process stream; instead, the hollow ball rotates into the open or closed position. Ball valves provide very little restriction to flow and can be opened 100% with a quarter turn on the valve handle. In the closed position the port is turned away from the process flow. In the open position the port lines up perfectly with the inner diameter of the pipe. Larger valves require hand wheels and gearboxes to be opened but most only require a quarter turn on a handle.

Plug Valves
Quick opening, one-quarter turn plug valves are very popular in the manufacturing industry. The plug valve takes its name from a plug shaped flow control element it uses to regulate flow. Plug valves provide very little restriction to flow, and can be opened 100% with a quarter turn on the valve handle. In the closed position the port is turned away from the process flow. In the open position the port lines up perfectly with the inner diameter of the pipe.

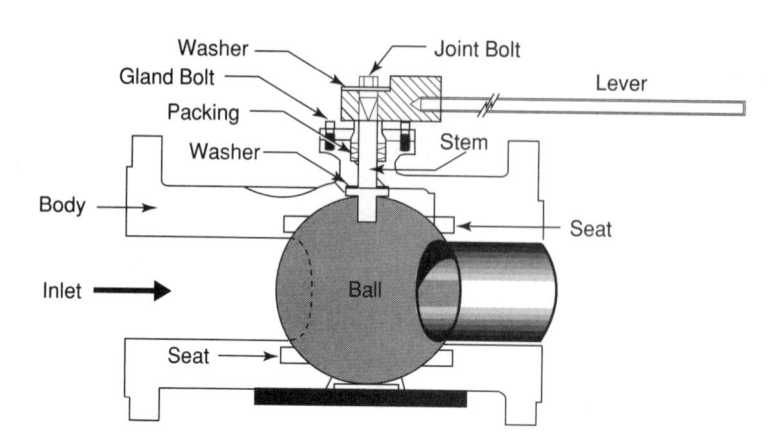

Figure 5.3-3 *Ball Valve*

Check Valves

A check valve, Figure 5.3–4, is a type of automatic valve designed to control flow direction and prevent possible contamination or damage to equipment. The check valve will prevent back flow as long as the device is operating properly. Check valves come in a variety of designs and applications. Typical designs include:

Swing check—this type of valve has a hinged disk that slams shut when flow reverses. Flow lifts the disc and keeps it lifted until flow stops or reverses. The body of the check valve has a cap for easy access to the flow control element.

Lift check—has a disc that rests on the seat when flow is idle and lifts when flow is active. Special guides keep the disc in place. Like the swing check it is designed to close when flow reverses. Lift checks are ideal for systems where flow rates fluctuate. The lift check is more durable than a swing check.

Ball check—has a ball-shaped disc that rests on a beveled, round seat. The ball is down when flow is idle and up when flow is active. Special guides keep the ball disc in place. Like the swing check it is designed to close when flow reverses. Ball checks are ideal for systems where flow rates fluctuate. The ball check is as durable as a lift check and more durable than a swing check.

Stop check—has design characteristics of a lift check and a globe valve. In the closed position the stop check disc is firmly seated. In the open position the stem rises out of the body of the flow control element and acts like a guide for the disc. In the open position the stop check functions like a lift check with one exception: the degree of lift can be controlled.

Butterfly Valve

A common valve used for throttling and on-off service is a butterfly valve. The body of this type of valve is relatively small when compared with other valves and therefore occupies much less space in a pipeline. The flow control element closely resembles a well-worn catcher's mitt. A metal shaft extends through the center of the mitt and allows the disc to rotate one quarter turn. A quarter turn is all it takes to open the valve 100%.

90° turn or quarter turn

Chapter 5 • *Equipment One*

BALL CHECK VALVE

LIFT CHECK VALVE

SWING CHECK VALVE

STOP CHECK VALVE

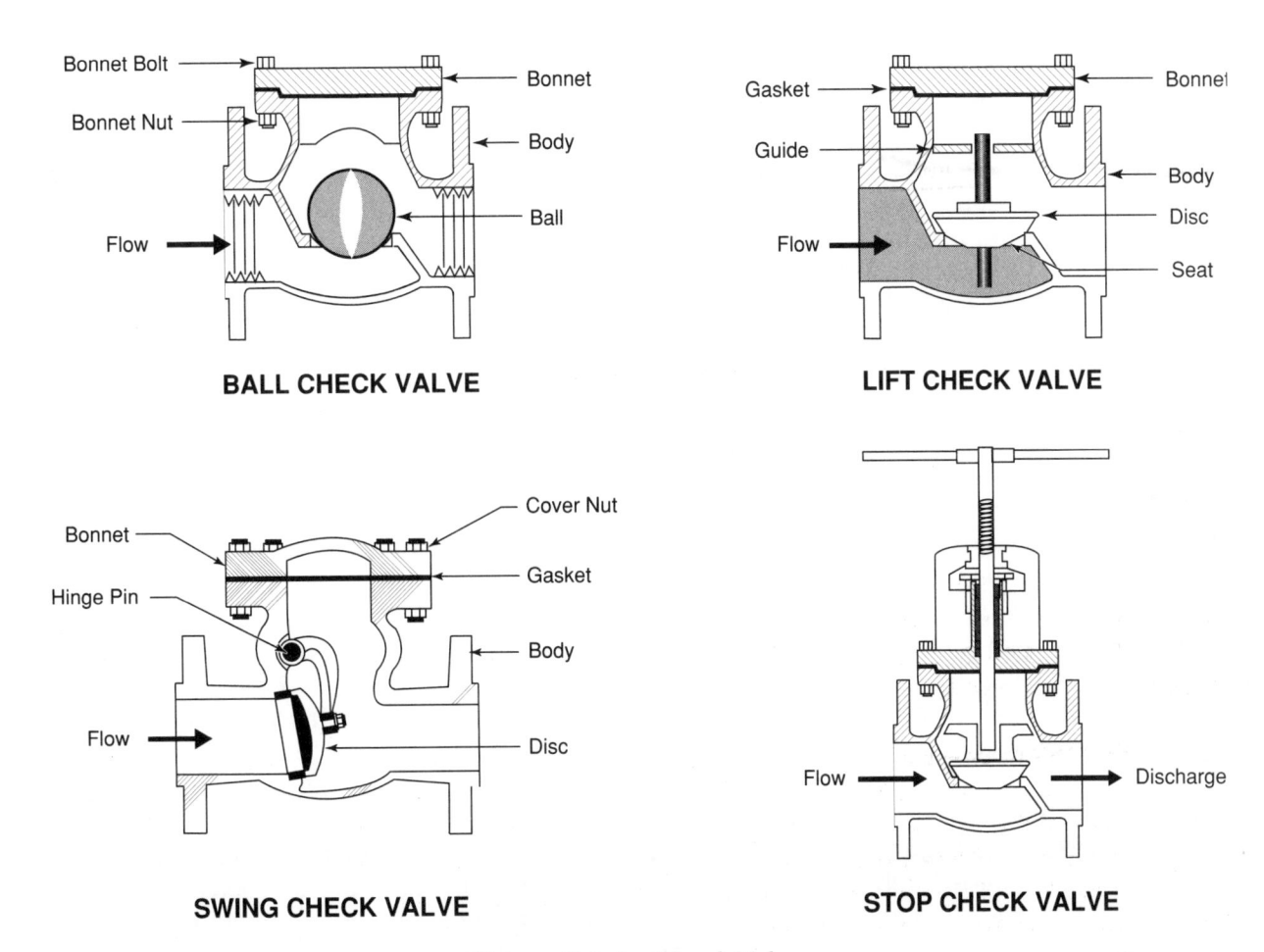

Figure 5.3-4 *Check Valves*

Diaphragm Valve

In a chemical plant a variety of corrosive or sticky substances are transferred from place to place. Standard valves would have a difficult time with this type of product but diaphragm valves are specifically designed for the job. Diaphragm valves use a flexible diaphragm and seat to regulate flow. The hand wheel operates just like a gate or globe valve. The stem is attached to the center of the flexible diaphragm. The diaphragm rests on the seat. The internal parts of the valve never come in contact with the process. The diaphragm forms a seal and holds the seal until the process pressure overcomes the control pressure. Diaphragm valves are typically used in low pressure applications. Diaphragm valves come in two designs:

- Weir diaphragm valve—has a weir located in the body of the valve. Flow must go over the top of the weir and lift the compressor to exit. There is a large pressure drop across the body of the valve. This valve uses thicker, more durable diaphragm material.
- Straight-through diaphragm valve—flexible diaphragm extends across pipe. There is very little pressure drop across this type of valve.

handles corrosive

91

Diaphragm valves handle corrosive fluids, have good throttling capability, and are used in low pressure applications. These valves are used in operations that have moderate temperature and pressure fluctuations.

Relief Valves
Relief valves are designed to automatically respond to sudden increases in pressure. A relief valve opens at a predetermined pressure. In a relief valve a disc is held in place by a spring that will not open until system pressure exceeds its operating limits. Tremendous pressures can be generated in process units. When a system overpressures, safety valves respond to allow excess pressure to be vented to the flare header or atmosphere. This prevents damage to equipment and personnel. Relief valves are designed for pressurized liquid service. They do not respond well in gas service. Relief valves are designed to open slowly. This is a poor feature for gas service.

Safety Valve
Safety valves are considered to be a process system's last line of defense. They are designed to respond quickly to excess vapor or gas pressures. This type of valve is very similar in design to a relief valve. The three major differences between a relief valve and a safety valve are (1) liquid versus gas service, (2) the pressure response time, and (3) a larger exhaust port. Relief valves are designed to lift slowly, while safety valves tend to pop-off. Because the exhaust port is much larger in a safety valve it can release more flow at much lower velocities. This keeps the trim from being damaged. Figure 5.3–5 is an illustration of a safety valve.

Automatic Valves
The chemical processing industry uses a complex network of automated systems to control its processes. The smallest unit in this network is called a control loop. Control loops usually have (1) a sensing device, (2) a transmitter, (3) a controller, (4) transducer, and (5) an automatic valve. Automatic valves, Figure 5.3–6, can be controlled from remote locations making them invaluable in modern processing. Any of the valves studied in this chapter can be automated. To automate

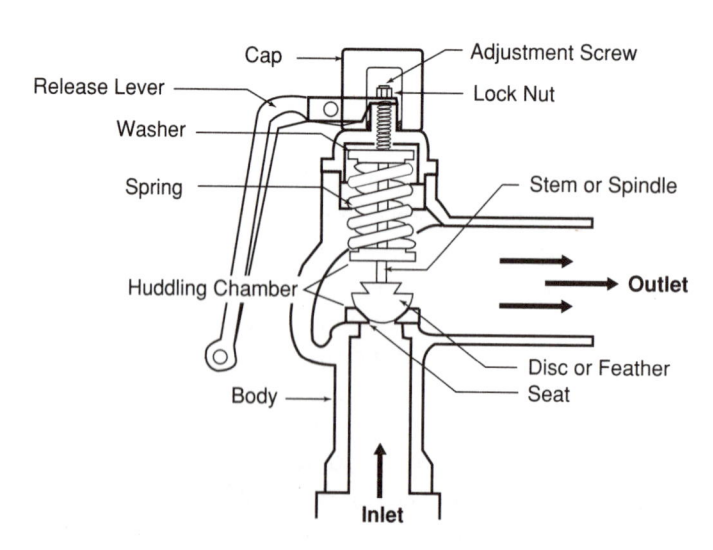

Figure 5.3-5 *Safety Valve*

Chapter 5 • *Equipment One*

Figure 5.3-6 *Automatic Valves*

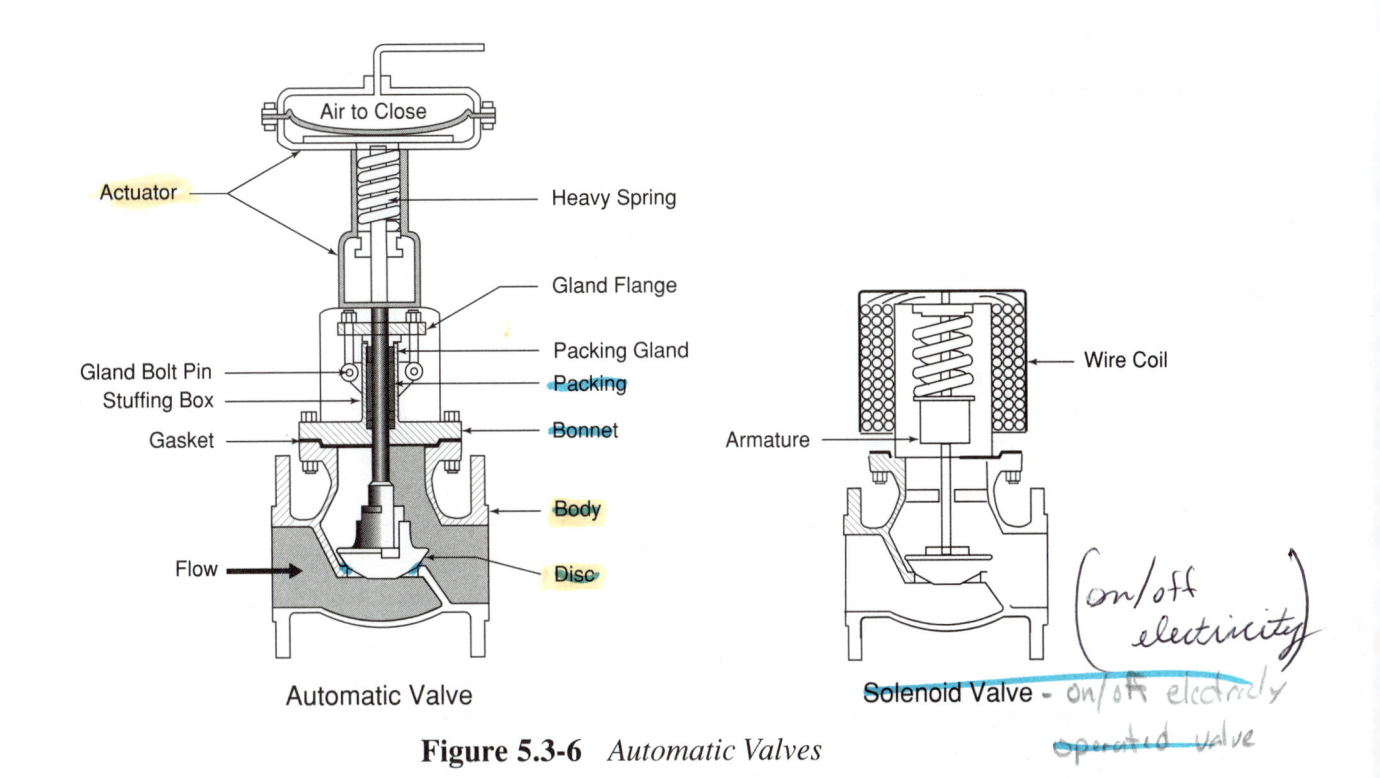

a valve, a device known as an actuator is installed. The actuator controls the position of the flow control element by moving and controlling the position of the valve stem. Actuators can be classified as pneumatic, hydraulic, or electric.

5.4 Piping and Storage Tanks

Industrial piping is composed of a variety of shapes, designs, and metals to safely contain and transport chemicals. The engineering design team carefully selects the type of materials that are compatible with the chemicals and operational conditions. Piping can be composed of stainless steel, carbon steel, iron, plastic, or specialty metals. Individual joints can be threaded on each end, flanged, welded, or glued.

To connect the piping a wide array of fittings are used. The various types of fittings include couplings, unions, elbows, tees, nipples, plugs, caps, and bushings. Figure 5.4–1 illustrates the various types of fittings and piping.

The chemical processing industry uses a variety of tanks, drums, bins, and spheres to store chemicals. The most popular designs are illustrated in Figure 5.4–2. The materials used in these designs include carbon steel, stainless steel, iron, specialty metals, and plastic. Each vessel includes a code stamp that indicates high pressure and temperature ratings, manufacturer, date, type of metal, storage capacity, and special precautions. Most vessels include strapping tables that allow a technician access to data that can be used to identify capacity.

93

Equipment One • Chapter 5

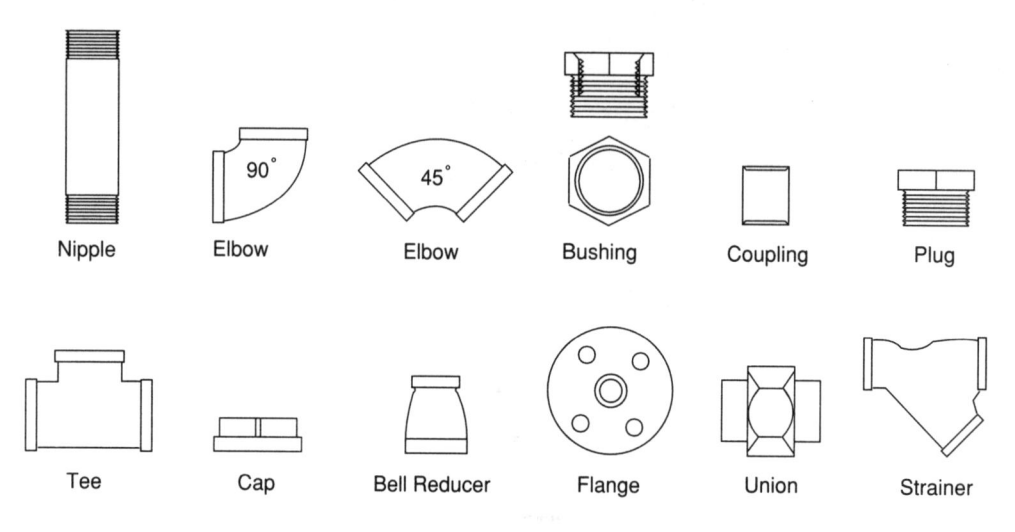

Figure 5.4-1 *Pipe Fittings*

Aboveground storage vessels that have pressures greater than 15 psig are governed by the ASME Code, Section V111. Common storage designs include spheres, spheroids, horizontal cylindrical tanks (drums), bins, and fixed and floating roofs. Tanks, drums, and vessels are typically classified as low pressure, high pressure, liquid service, gas service, insulated, steam traced, or water cooled. Wall thickness and shape often determine the service for a stationary vessel. Some tanks are designed with internal or external floating roofs, double walls, dome or

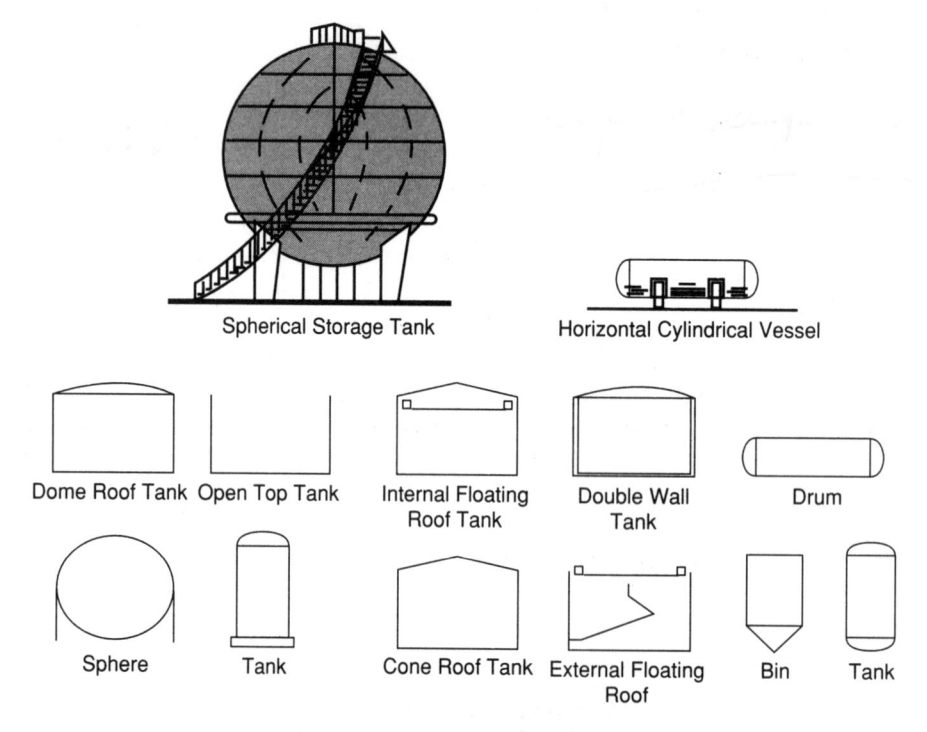

Figure 5.4-2 *Tank Designs*

cone roofs, or open top. Earthen or concrete dikes often surround a tank and are designed for containment in the event of a spill.

Spherical and spheroidal storage tanks are designed to store gases or pressures above 5 psi. Spheroid tanks are flatter than spherical tanks. Figure 5.4–2 illustrates each of these designs. Horizontal cylindrical tanks or drums can be used for pressures between 15 and 1000+ psig. Floating roof storage tanks are used for materials near atmospheric pressure. In the basic design a void forms between the floating roof and the product, forming a constant seal. The primary purpose of a floating roof is to reduce vapor losses and contain stored fluids. In areas of heavy snowfall an internal floating roof is used with an external roof since the weight of the snow would effect the seal.

Process technicians often inspect their stationary vessels using the following methods: listen, touch, look, feel, and smell. An experienced technician can identify a problem by listening for abnormal sounds and vibrations. Touching the equipment allows a technician to identify unusual heat patterns. Visually inspecting tank and sump levels allows a technician to look at and determine corrective action. Figure 5.4–3 shows a typical tank arrangement.

Figure 5.4-3 *Tank Storage*

Filters
The chemical processing industry has adopted the practice of using surface water for industrial applications instead of well water. When large quantities of water are pulled out of the ground the upper layers of soil drop. Some residences in heavily industrialized areas have seen the ground level drop so rapidly that their homes and businesses have been dropped below sea level and flooded. In higher locations this process can cause foundations to shift or crack, damaging the overall structure. Because of this problem, chemical manufacturers bring water in from local rivers and lakes. The water is initially brought into a large water basin where sediments are allowed to settle. Several large pumps take suction off the basin and transfer the water to filters designed to remove suspended solids. Figure 5.4–4 illustrates a typical industrial filter. Demineralizers are used to remove dissolved impurities in the water.

Equipment One • Chapter 5

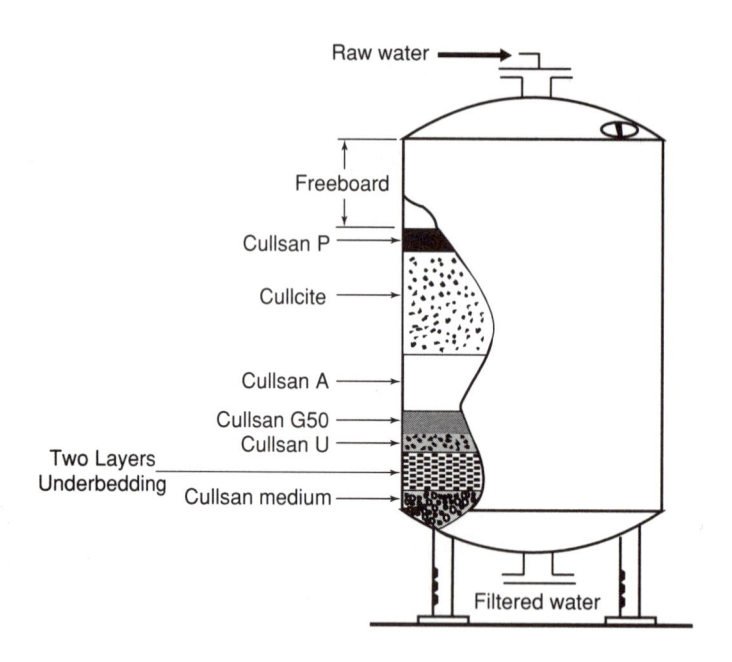

Figure 5.4-4 *Filter*

5.5 Pumps

Dynamic pumps can be classified as either centrifugal or axial, Figures 5.5–1 and 5.5–2. Centrifugal pumps move liquids by centrifugal force. The primary principle involves spinning the liquid in a circular rotation that propels it outward and into a discharge chute known as a volute. Centrifugal force and the design of the volute adds energy or velocity to the liquid. As the liquid

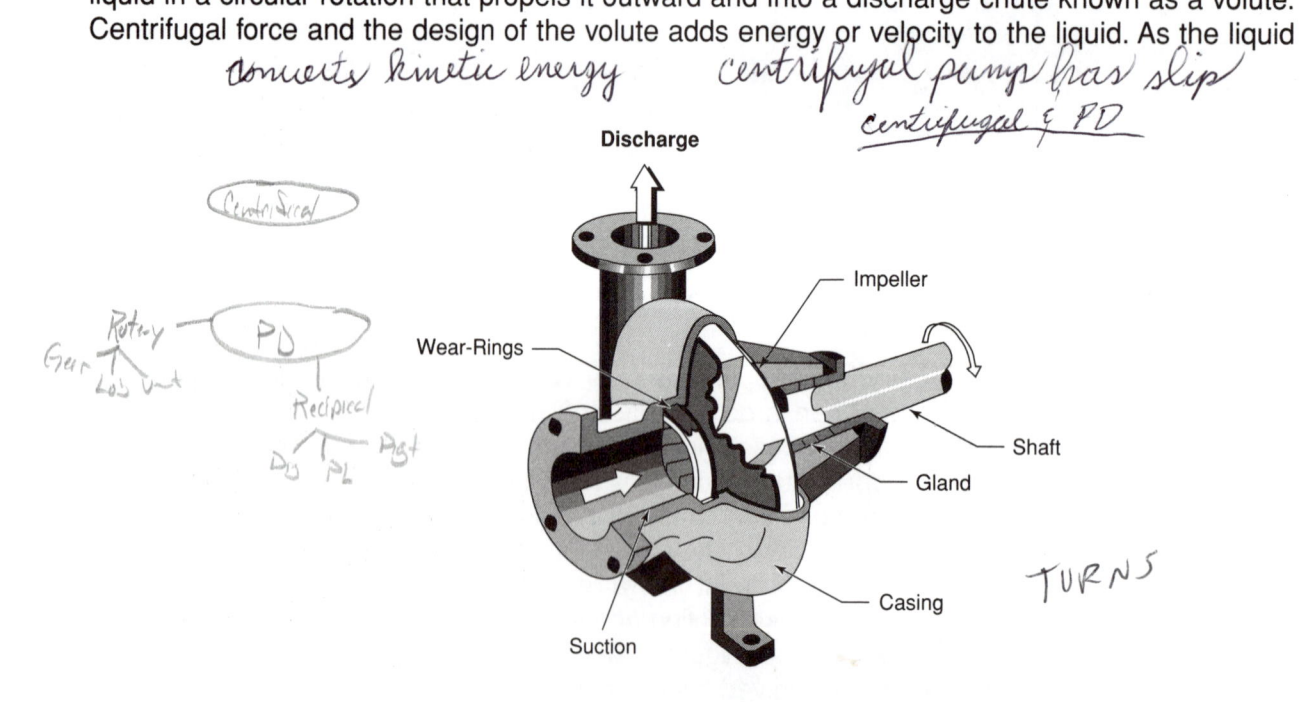

Figure 5.5-1 *Centrifugal Pump*

96

Chapter 5 • **Equipment One**

Figure 5.5-2 *Axial Pump*

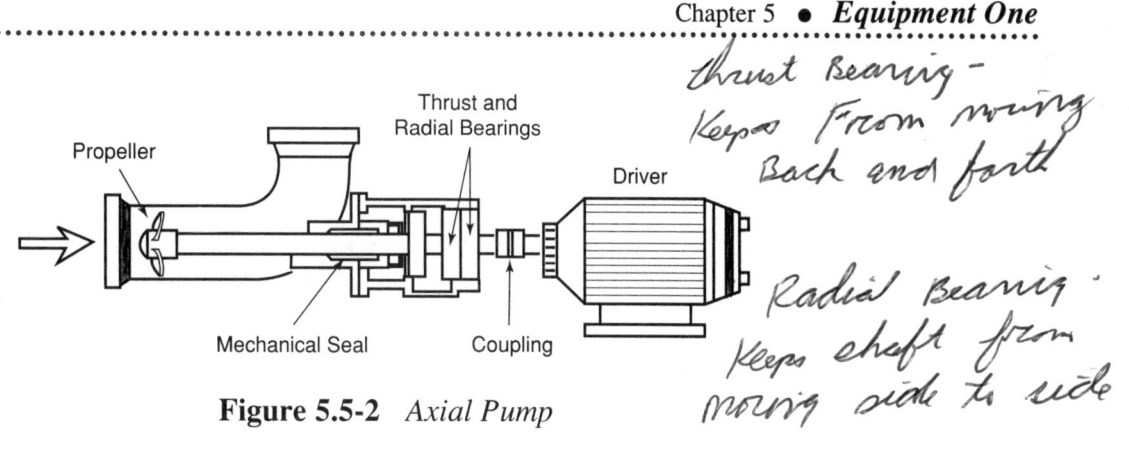

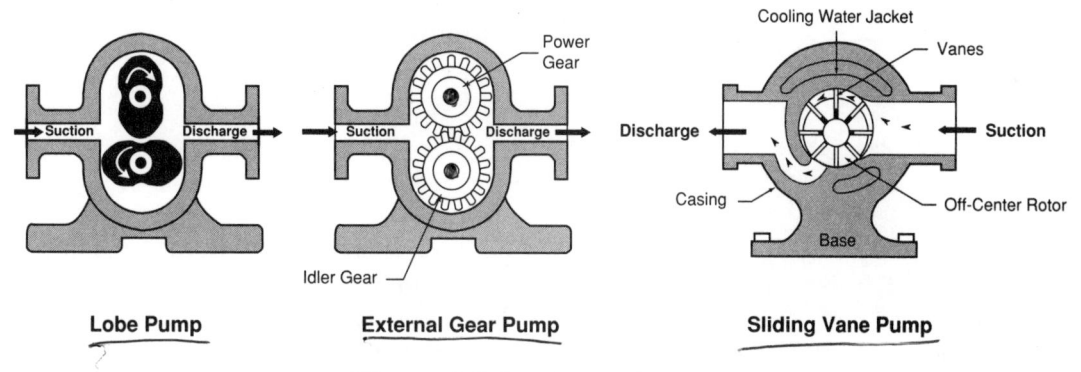

Figure 5.5-3 *Rotary Pumps*

leaves the volute it begins to slow down, creating pressure. Fluid pressure moves the process through the pipes. The basic components of a centrifugal pump include the casing, motor or driver, coupling, volute, suction eye or inlet, impellers, wear rings, seals, bearings, discharge port, suction gauge, and discharge gauge. Axial pumps are composed of a motor, coupling, bearings, seals, propeller, and shaft. As the propeller turns, fluids are propelled axially along the shaft. This feature is similar to the way a ceiling fan moves air around a room.

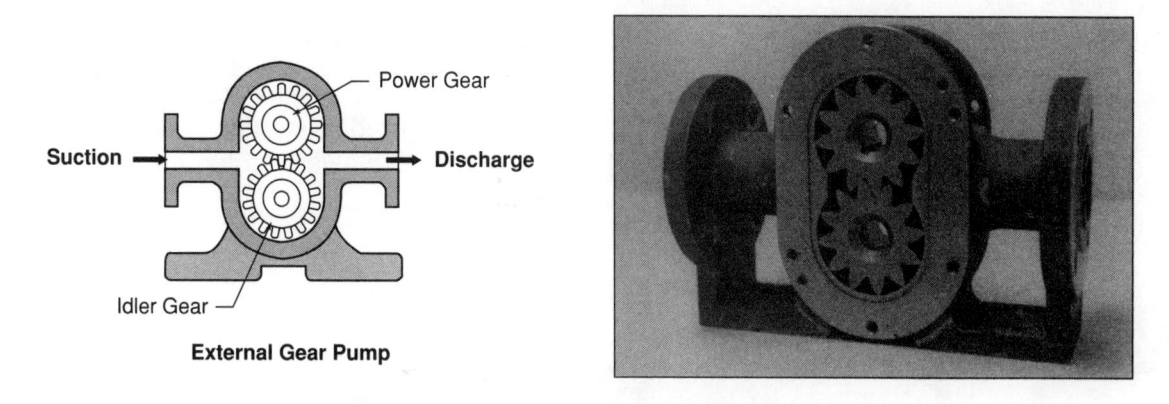

Figure 5.5-4 *External Gear Pump*

Equipment One • Chapter 5

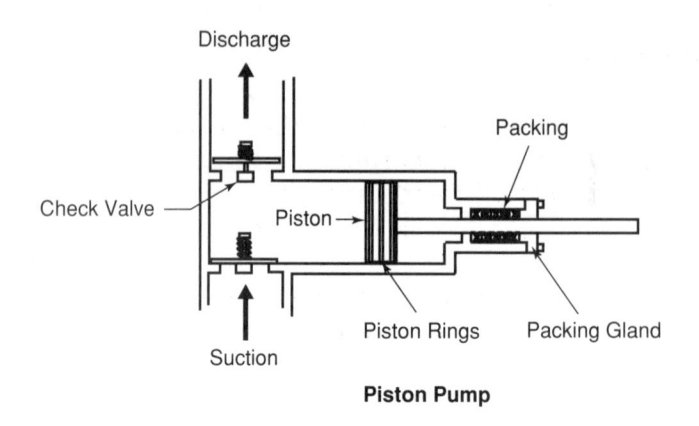

Figure 5.5-5 *Reciprocating Pumps*

Positive displacement pumps displace a specific volume of fluid on each stroke or rotation. These pumps can be classified as rotary or reciprocating. Rotary pump designs include screw pumps (progressive cavity and screw), gear pumps (internal and external), vane pumps (sliding and flexible), and lobe pumps, see Figures 5.5–3 and 5.5–4. Rotary pumps displace fluids with gears and rotating screw elements. Reciprocating pumps move fluids by drawing them into a chamber on the intake stroke and positively displacing them with a piston or diaphragm on the discharge stroke.

Reciprocating pump designs include diaphragm, piston (Figures 5.5–5 and 5.5–6), and plunger. Each of these designs is characterized by a back and forth motion. The basic components of a reciprocating pump include a connecting rod, piston/plunger or diaphragm, seals, check valves, motor, casing, and bearings.

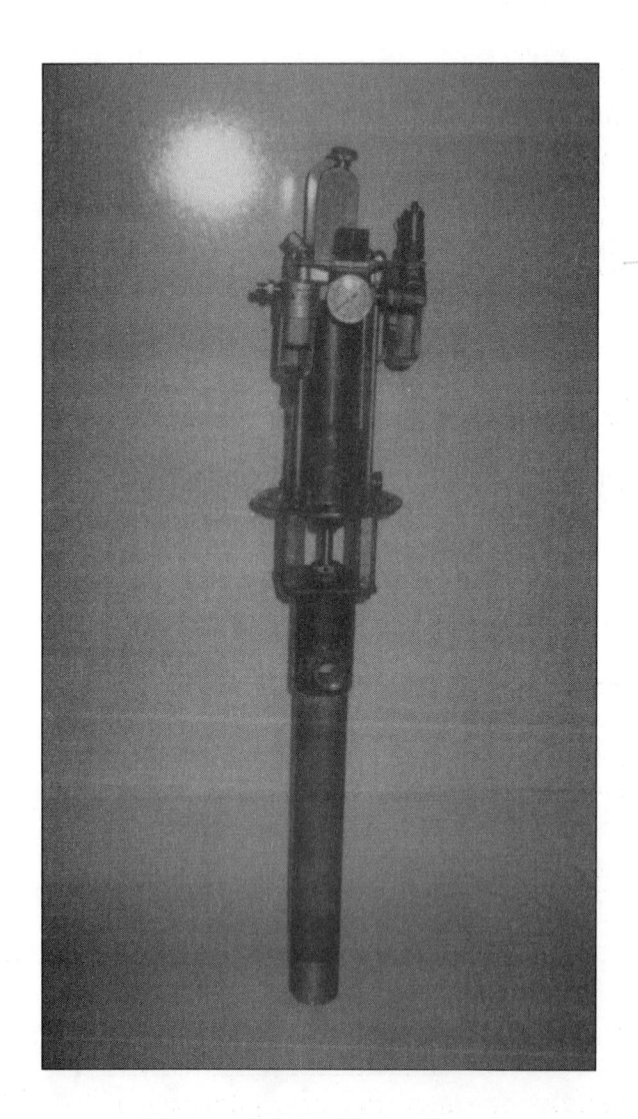

Figure 5.5-6 *Piston Pump*

Chapter 5 • *Equipment One*

5.6 Compressors

The operation and design of a compressor can usually be classified in two groups: positive displacement and dynamic. Dynamic compressors operate by accelerating the gas and converting the energy to pressure. This type of compressor can be either centrifugal or axial. Centrifugal compressors, Figure 5.6–1, operate by adding centrifugal force to the product stream. The design and application of centrifugal compressors accelerate the velocity of the gases. This velocity or kinetic energy is converted to pressure as the gas flow leaves the volute and enters the discharge pipe. Centrifugal compressors can deliver much higher flow rates than positive displacement compressors.

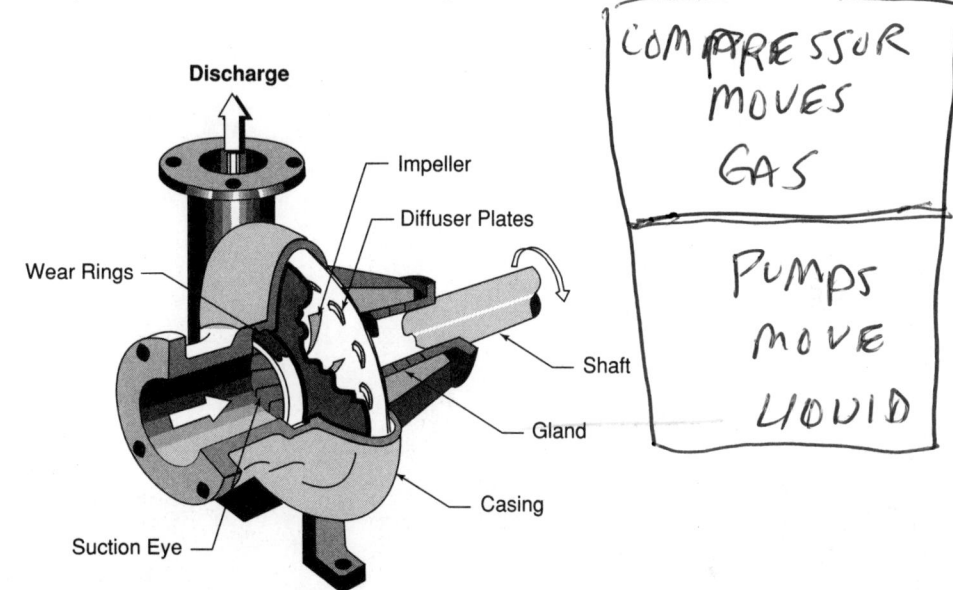

Figure 5.6-1 *Centrifugal Compressors*

The basic components of a centrifugal compressor include the casing, motor or driver, coupling, volute, suction eye or inlet, impellers, wear rings, seals, bearings, discharge port, suction gauge, and discharge gauge. An axial flow compressor is composed of a rotor that has rows of fan-like blades. Unlike centrifugal compressors, axial compressors do not use centrifugal force to increase gas velocity. Airflow is moved axially along the shaft. Rotating blades attached to a shaft push gases over stationary blades called stators. The stators are mounted or attached to the casing. As the gas velocity is increased by the rotating blades, the stator blades slow it down. As the gas slows, kinetic energy is released in the form of pressure. Gas velocity increases as it moves from stage to stage until it reaches the discharge port. Figure 5.6–2 illustrates a single-stage centrifugal blower.

Positive Displacement Compressors
Positive displacement compressors, Figure 5.6–3, operate by trapping a specific amount of gas and forcing it into a smaller volume. They are classified as rotary or reciprocating. Positive displacement compressors and positive displacement pumps operate under similar conditions. The primary difference is that compressors are designed to transfer gases while pumps move liquids.

Equipment One • Chapter 5

Figure 5.6-2 *Blower*

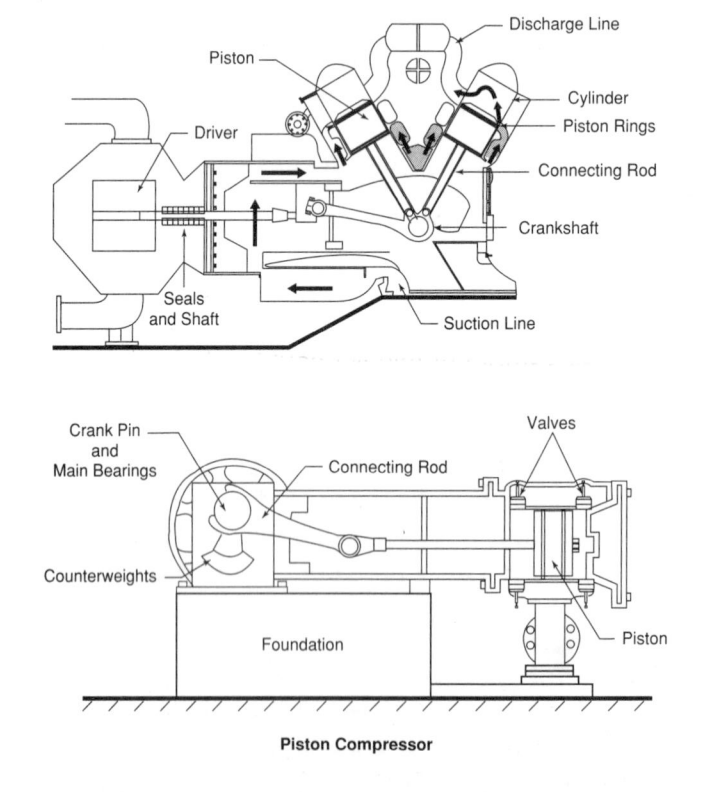

Figure 5.6-3 *PD Compressor*

Figure 5.6-4 *Piston Compressor*

Rotary compressor design includes a rotary screw, sliding vane, lobe, and liquid ring. Reciprocating compressors include piston (Figure 5.6–4) and diaphragm.

Typical Compressor System

In a refinery or chemical plant compressors are used to compress gases like nitrogen, hydrogen, carbon dioxide, and chlorine. These gases are sent to headers where they are distributed to a variety of applications. Compressors also provide clean, dry air for instruments and control devices. When compressors are used in a process system a wide assortment of supporting equipment is required. A small sample of this list could include the compressor, receiver, safety valves, heat exchangers, motor, lubrication systems, control instruments, valves, dryers, demister, regulators, and pipe header.

5.7 Steam Turbines

A steam turbine is a device "driver" that converts kinetic energy (steam energy of movement) to mechanical energy. Steam turbines have a specially designed rotor that rotates as steam strikes it. This rotation is used to operate a variety of shaft driven equipment. The steam used to operate a steam turbine is produced in a boiler. Boilers produce steam that can enter a turbine at temperatures as high as 1000 degrees F, and pressures as high as 3500 psi inlet and 200 psi outlet. High pressure steam is slowly admitted into a turbine to warm it up and remove the condensate.

Equipment One • Chapter 5

Steam enters the turbine through the steam chest. The steam chest typically has a strainer on the inlet side to remove solids. Inside the steam chest is a device called the governor valve. The governor valve opens and closes to admit steam into the turbine. A governor system controls the position of the governor valve. An overspeed trip mechanism is attached to the rotor that will shut off the flow of steam into the turbine when it reaches 115% of its design limit. The shutoff valve is typically located in front of the governor valve.

As steam leaves the steam chest it is directed into the nozzle block. The nozzle directs the steam onto the blading which is attached to the shaft. The blading and shaft make up the rotor. Impulse or reaction movement occurs as the steam strikes the rotor, converting the steam energy into mechanical energy. Each stage consists of a set of moving and stationary blades. The curved blades of each stage are designed so the spaces between the blading acts like the nozzle and increases steam velocity. As the steam zigzags between the stationary and moving blades it begins to expand as much as 1000 times its original volume. Modern turbine design increases the size of each stage giving the turbine a conical shape.

Steam turbines are typically classified as condensing, noncondensing, impulse, or reactive. In the condensing design a heat exchanger is used to condense the steam, while the noncondensing design utilizes the exhaust as low pressure steam. Impulse and reactive movement describe how the steam acts upon the rotor. In the reactive design, the nozzle is mounted on the rotor, while the impulse design allows the steam to blow against the rotor. Reactive movement is a reactive response to the release of steam. Steam turbines are used primarily as drivers for pumps, compressors, and electric power generation. Figure 5.7–1 illustrates the internal components of an impulse steam turbine.

5.8 Steam Traps

Steam traps are used to eliminate condensate from industrial steam systems. Condensate can cause a lot of serious problems as it flows with the steam. Slugs of water can damage equipment and lead to a condition known as water hammer. To eliminate this problem, steam traps are used to remove condensate. Figure 5.8–1 illustrates two different steam trap designs. Steam traps are grouped as either mechanical or thermostatic. Mechanical steam traps include floats and inverted buckets. Thermostatic traps include bellows type traps.

5.9 Electricity and Motors

The majority of electrical power produced in the world is alternating current (AC). Alternating current is defined as current that reverses direction at regular intervals. Most industrial motors use alternating current. Alternating current can be transformed using a step-down or step-up trans-

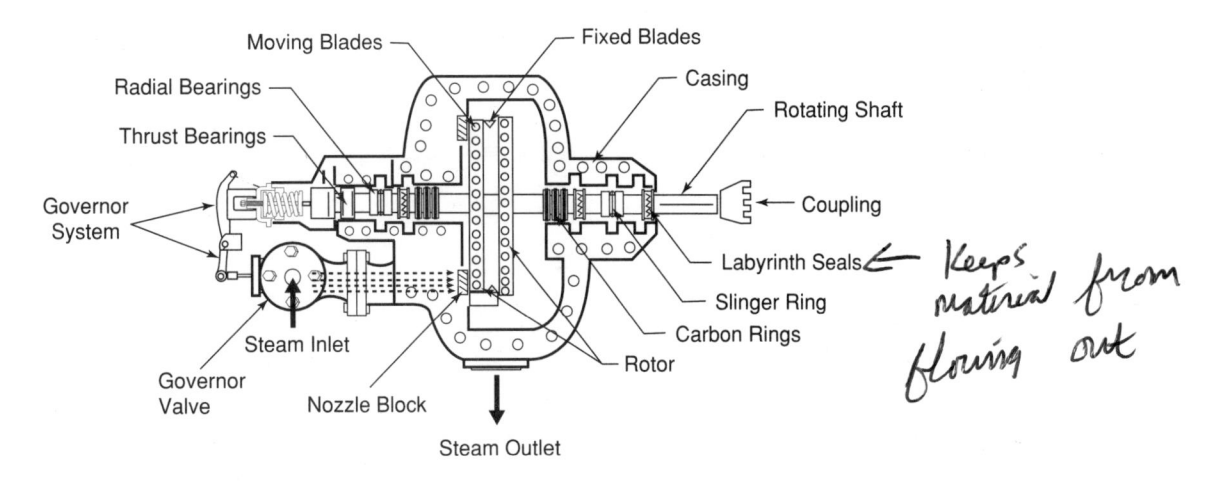

Figure 5.7-1 *Steam Turbine*

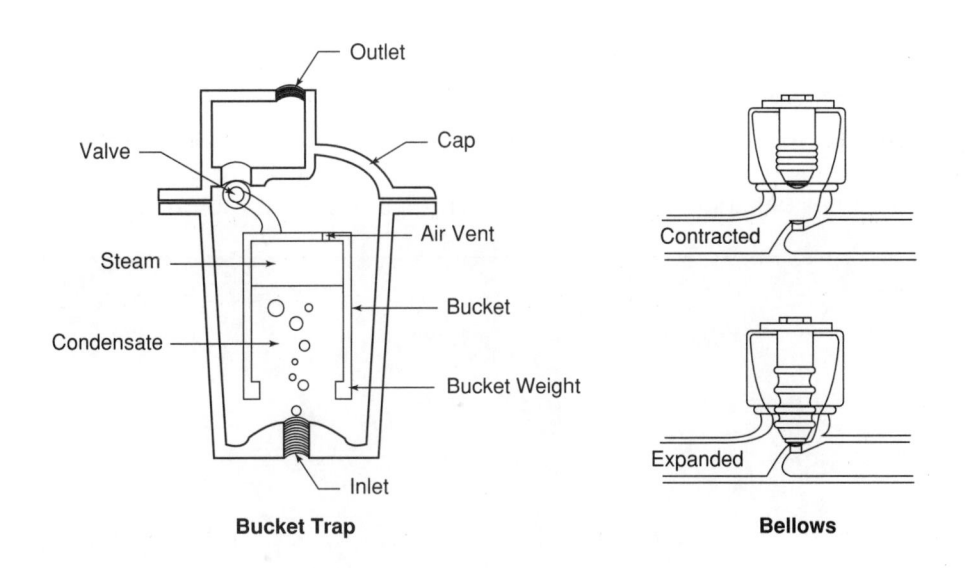

Figure 5.8-1 *Steam Traps*

former. Voltage can be increased for the purpose of transmission and then stepped down as it nears the electrical equipment. Voltages between 69kV, 138kV, and 345kV are frequently used. Direct current (DC) cannot be used the same way as alternating current. Direct current does not change flow direction.

During the 1904 World's Fair, Thomas Edison attempted to demonstrate that low-voltage direct current could light the fair more economically than George Westinghouse and Nikola Tesla's alternating current. Under Edison's plan it would have cost $1.00 for every light bulb vs.

Equipment One • Chapter 5

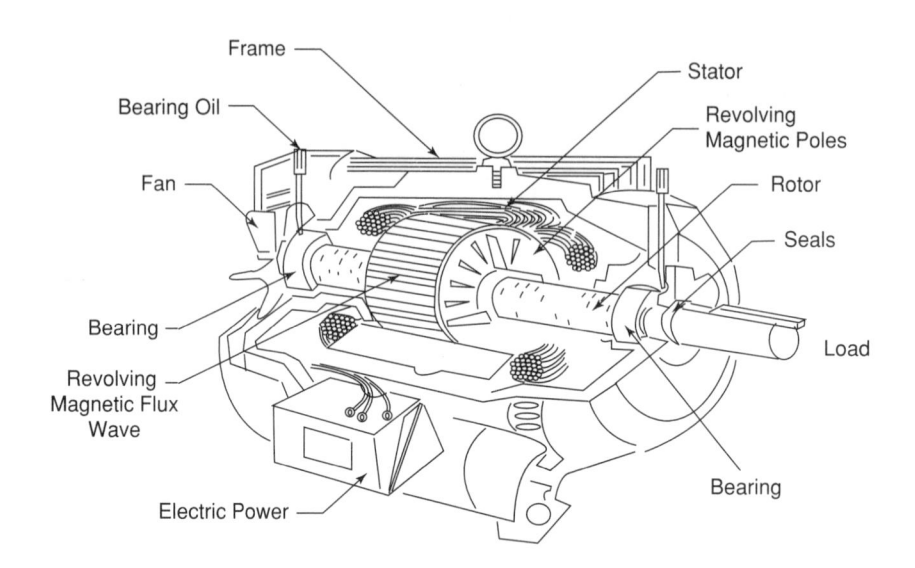

Figure 5.9-1 *Typical Motor*

Westinghouse's bid of 25 cents per light bulb. Alternating current easily won the contest and has remained the most popular option.

The chemical processing industry uses three-phase motors to operate pumps, compressors, fans, blowers, and other electrically driven equipment. Three-phase motors come in three basic

Figure 5.9-2 *AC Motor*

designs: squirrel-cage induction motors, wound rotor induction motors, and synchronous motors. The primary difference is in the rotor. The direction of rotation in a motor is determined by strong magnetic fields. A typical motor is composed of stator windings, rotor and shaft, bearings and seals, conduit box, frame, fan, lubrication system, and shroud. Figure 5.9–1 illustrates the location of these components. Figure 5.9–2 shows an AC motor.

5.10 Equipment Lubrication, Bearings, and Seals

One of the primary functions a process technician performs is periodic equipment checks. During these routine checks equipment oil levels and operating conditions are closely inspected. High temperatures, unusual noises or smells, and erratic flows are all signs that a problem has developed. To ensure the good operation of process equipment, proper lubrication must be maintained. Lubrication protects the moving parts of equipment by placing a thin film of protection between surfaces that come into contact with each other. Under a microscope the smooth surface of a gear may appear very rough. Without lubrication a tremendous amount of friction would develop. Lubrication helps remove heat generated by friction and provides a fluid barrier between the metal parts to reduce friction. Loss of lubrication causes severe damage to compressors, steam turbines, pumps, generators, and engines. Most rotary equipment requires some type of lubrication.

Bearings
Radial and axial bearings can be found in most rotating equipment and require lubrication to operate properly. Radial bearings are designed to prevent up and down and side to side movement of the rotating shaft. Axial bearings are designed to prevent back and forth movement of the shaft. Radial bearings can be found in a variety of designs. Some of these designs include ball bearings, friction or sleeve bearings, rolling element bearings, and shielded bearings.

Seals
Shaft seals are designed to prevent leakage between internal compartments in a rotating piece of equipment. Shaft seals come in a variety of shapes and designs. Typical designs include labyrinth seals, carbon seals, packing seals, and mechanical seals. Labyrinth seals trap lubrication and fluids between a maze of ridges. Segmental carbon seals are mounted in a ring-shaped design around the rotating shaft. A spring holds the soft graphite seal in place and allows it to wear evenly. Mechanical seals come in a modular kit that is slid into place as one unit. Mechanical seals provide a stationary seat and a moving seal face. Mechanical seals are designed to withstand high pressure situations where carbon seals and labyrinth seals do not. Shaft seals minimize air leakage into and out of the equipment; keep dirt, chemicals, and water out of the lubricant; and keep the clean lubricant in the chamber where the bearings and moving components are located. Seals and bearings are illustrated in Figure 5.10–1.

Summary

Tanks and pipes store and contain fluids. Tank designs vary depending upon their service. Pipe size and design determine flow rates, pump and valve sizes, turbulent or laminar flow, instrument type, and automation. Valves control the flow of fluids. Valves come in a variety of shapes, sizes,

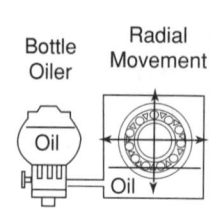

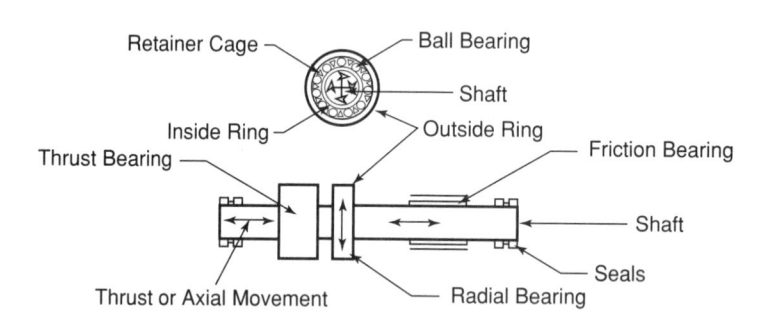

Figure 5.10-1 *Seals and Bearings*

and designs that throttle, stop, or start flow. The more common designs include gate, globe, ball, plug, check, and butterfly. Filters remove solids from fluids. Strainers remove solids from a process before they can enter a pump and damage it. A cyclone is used to remove solids from a gas stream. A typical cyclone is shaped like a V-bottom tank with a port in the top, bottom, and upper side. Gases and solids enter the top upper side of the tank and are swirled around the tank. Solids drop to the bottom of the cone while gases escape out the top of the tank. Demineralizers remove dissolved substances from a fluid.

Pumps are primarily used to move liquids from one place to another. They come in two basic designs, positive displacement and dynamic. Positive displacement pumps can be classified as rotary or reciprocating. Reciprocating pumps are characterized by a back and forth motion, while rotary pumps move in a circular rotation. Dynamic pumps can be classified as centrifugal or axial. The centrifugal pump uses centrifugal force to move liquids while the axial pumps push liquids along a straight line. Compressors come in two basic designs, positive displacement (rotary and reciprocating) and dynamic (axial or centrifugal). A compressor is designed to accelerate or compress gases. Compressors are closely related to pumps.

Steam turbines are used as drivers to turn pumps, compressors, and electric generators. High pressure steam is directed into buckets designed to turn a rotor and provide rotational energy. Steam turbines provide the same feature as electric motors. A typical motor is composed of stator windings, rotor and shaft, bearings and seals, conduit box, frame, fan, lubrication system, and shroud. Steam turbines and motors are two of the most popular devices used by industry as drivers.

Shaft seals are designed to prevent leakage between internal compartments in a rotating piece of equipment. Shaft seals come in a variety of shapes and designs. Typical designs include labyrinth seals, carbon seals, packing seals, and mechanical seals. Radial and axial bearings can be found in most rotating equipment and require lubrication to operate properly. Radial bearings are designed to prevent up and down and side to side movement of the rotating shaft while axial bearings are designed to prevent back and forth movement of the shaft.

Chapter 5

Review Questions

1. Draw and label a gate valve.
2. Draw and label a centrifugal pump.
3. What is the primary difference between a pump and a compressor?
4. Describe how a steam turbine works. Sketch a simple drawing if needed.
5. Describe alternating current.
6. Draw and label a rotary pump. Show rotation.
7. Explain the purpose of bearings and seals.
8. What are the basic components of an electrical motor?
9. List the basic hand tools used by process technicians.
10. Describe how a steam trap operates.
11. Draw and label a globe valve.
12. What is the primary purpose of a floating roof?
13. List the standard pipe fittings used to connect pipe.
14. What type of materials are used in the manufacture of storage tanks?
15. How much pressure can a typical horizontal cylindrical tank hold?
16. Draw and label the type of valve used to relieve pressure.
17. Describe how an industrial motor works.
18. Describe centrifugal movement.
19. Draw and label a reciprocating pump. Show rotation.
20. Explain the importance of lubrication for a pump, compressor, or turbine.

Chapter 6

Equipment Two

OBJECTIVES

After studying this chapter, the student will be able to:

- *Describe the major components and operation of a heat exchanger.*

- *Describe the key components and operation of a cooling tower.*

- *List the primary components of a boiler.*

- *Describe the primary components of a furnace.*

- *List the key components of a mixing reactor.*

- *List the primary components of a plate distillation column.*

- *Describe the primary components of a packed distillation column.*

- *Explain how an extruder operates.*

Equipment Two • Chapter 6

KEY TERMS

Absorber—a device used to remove selected components from a gas stream by contacting it with a gas or liquid.

Adsorber—a device (reactor, dryer, etc.) filled with a porous solid designed to remove gases and liquids from a mixture.

Aerators—a device used to stir up and add oxygen to a microbiological system.

Boilers—provide steam for industrial applications. Boilers are classified into two groups: (1) water tube and (2) fire tube. A boiler is composed of a furnace, stack, economizer section, large upper drum, and smaller lower drum.

Cooling towers—consist of a box shaped collection of multi-layered wooden slats and louvers that directs airflow and breaks up water as it falls from the top of the tower or water distribution header. Cooling towers are classified by the way they produce airflow and by the way the air moves in relation to the downward flow of water.

Cyclone—a device used to remove solids from a gas stream.

Demineralizer—a filtering type device that removes dissolved substances from a fluid.

Distillation column—separate chemical mixtures by boiling points. A distillation column is a collection of stills stacked one on top of the other. Distillation columns fall into two distinct classes: plate or packed.

Extruder—a device used to melt and extrude granular or powdered plastic. Most extruders are composed of a single or twin screw design surrounded by a heated barrel. The molten polymer is forced or pumped through a die.

Flares—safely burn excess hydrocarbons. A flare system is composed of a flare, knock-out drum, flare header, fan (optional), steam line and steam ring, fuel line, and burner.

Fired heaters—consist of a battery of tubes that pass through a firebox. Fired heaters or furnaces are commercially used to heat up large volumes of crude oil or hydrocarbons.

Heat exchanger—an energy transfer device designed to convey heat from one substance to another.

Incinerator—a permitted device used to burn industrial wastes.

Chapter 6 • **Equipment Two**

Reactors—a device used to combine raw materials, heat, pressure, and catalysts in the right proportions. Reactors are classified as either batch or continuous and as fixed bed or fluidized bed design.

Scrubber—devices used to remove chemicals and solids from process gases.

6.2 Heat Exchangers

In unit operations, the transfer of heat from one process to another is typically accomplished with heat exchangers. A heat exchanger allows a hot fluid to transfer energy in the form of heat to a cooler fluid without physically coming into contact with each other. A heat exchanger can provide heat or cooling to a process. Heat exchangers can be found in the following categories:

- Shell and tube heat exchanger
- Plate and frame
- Spiral
- Air cooled

Advanced classes in the process technology curriculum will go into greater detail in each of these areas. For example, shell and tube heat exchangers can be broken down into (1) double-pipe heat exchanger, (2) fixed head, single pass, (3) fixed head, multi-pass, (4) floating head, multi-pass (u-tube), (5) kettle reboiler, (6) thermosyphon reboiler, (7) shell nomenclature, etc.

A typical shell and tube heat exchanger is composed of a series of tubes surrounded by a shell. The tubes are typically connected to a fixed tube sheet and are supported by a series of internal baffles. The shell and tube cylinder has a water box or head securely attached on the inlet and outlet side. A tube inlet and tube outlet admits fluid through the tubes. A shell inlet and outlet admits flow through the shell. Figure 6.2–1 illustrates the typical layout for a shell and tube heat exchanger.

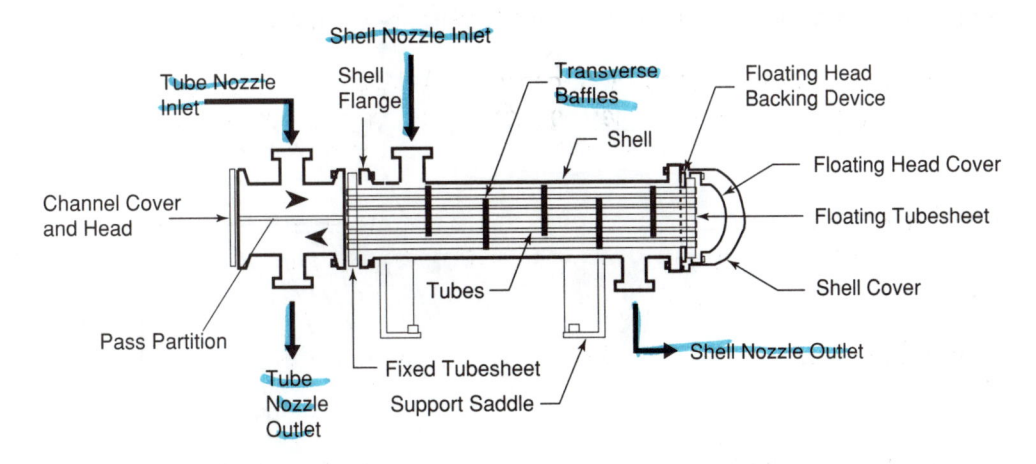

Figure 6.2-1 *Shell and Tube Heat Exchanger*

111

6.3 Cooling Towers

A cooling tower is a box-shaped collection of multi-layered wooden slats and louvers that directs airflow and breaks up water as it falls from the top of the tower or water distribution header. The internal design of the tower ensures good air and water contact. A cooling tower is a simple device used by industry to remove heat from water. Hot water transfers heat to cooler air as it passes in the tower. This type of heat is called sensible heat. Sensible heat can be measured or felt. Sensible heat accounts for 10 to 20% of the heat transfer in a cooling tower. The major portion of heat stripped from a tower is caused by evaporation. Evaporation accounts for 80 to 90% of the heat transfer in a cooling tower. When water changes to vapor, it takes heat energy with it, leaving behind the cooler liquid. The principle of evaporation is the most critical factor in cooling tower efficiency.

Cooling towers are classified by (1) how they produce airflow and (2) the direction the airflow takes in relation to the downward flow of water. Airflow is produced naturally or mechanically. Mechanical drafts are created by fans located on the side or top of the cooling tower. Flow direction into a tower is either crossflow or counterflow. Crossflow goes horizontally across the fill while counterflow is directed vertically through the tower. Cooling towers are categorized as:
- Natural draft—counterflow
- Forced draft—counterflow
- Induced draft—counterflow or crossflow
- Hyperbolic—counterflow or crossflow

The basic components of a cooling tower include a water basin, pump, and water make-up system at the base of the cooling tower. The internal frame is made of pressure-treated wood or plastic and is designed to support the internal components of the tower. Some of these components include the fill or splash boards and drift eliminators. The fill or splash boards enhance liquid air contact while the drift eliminators reduce the amount of water lost from the tower because of excess airflow. Louvers on the side of the cooling water tower admit air into the device. A hot water distribution system is typically located on the top of the cooling tower fill. A fan may be used to enhance airflow through the cooling tower. Fan location determines whether airflow is induced

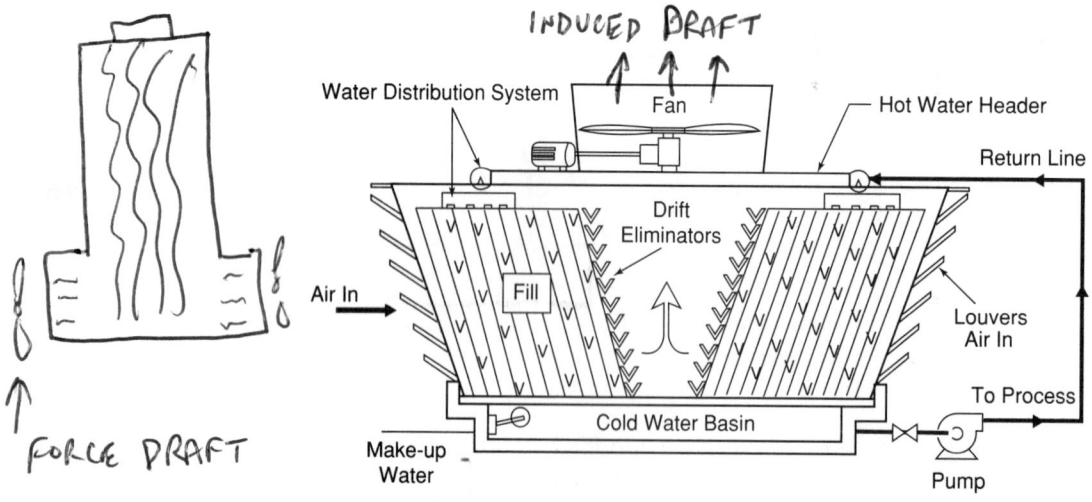

Figure 6.3-1 *Cooling Tower*

(drawn in) or forced (pushed in.) Figure 6.3–1 shows a typical cooling tower. Additional information about cooling towers will be covered in the equipment and systems course.

Heat Exchangers and Cooling Towers

Heat exchangers and cooling towers often team up to form industrial cooling systems. The system consists of a cooling tower, heat exchanger, and pump. During operation cooling water is pumped into the shell-side of a heat exchanger and returned (much hotter) to the top of the cooling tower. As the hot water goes into the top of the cooling tower it enters a water distribution header where it is sprayed over the internal components (fill) of the tower. As the water falls on the splash bars, cooler air and water contact occurs. This process removes 10 to 20% of the sensible heat. Another 80 to 90% of the heat energy is removed through evaporation. The cooled water collects in a basin at the foot of the cooling tower where a recirculation pump sends it back to the heat exchanger.

6.4 Boilers

Steam generators, also called boilers, are used by industrial manufacturers to produce steam. Steam is used to drive turbines and provide heat to process equipment. Steam generators are classified as fire tube and water tube boilers.

Fire tube heaters contain the combustion gases in tubes that occupy a small percentage of the overall volume of the heater. The heated tubes run through a shell that contains the heated medium. A fire tube boiler, Figure 6.4–1, resembles a multi-pass, shell and tube heat exchanger. This

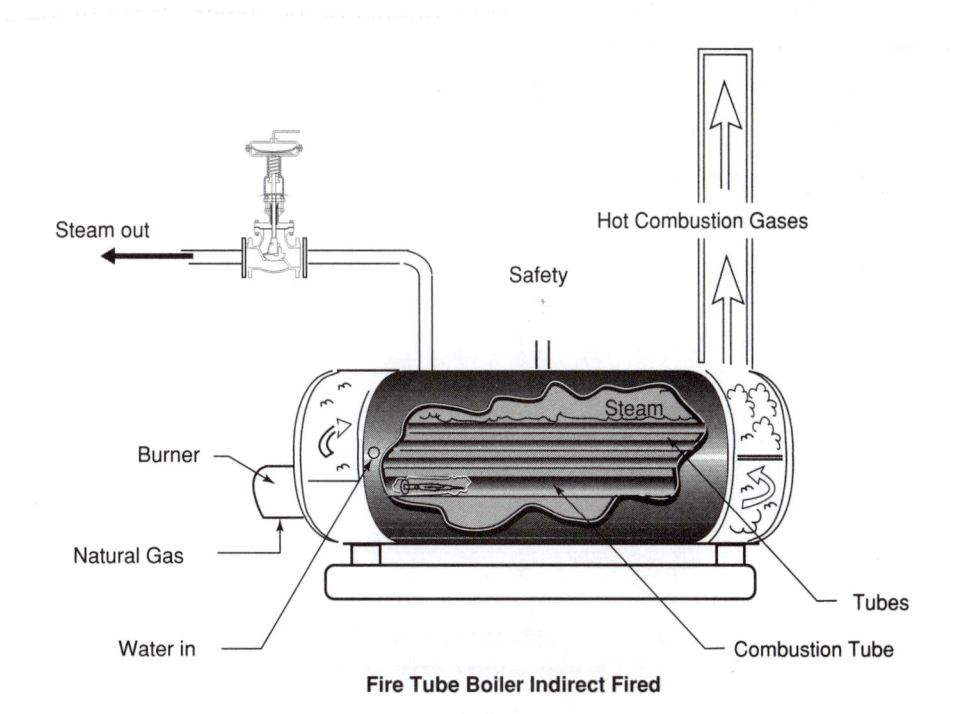

Figure 6.4-1 *Fire tube boiler*

Equipment Two • Chapter 6

type of boiler is composed of a shell and a series of steel tubes designed to transfer heat through the combustion chamber (tube) and into the horizontal fire tubes. Exhaust fumes exit through a chamber similar to an exchanger head and pass safely out of the boiler. The water level in the boiler shell is maintained above the tubes to protect them from overheating. The term "fire tube" comes from the way the boiler is constructed.

The basic components of a fire tube boiler include a large shell that surrounds a horizontal series of tubes. A large, lower combustion tube is attached to a burner that admits heat into the tubes. The upper tubes transfer hot combustion gases through the system and out the stack. Airflow is closely controlled with the inlet air louvers and the stack damper. Water level in the shell is maintained slightly above the tubes. As heat energy is transferred into the water the temperature rises until the fluid boils. A pressure control valve maintains the correct operating pressure on the vessel. Every fire tube boiler is equipped with a pressure relief system. A series of safety valves may be located on the discharge side of the shell. Low pressure steam is discharged into a common steam header that is connected to various locations in the facility. A condensate return line admits the condensed steam into a deairator drum and the water make-up system.

The chemical processing industry uses large industrial boilers commonly called water tube boilers, Figure 6.4–2. A water tube boiler consists of an upper and lower drum connected by tubes. The lower drum and water tubes are completely filled with water while the upper drum is only partially full. This vapor cavity allows steam to collect and pass out of the header. Water is car-

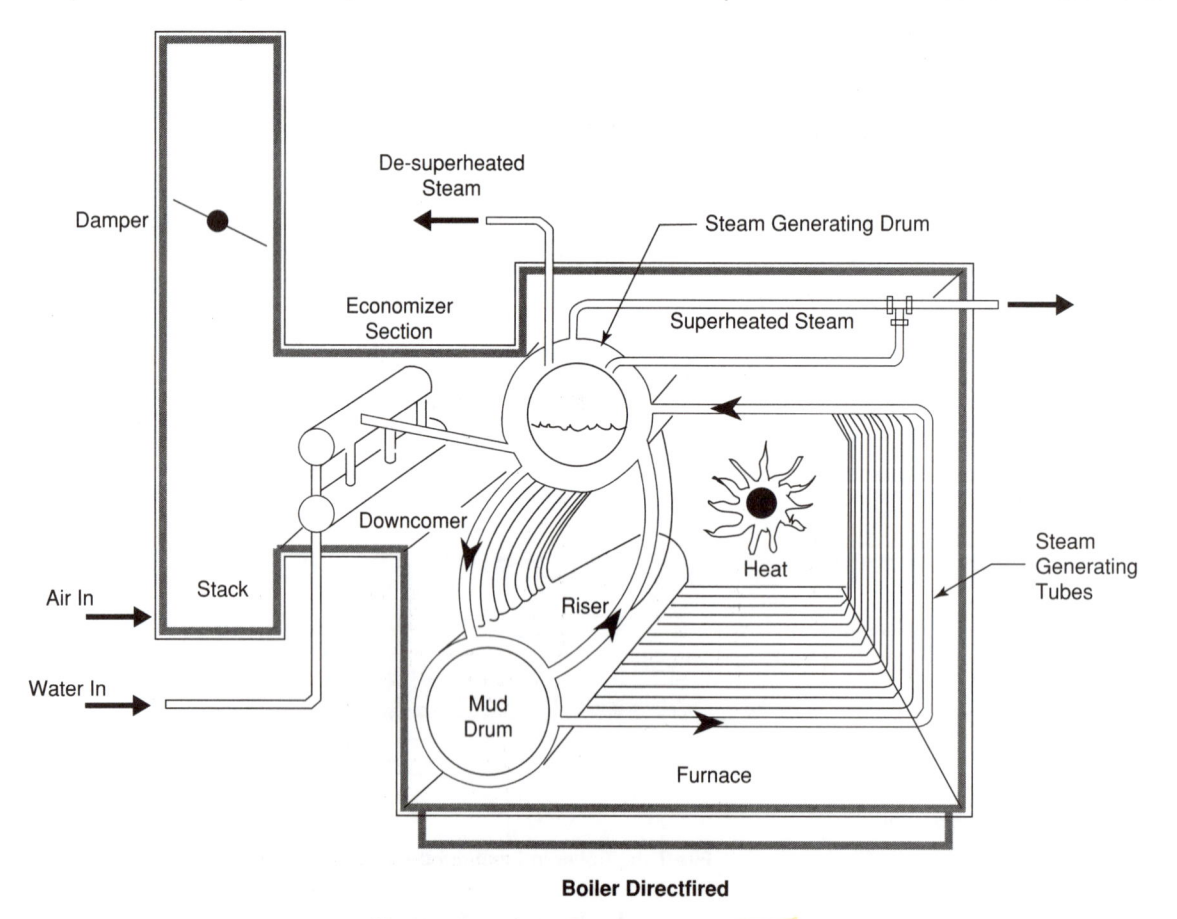

Boiler Directfired

Figure 6.4-2 *Water tube boiler*

ried through tubes that flow into a heated chamber. As heat is applied to the water generating tubes and drums the water circulates around the boiler, down the downcomer tube, into the lower drum, and back up the riser tube and steam generating tubes of the furnace. During normal operating conditions steam rises into the upper drum and out the steam header. Lost water in the boiler is replaced by the make-up water line. Additional information about boilers will be covered in the equipment and systems courses.

6.5 Furnaces

Furnaces are classified as (1) direct fired or (2) indirect fired. Direct fired furnaces can be identified by the amount of volume the combustion gases occupy inside the furnace. Furnaces are used in many processes including distillation, reactors, olefins, and hydrocracking. Furnaces heat up raw materials so they can produce products like gasoline, oil, kerosene, plastic, and rubber. Furnaces consist essentially of a battery of pipes or tubes that pass through a fire box. These tubes run along the inside walls and roof of a furnace. The heat released by the burners is transferred through the tubes and into the process fluid. The fluid remains in the furnace just long enough to reach operating conditions before exiting and being shipped to the processing unit. Furnace designs include:
- Cylindrical—direct fired
- Cabin—direct fired
- Box—direct fired
- A-frame—direct fired
- Fire tube—indirect

Figure 6.5–1 illustrates the basic layout of a furnace.

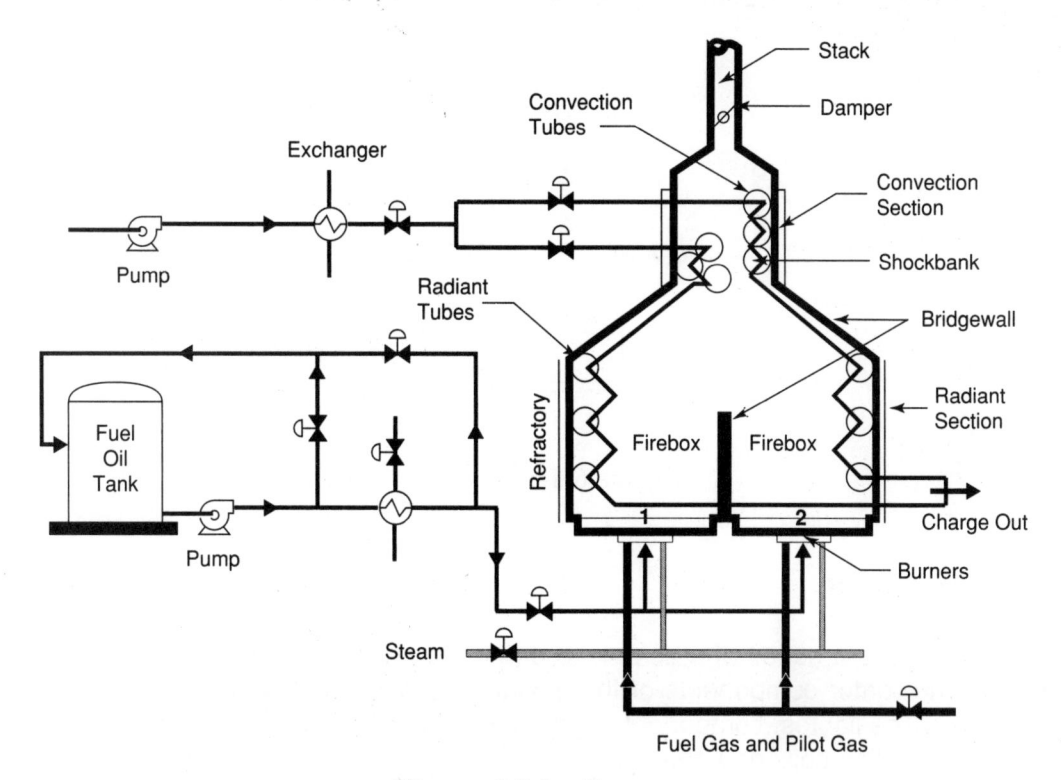

Figure 6.5-1 *Furnace*

Equipment Two • Chapter 6

The basic components of a fired heater include a tough metal shell surrounding a fire box, convection section, and stack. The inside of the furnace is lined with a special refractory (brick, blocks, peep stones, gunite) designed to reflect heat. A battery of tubes pass through the convection and radiant sections and into a common insulated header that passes out of the furnace. A series of burners are located on the bottom of the furnace or on the sides. Fluid flow is carefully balanced through the tubes to prevent equipment or product damage. Airflow and oxygen content are controlled through primary, secondary, and damper adjustments.

6.6 Reactors

A reactor is a device used to convert raw materials into useful products through chemical reactions. Reactors are designed to operate under a variety of conditions. These devices combine raw materials with catalyst, gases, pressure, or heat. A catalyst is a material designed to increase or decrease the rate of a chemical reaction without becoming part of the final product. The shape and design of a reactor are dictated by its application. Reactors are used in a variety of processes and systems:

- Alkylation
- Fluid catalytic cracking
- Fixed and fluidized bed reactors
- Hydrodesulfurization
- Fluid coking
- Chemical synthesis
- Batch and continuous
- Hydrocracking

The basic components of a reactor include a shell, a heating or cooling device, two or more product inlet ports, and one outlet port. A mixer may be used to blend the materials together. Figure 6.6–1 is an illustration of a simple mixing reactor.

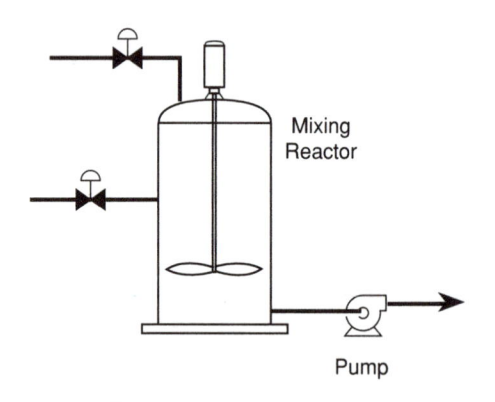

Figure 6.6-1 *Reactors*

6.7 Distillation

A distillation column is a series of stills placed one on top of the other, Figure 6.7–1. As vaporization occurs the lighter components of the mixture move up the tower and are distributed on the various trays. The lightest component goes out the top of the column in a vapor state and is passed over the cooling coils of a shell and tube condenser. As the hot vapor comes in contact

116

Chapter 6 • *Equipment Two*

with the coils it condenses and is collected in the overhead accumulator. Part of this product is sent to storage while the other is returned to the tower as reflux.

Distillation is a process that separates a substance from a mixture by its boiling point. During the distillation process a mixture is heated until it vaporizes, then is recondensed on the trays or at various stages of the column where it is drawn off and collected in a variety of overhead, side-stream, and bottom receivers. The condensed liquid is referred to as the distillate while the liquid that does not vaporize in a column is called the residue. During tower operation, raw materials are pumped to a feed tank and mixed thoroughly. Mixing is usually accomplished with a pump around loop or a mixer. This mixture is pumped to a feed preheater or furnace where the temperature of the fluid mixture is brought up to operating conditions. Preheaters are usually shell and tube heat exchangers or fired furnaces. This fluid enters the feed tray or section in the tower. Part of the mixture vaporizes as it enters the column while the rest begins to drop into the lower sections of the tower.

Heat balance on the tower is maintained by a device known as a reboiler. Reboilers take suction off the bottom of the tower. The heaviest components of the tower are pulled into the reboiler and stripped of smaller molecules. The stripped vapors are returned to the column and allowed to separate in the tower. Distillation towers are classified as plate or packed. Plate columns could

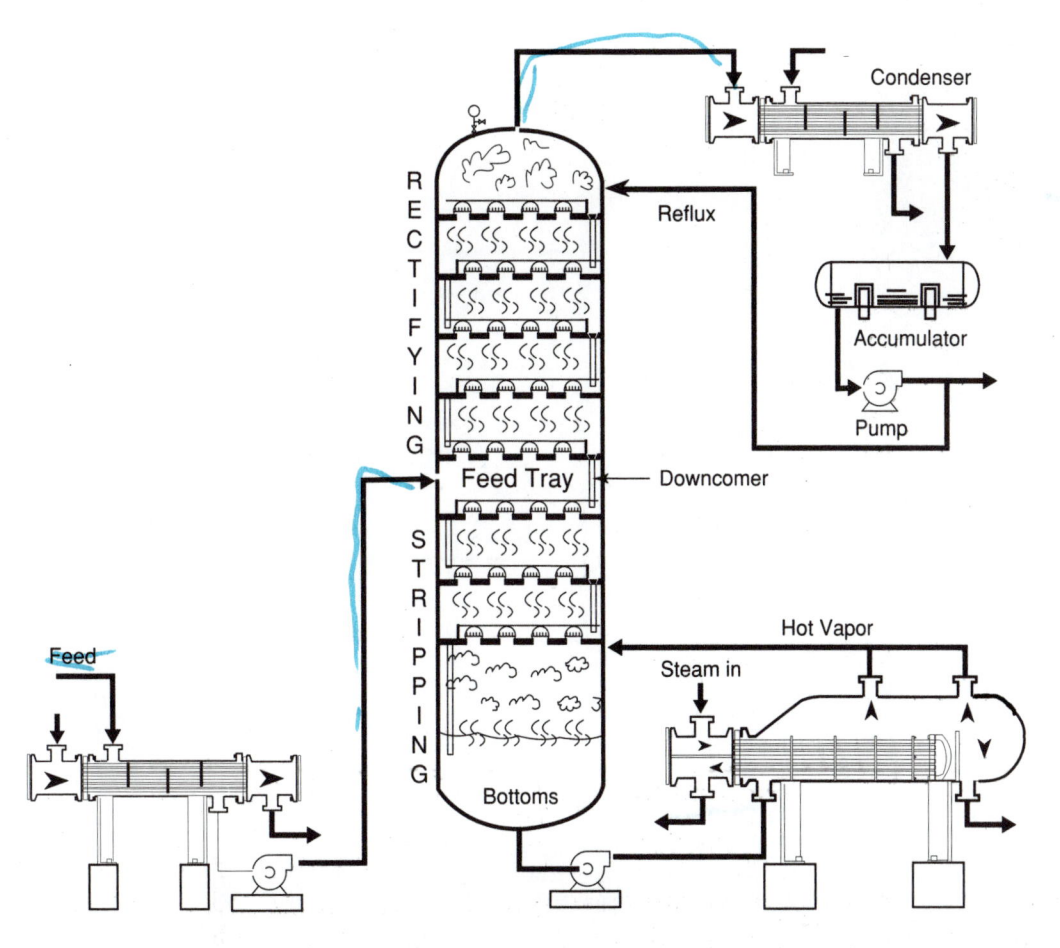

Figure 6.7-1 *Distillation Column*

117

have bubble-cap trays, valve trays, or sieve trays. Packed towers could be filled with sulzer packing, rasching ring, flexiring, pall ring, intalox saddle, berl saddle, metal intalox, teller rosette, or mini-ring packing.

Absorption, Stripping, and Scrubbing Columns

An absorber column is a device used to remove selected components from a gas stream by contacting it with a gas or liquid. Absorption can roughly be compared to fractionation. A typical gas absorber is a plate or packed distillation column that provides intimate contact between raw natural gas and an absorption medium. Absorption columns work differently than typical fractionators because during the process the vapor and liquid do not vaporize to any degree. Figure 6.7–2 illustrates the scientific principles involved in absorption. Product exchange takes place in one direction, vapor phase to liquid phase. The absorption oil gently tugs the pentanes, butanes, etc. out of the vapor. In an absorber the gas is brought into the bottom of the column while lean oil is pumped into the top of the column. As the lean oil moves down the column it absorbs elements from the rich gas. As the raw, rich gas moves up the column it is robbed of specific hydrocarbons and exits as lean gas.

Stripping columns are used with absorption columns to remove liquid hydrocarbons from the absorption oil. To the untrained eye a stripping and absorption column are identical. As rich oil leaves the bottom of the absorber it is pumped into the midsection of a stripping column. Figure 6.7–2 illustrates how steam is injected directly into the bottom of the stripper allowing for 100% conversion of BTUs. As the hydrocarbons break free from the absorption oil, they move up the column while the lean oil is recycled back to the absorber.

Adsorption

Absorption Column

The liquid phase removes lighter components from the vapor phase. One direction component removal.

Stripping Column

Reverses absorption process. Strips out hydrocarbons from absorption oil.

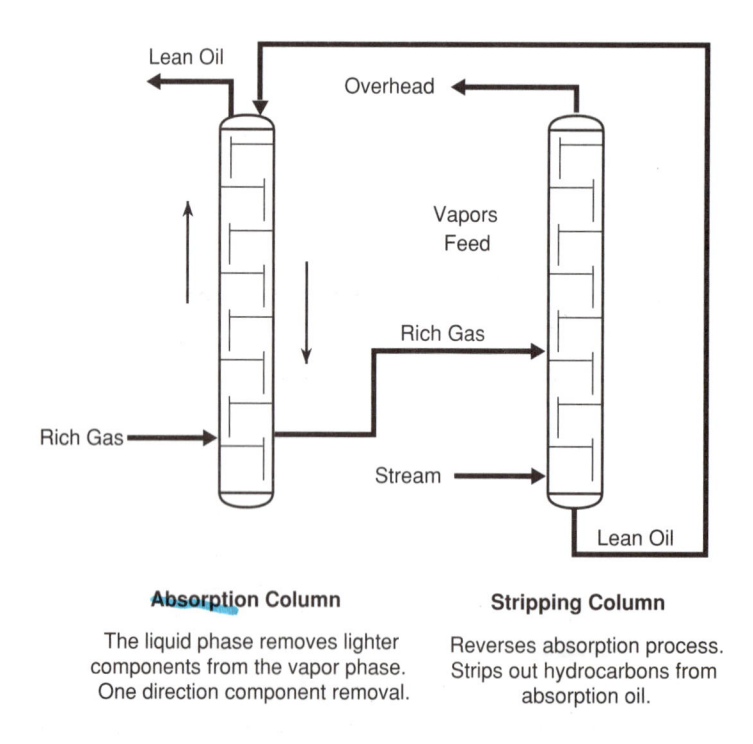

Figure 6.7-2 *Absorption and Stripping*

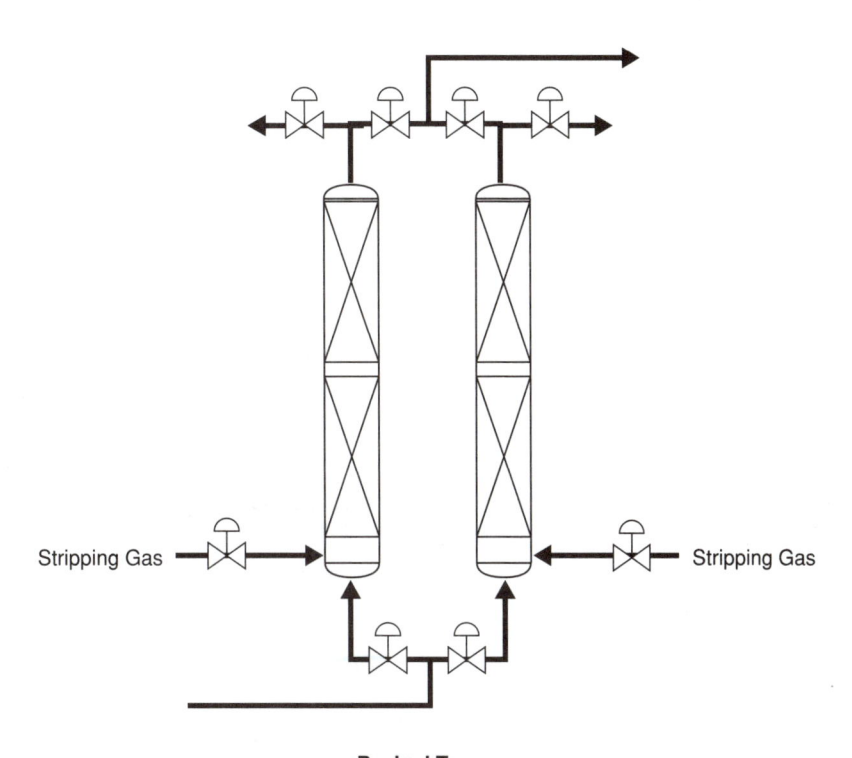

Packed Tower
Activated Alumina or Charcoal

Figure 6.7-3 Adsorption

During the adsorption process a device is filled with a porous solid designed to remove gases or liquids from a mixture. Typically the process is run in parallel with a primary and secondary vessel. The adsorption material can be activated alumina or charcoal. A variety of adsorption materials can be used. The adsorption material has selective properties that remove specific components of the mixture while passing over it. A stripping gas is used to remove the stripped components from the adsorption material.

During the adsorption process the mixture to be separated is passed over the fixed bed medium (adsorbent) in the primary device. Figure 6.7–3 illustrates this process. At the conclusion of the cycle time the process flow is transferred to the secondary device. A stripping gas is admitted into the primary device. The stripping gas is designed to remove or separate the selected chemical from the adsorption material. At the conclusion of this cycle the valves switch and the stripping gas stops as the process switches back and repeats.

Scrubber

A scrubber is an environmental device used to remove chemicals and solids from process gases. Scrubbers are cylindrical and can be filled with packing material or left empty. As dirty gases enter the lower section of a scrubber they begin to rise. As these dirty vapors rise they encounter a liquid chemical wash that is being sprayed downward. As the vapors and liquid come into contact the undesirable products entrained in the stream are removed. As the dirty material is absorbed into the liquid medium they fall to the bottom of the scrubber where they are mechan-

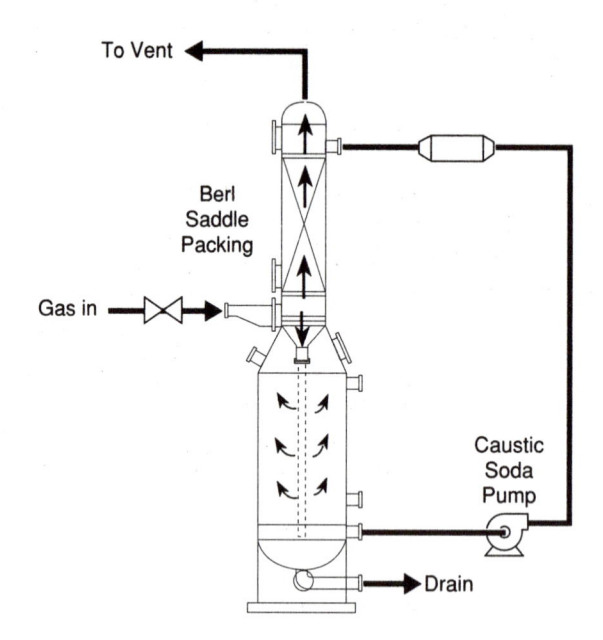

Figure 6.7-4 Scrubber

ically removed. Clean gases flow out of the top of the scrubber and on for further processing. Figure 6.7–4 is an illustration of a simple scrubber.

6.8 Plastics Plant Equipment

An extruder is a complex device composed of a heated jacket, one screw or a set of screws, a heated die, large motor, gear box, and a pelletizer, Figure 6.8–1. The purpose of the extruder is to melt the granule mixture, quickly quench it and cut it into small pellets that are easier to handle. Melting the granules encapsulates the additives in the polypropylene. In its granular form the molecular weight distribution and swell are too broad. Various additives such as peroxide help narrow this down. Customers will not accept raw granules because of the danger of dust explosions.

Solids feeders are composed of single or multiple screws that rotate inside a sleeve. Granules and additives from the feed and additive tanks are conveyed to the homogenizer by the feeders. Most solids feeders are single screws that deliver solids at a specific rate. Figure 6.8–2 illustrates how a solids feeder operates. Granules leave the feed tank continuously when the extruder is in operation.

Some solids feeders have two-stage designs that allow granules to drop into a variable speed screw mounted on a scale. From the scale, a constant speed screw delivers granules to a discharge line. The speed of the first screw is adjusted to maintain a constant scale weight. The result is a feed that is constant by weight.

Chapter 6 • **Equipment Two**

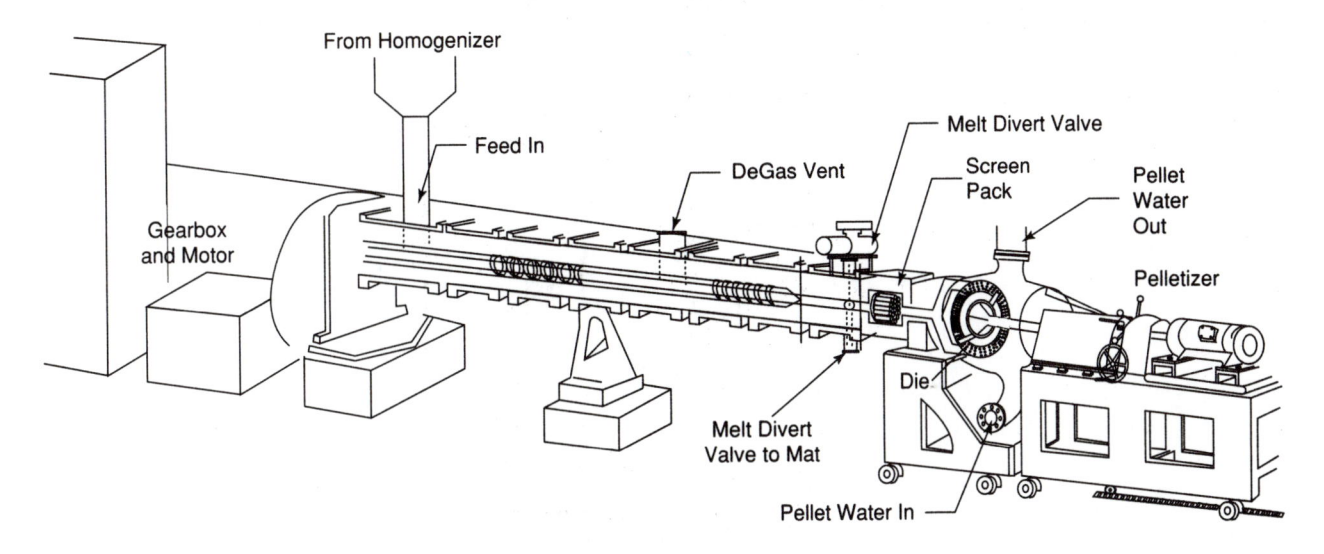

Figure 6.8-1 *Extruder*

The classifier is a vibrating tub with screens in two stages that permit the desired size of pellets to pass through. Larger pellets or clumps that have managed to pass through the scalping box are eliminated from the product flow here. Between the two screens is a cleaning kit that is used to prevent the lodging of pellets in the individual holes of the classifier. Figure 6.8–3 illustrates this device.

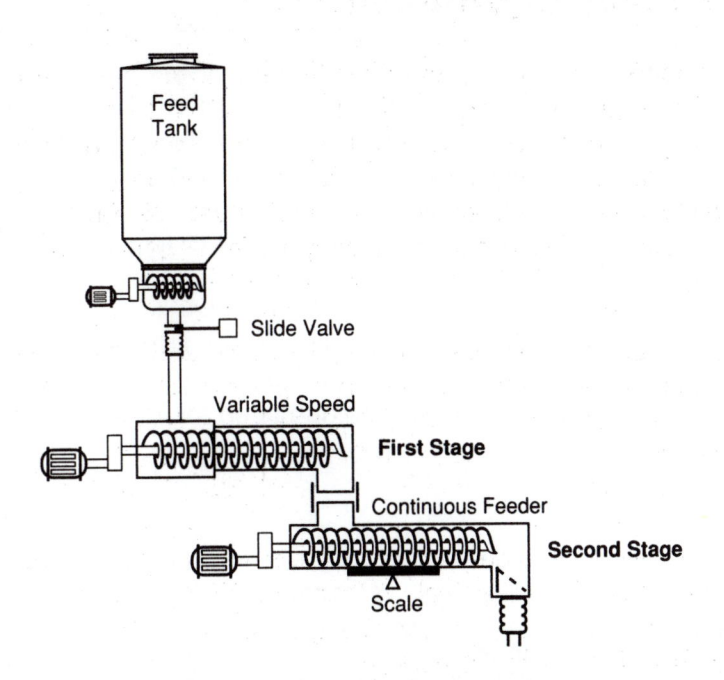

Figure 6.8-2 *Solids Feeders*

121

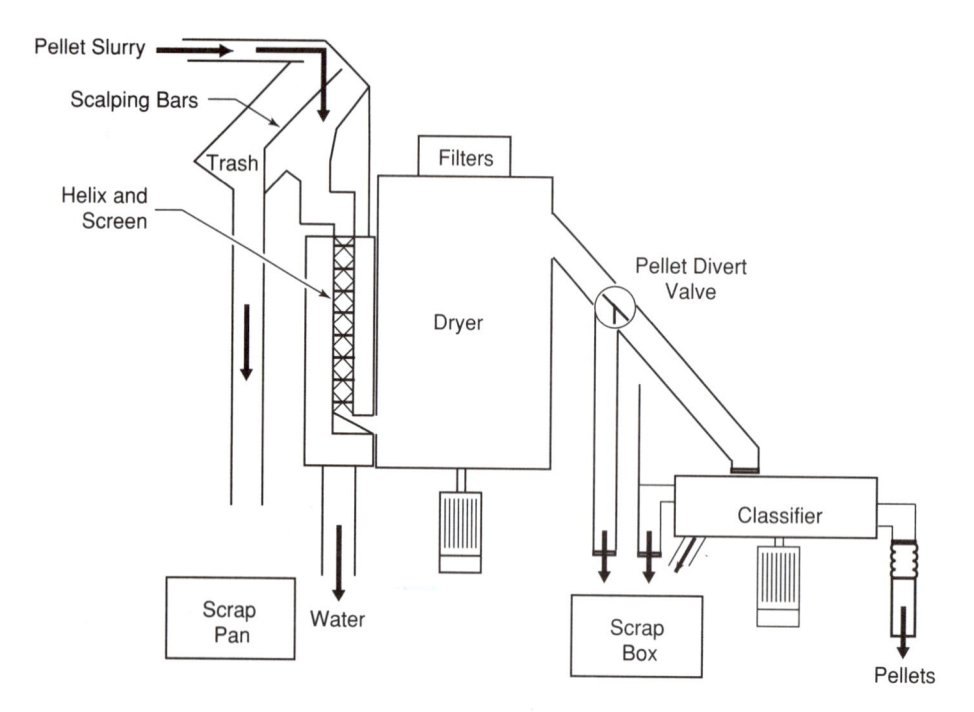

Figure 6.8-3 Classifier

6.9 Pressure Relief Equipment

Flare systems are used to safely remove excess hydrocarbons from a variety of plant processes, Figure 6.9–1. Flare systems are connected by a complex network of pipes and headers to a knock-out drum and flare. Governmental laws and regulations require the flare to be located a safe distance from the operating units and populated areas. Flare systems are part of a plant's safety system. Most process units are lined up to safety relief valves that lift when specified pressures are exceeded. These safety valves discharge into the flare header. Unexpected process upsets are dumped to the flare system as a last course of action. A typical flare system includes:

- Flare—a long, narrow pipe mounted vertically
- Steam ring—mounted at the top of the flare; used to disperse hydrocarbon vapors
- Ignition source—located at the top of the flare
- Fan—mounted at the base of the flare and used for forced draft operation
- Knock-out drum with water seal
- Flare header

Summary

Heat exchangers transfer energy in the form of heat between two fluids without physically coming into contact with each other. A shell and tube heat exchanger is characterized by two shell ports and two tube ports. A standard exchanger has a shell, tubes, tube sheet, shell inlet and

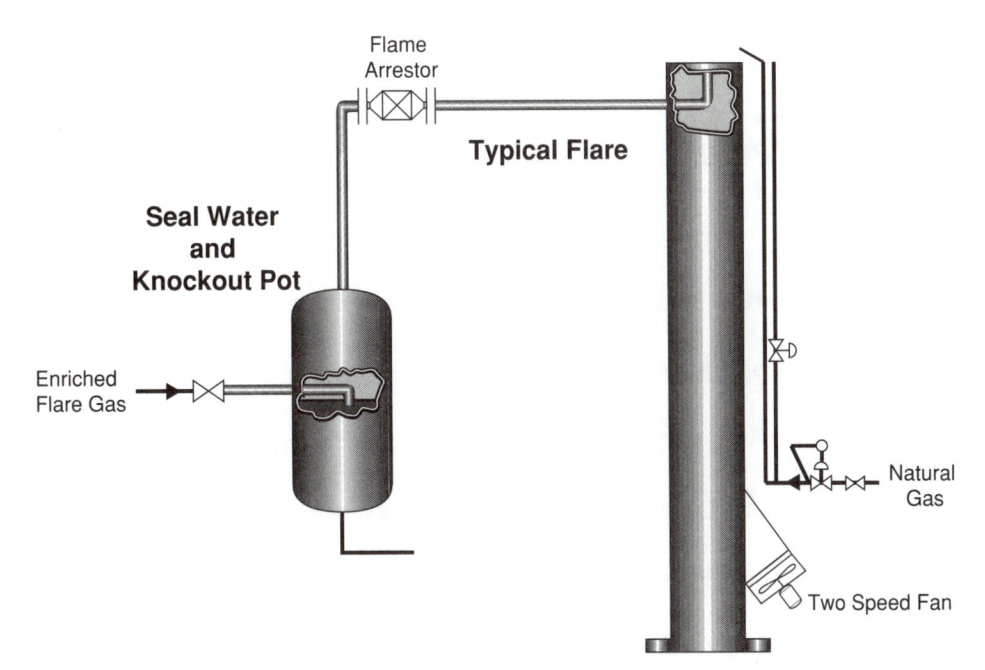

Figure 6.9-1 Flare System

outlet, tube inlet and outlet, and baffles. Cooling towers consist of a box-shaped collection of multi-layered wooden slats and louvers that directs airflow and breaks up water as it falls from the top of the tower or water distribution header. Cooling towers are classified by the way they produce airflow and by the way the air moves in relation to the downward flow of water.

Fired heaters consist of a battery of tubes that pass through a firebox. Fired heaters or furnaces are commercially used to heat up large volumes of crude oil or hydrocarbons. Boilers are classified into two groups, water tube and fire tube. Boilers provide steam for industrial applications. Steam traps remove condensate from steam systems. An incinerator is a permitted device used to burn industrial wastes.

Reactors are used to combine raw materials, heat, pressure, and catalysts in the right proportions. Reactors are classified as either batch or continuous and as fixed bed or fluidized bed design. An absorber is a device used to remove selected components from a gas stream by contacting it with a gas or liquid. An adsorber is a device (reactor, dryer, etc.) filled with a porous solid designed to remove gases and liquids from a mixture. Distillation towers separate chemical mixtures by boiling points. A distillation tower is a collection of stills stacked one on top of the other. As chemical mixtures enter the tower they are separated by boiling point and distributed on different trays or sections of the tower. Distillation columns fall into two distinct classes: plate and packed. A scrubber is a device used to remove chemicals and solids from process gases.

Flares are designed to safely burn excess hydrocarbons. A flare system is composed of a flare, knock-out drum, flare header, fan (optional), steam line and steam ring, fuel line, and burner. A flare is a tall pipe located a specified distance from the facility.

Chapter 6

Review Questions

1. Draw a shell and tube heat exchanger. Label and show flows with a red pen.

2. Draw and label a cooling tower. Label and show flows with a red pen.

3. Compare heat transfer in a cooling tower and a heat exchanger.

4. Draw and label a furnace. Label and show flows with a red pen.

5. Describe how a typical distillation column works.

6. What is the primary function of a reactor?

7. Draw and label a boiler. Illustrate flows with a red pen.

8. Draw and label an absorption column.

9. Draw a simple flare system.

10. What is the purpose of an industrial filter?

11. Draw and label a plate distillation column. Label and show flows with a red pen.

12. Draw and label a packed column.

13. Draw and label a mixing reactor.

14. Sketch a simple stripping column.

15. Describe a scrubber's primary function.

16. Describe an absorption column's primary function.

17. Describe the function and operation of an extruder.

18. Describe the function and operation of a solids feeder.

19. Compare water tube and fire tube boilers.

20. List the four basic furnace designs discussed in this chapter.

Chapter 7

Process Instrumentation

OBJECTIVES

After studying this chapter, the student will be able to:

- *Draw the basic symbols for equipment used in the chemical processing industry.*

- *Describe the basic equipment found in the chemical processing industry.*

- *Identify and draw standard instrument symbols.*

- *Describe the standard instruments used in industry.*

- *Draw typical line symbols used in industry.*

- *Describe the various process equipment relationships.*

- *Draw a simple flow diagram.*

- *Draw a process and instrument drawing (P&ID).*

- *Identify the five elements of a control loop.*

- *Describe process legends, foundation, elevation, electrical, and equipment location drawings.*

KEY TERMS

Control loop—a collection of instruments that work together to automatically control a process.

Elevation drawing—a graphical representation that shows the location of process equipment in relation to existing structures and ground level.

Electrical drawings—symbols and diagrams that depict an electrical process system.

Equipment location drawings—show the exact floor plan location of equipment in relation to the plant's physical boundaries.

Flow diagram—a simplified diagram that uses process symbols to describe the primary flow path through a unit.

Foundation drawings—concrete, wire mesh, and steel specifications that identify width, depth, and thickness of footings, support beams, and foundation.

Legends—describe symbol meanings, abbreviations, prefixes, and other specialized equipment.

P&ID—Piping and instrument drawing; a complex diagram that uses process symbols to describe a process unit.

Process equipment—piping, tanks, valves, pumps, compressors, steam turbines, heat exchangers, cooling towers, furnaces, boilers, reactors, distillation towers, etc.

Process instrumentation—transmitters, controllers, transducers, primary elements and sensors, etc.

Process symbols—graphically depict process equipment, piping, and instrumentation.

Chapter 7 • *Process Instrumentation*

7.2 PFDs and P&IDs

Process flow diagrams (PFDs) and process instrumentation drawings (P&IDs) are used to outline or explain the complex flows, equipment, instrumentation, electronics, elevations, footings, and foundations that exist in a process unit. New technicians are required to study a simple flow diagram of their assigned unit. Process flow diagrams typically include the major equipment and piping path the process takes through the unit. As operators learn more about symbols and diagrams they graduate to the much more complex process instrumentation drawing.

Some symbols are common between plants while others change depending upon the company. In other words, there may be two different symbols used to identify a centrifugal pump or a valve. Some standardization of process symbols and diagrams is taking place. The symbols used in this chapter reflect a wide variety of petrochemical and refinery operations.

A process instrumentation and piping drawing (P&ID) is a complex representation of the various units found in a plant. A simple process flow diagram (PFD) is typically used to describe the pri-

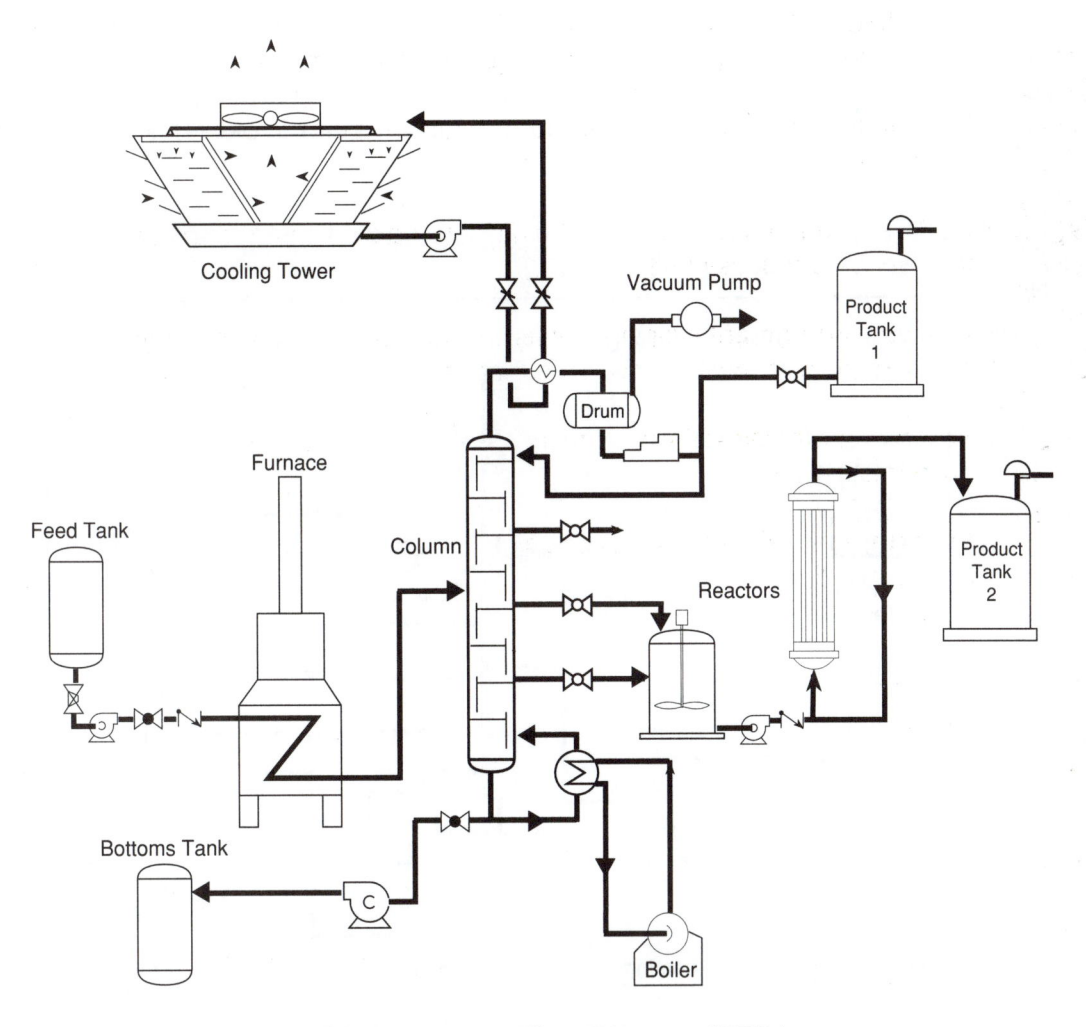

Figure 7.2-1 *Process Flow Diagram (PFD)*

127

Process Instrumentation • Chapter 7

mary flow path through a unit. Like a well illustrated road map a P&ID can show intricate details of a unit that cannot easily be noticed during a walk-through. Process technicians are expected to read simple flow diagrams within hours of starting their initial training. Technicians will graduate to complex P&IDs over the course of their training.

In order to read a P&ID an understanding of the equipment, instrumentation and technology is needed. Some of this equipment includes piping, valves, pumps, tanks, compressors, steam turbines, process instrumentation, heat exchangers, cooling towers, furnaces, boilers, reactors and distillation columns. The next step in using a P&ID is to memorize your plant's process symbol list. This information can be found on the process legend. Process and instrumentation drawings have a variety of elements. Some of these elements include flow diagrams, equipment layouts, elevation plans, electrical layouts, title blocks and legends, footings, and foundation drawings.

Process diagrams can be broken down into two major categories: flow diagrams and P&IDs. A flow diagram is a simplified illustration that uses process symbols to describe the primary flow path through a unit. A process and instrument drawing (P&ID) is a complex diagram that uses process symbols to describe a process unit.

Figure 7.2–1 on page 127 shows the basic relationships and flow paths found in a process unit. It is easier to understand a simple flow diagram if it is broken down into four different sections: feed, pre-heating, the process, and the final products, Figure 7.2–2. This simple left to right

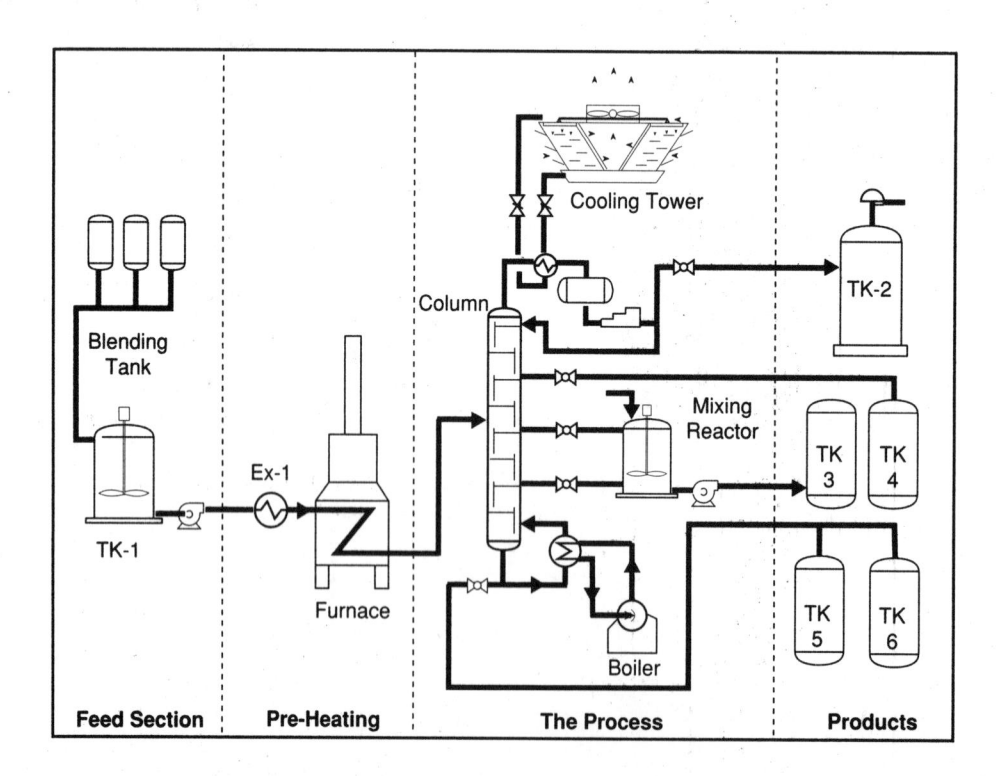

Figure 7.2-2 *4 Section Flow Diagram*

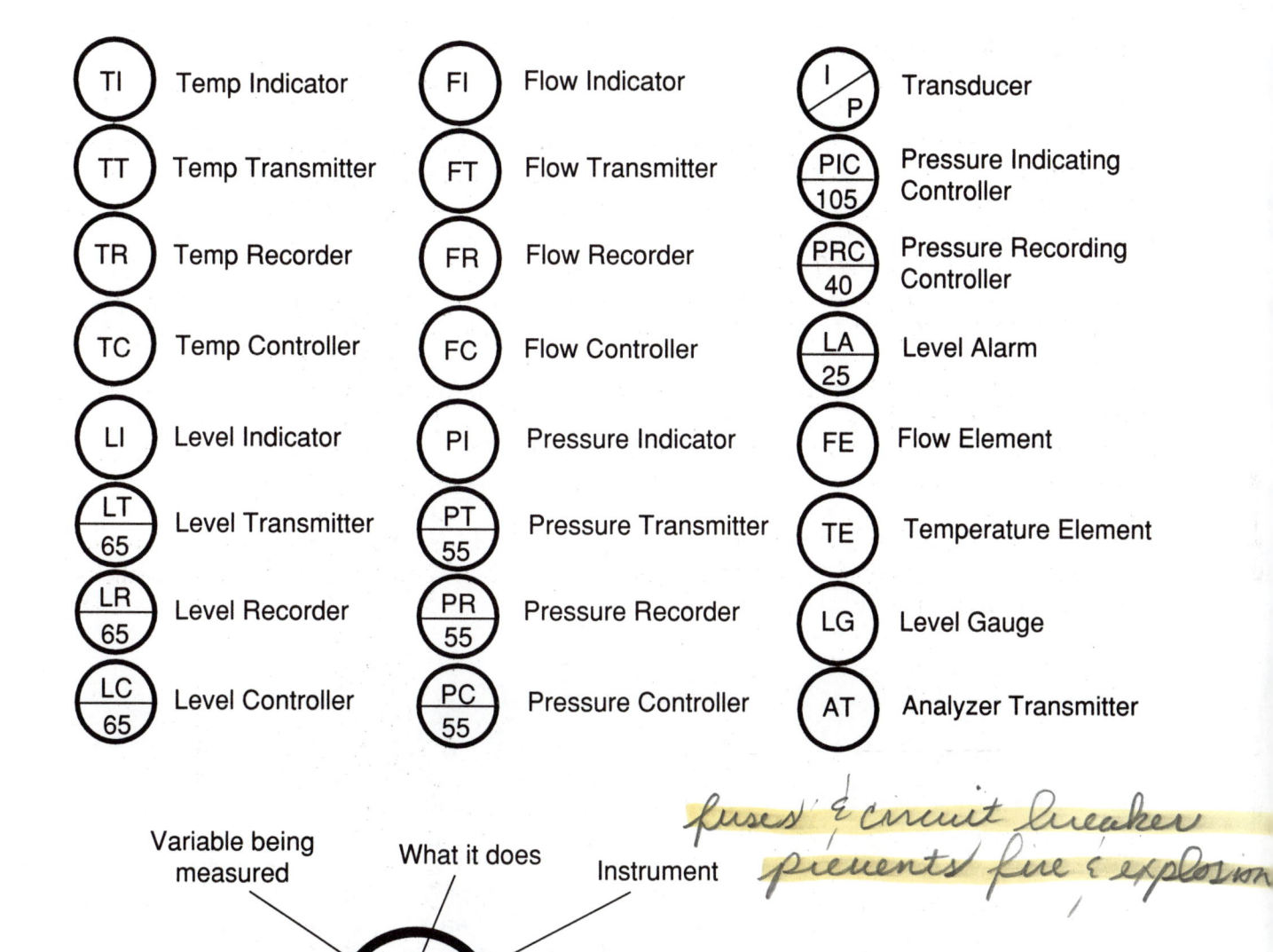

Chapter 7 • **Process Instrumentation**

Symbol	Name	Symbol	Name	Symbol	Name
TI	Temp Indicator	FI	Flow Indicator	I/P	Transducer
TT	Temp Transmitter	FT	Flow Transmitter	PIC 105	Pressure Indicating Controller
TR	Temp Recorder	FR	Flow Recorder	PRC 40	Pressure Recording Controller
TC	Temp Controller	FC	Flow Controller	LA 25	Level Alarm
LI	Level Indicator	PI	Pressure Indicator	FE	Flow Element
LT 65	Level Transmitter	PT 55	Pressure Transmitter	TE	Temperature Element
LR 65	Level Recorder	PR 55	Pressure Recorder	LG	Level Gauge
LC 65	Level Controller	PC 55	Pressure Controller	AT	Analyzer Transmitter

fuses & circuit breaker prevents fire & explosion

FIC / 55
- Variable being measured
- What it does
- Instrument
- Control Loop
- Remote Location (board mounted)
- Remote Location (behind control panel)
- Field Mounted

Figure 7.2-3 *Instrument Symbols*

Control Loop
Transmitter, Controller, Transducer & Valve

129

Process Instrumentation • Chapter 7

Figure 7.2-4 P&ID

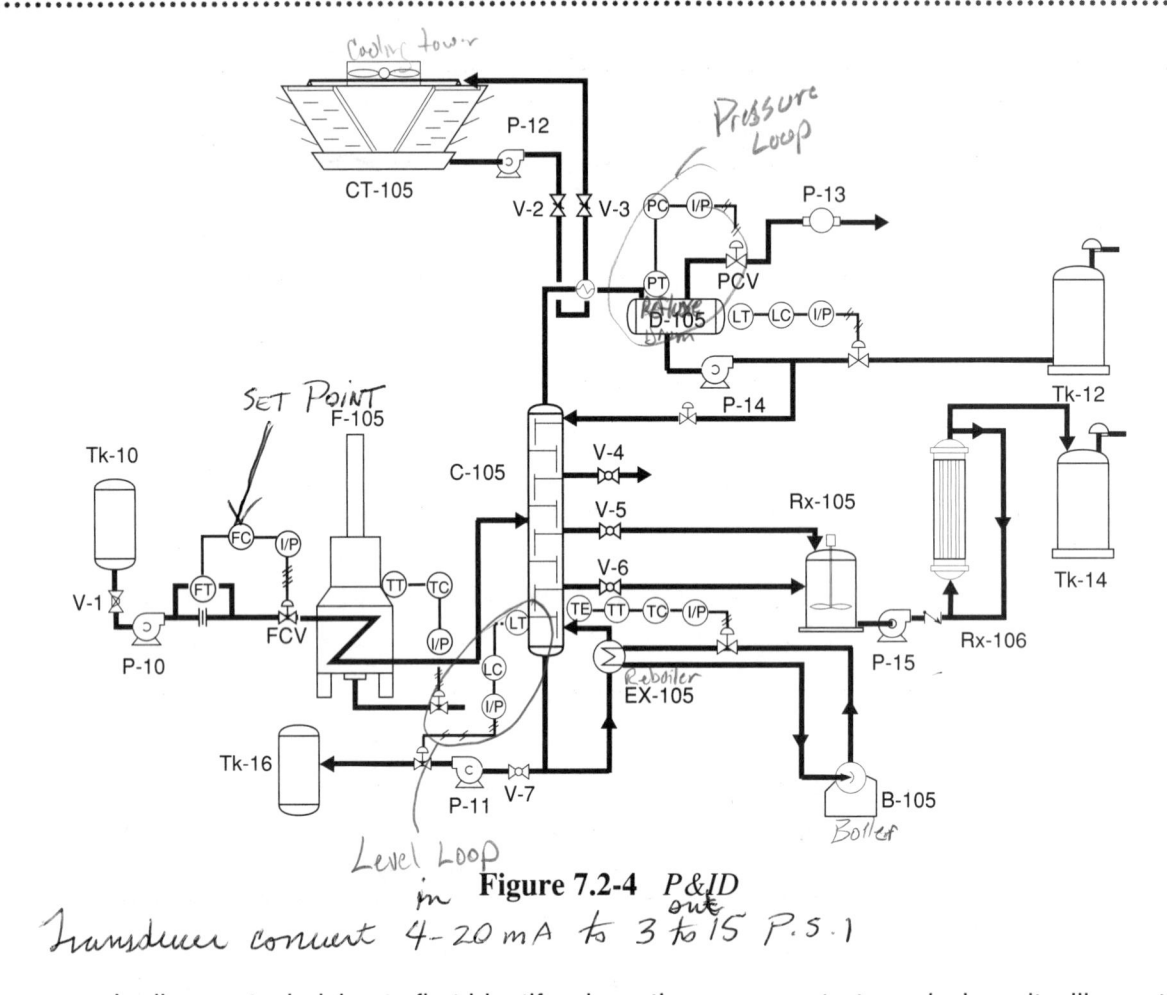

approach allows a technician to first identify where the process starts and where it will eventually end. The feed section includes the feed tanks, mixers, piping, and valves. In the second step the process flow is gradually heated up for processing. This section includes heat exchangers and furnaces. In the third section the process is included. Typical examples found in the process section could include distillation or reaction. The process area is a complex collection of equipment that works together in a system. The process is designed to produce products that will be sent to the final section.

Instrumentation symbols are shown on a P&ID as a circle. Inside the circle information is included that tells the process technicians what type of instrument is represented. Figure 7.2–3 includes examples of typical instrument symbols.

The P&ID includes a graphic representation of the equipment, piping, and instrumentation, Figure 7.2–4. Modern process control is vividly illustrated in this type of drawing. Process technicians can look at their process and see how the engineering department has automated their unit. Pressure, temperature, flow, level, and analytical control loops are all included on the unit P&ID.

Symbols and diagrams have been developed for most pieces of industrial equipment, process flows, and instrumentation. The symbols covered in this chapter include those typically used with

Chapter 7 • Process Instrumentation

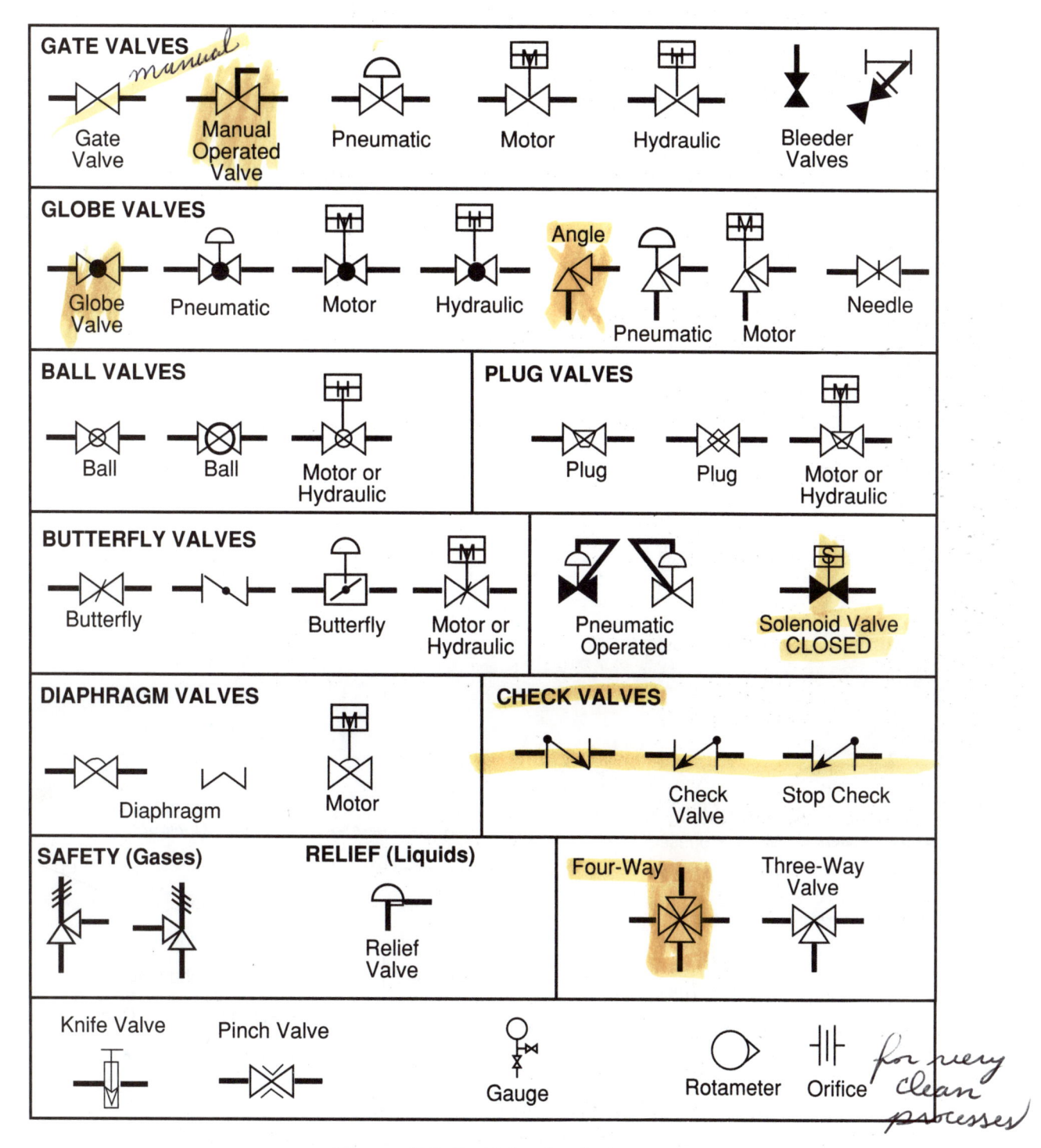

Figure 7.2-5 *Valve Symbols*

valves, pumps, compressors, steam turbines, heat exchangers, cooling towers, furnaces, boilers, distillation columns, instrumentation, and process flows. Figures 7.2–5 and 7.2–6 include many of the basic symbols found in the chemical processing industry.

Process Instrumentation • Chapter 7

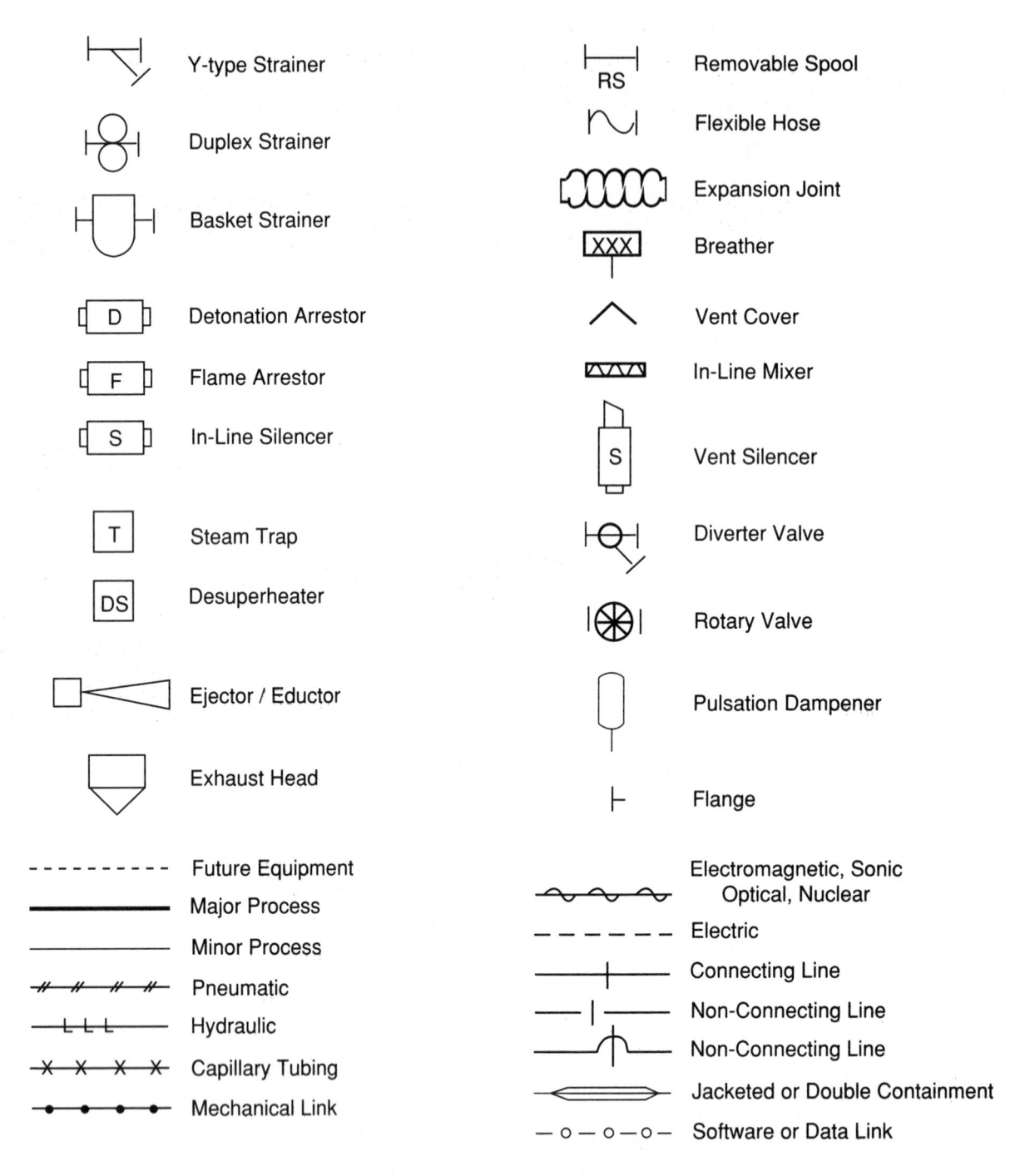

Symbol	Name	Symbol	Name
	Y-type Strainer		Removable Spool
	Duplex Strainer		Flexible Hose
	Basket Strainer		Expansion Joint
			Breather
	Detonation Arrestor		Vent Cover
	Flame Arrestor		In-Line Mixer
	In-Line Silencer		Vent Silencer
	Steam Trap		Diverter Valve
	Desuperheater		Rotary Valve
	Ejector / Eductor		Pulsation Dampener
	Exhaust Head		Flange
----------	Future Equipment		Electromagnetic, Sonic Optical, Nuclear
	Major Process	– – – – – –	Electric
	Minor Process		Connecting Line
	Pneumatic		Non-Connecting Line
	Hydraulic		Non-Connecting Line
	Capillary Tubing		Jacketed or Double Containment
	Mechanical Link	— o — o — o —	Software or Data Link

Figure 7.2-6 *Piping Symbols*

Each plant has a standardized file for their piping symbols. Process technicians should carefully review the piping symbols for major and minor flows, electric, pneumatic, capillary, hydraulic, and future equipment. The major flow path through a unit illustrates the critical areas of focus for a

Chapter 7 • *Process Instrumentation*

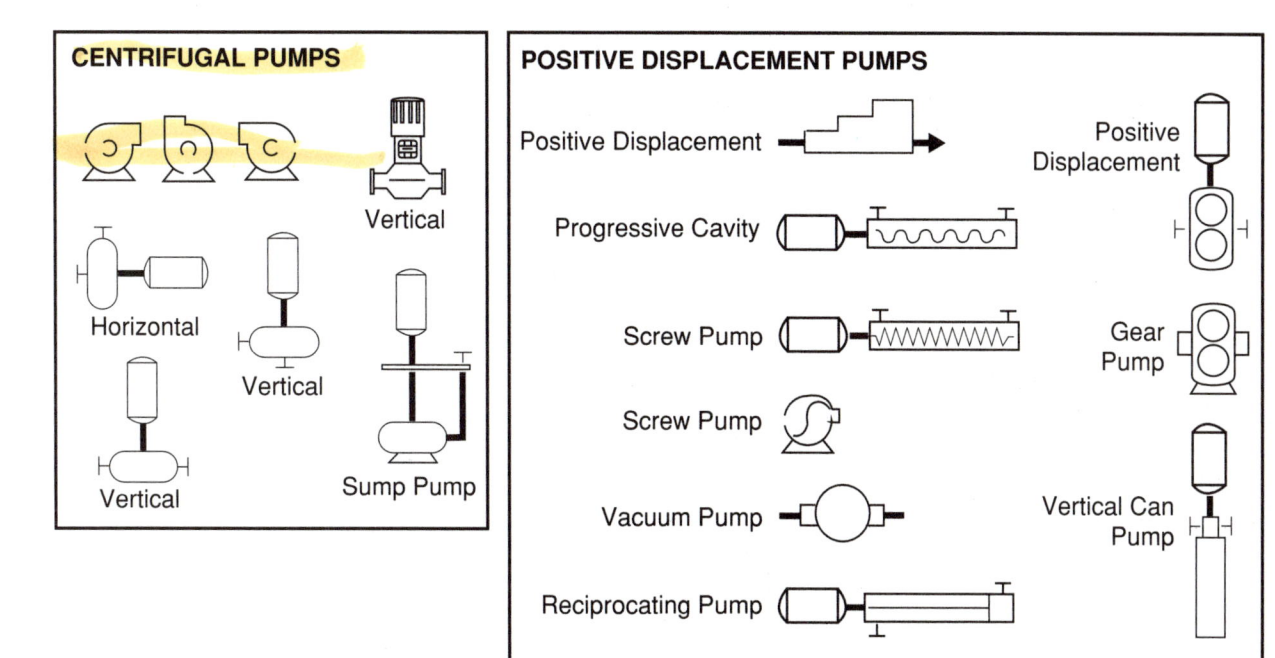

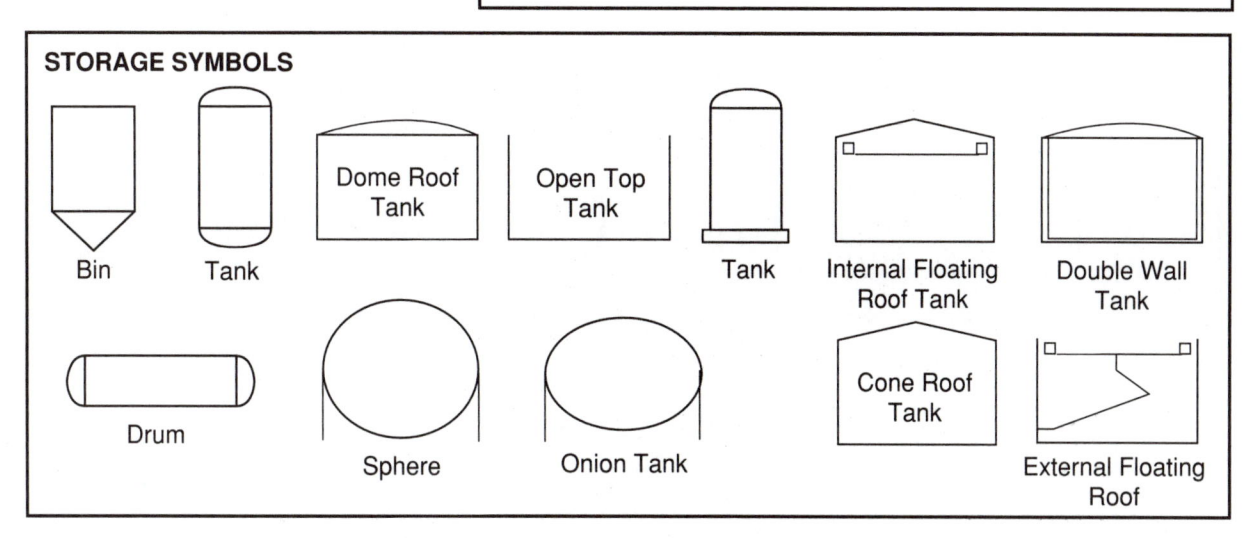

Figure 7.2-7 *Pump and Tank Symbols*

new technician. A variety of other symbols are laced into the piping symbols. Some of these devices include strainers, filters, flanges, spool pieces, and steam traps.

Pumps and tanks come in a variety of designs and shapes. Process symbols are designed to graphically display the process unit. Common pump and tank symbols can be found in Figure 7.2–7.

Compressors and pumps share a common set of operating principles. The dynamic and positive displacement families share common categories. The symbols for compressors may closely resemble a pump. In most cases the symbol is slightly larger in the compressor symbol file. In

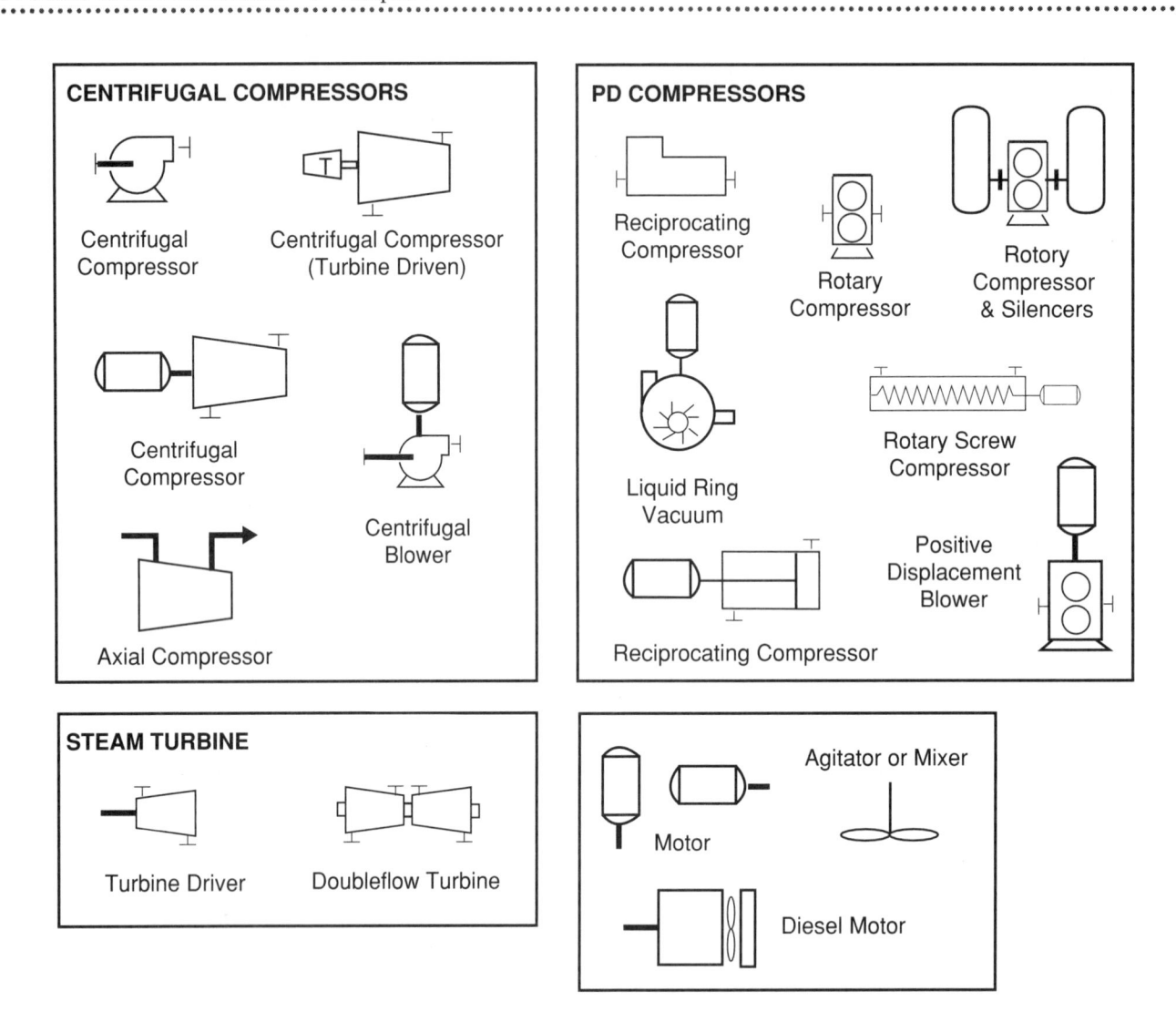

Figure 7.2-8 *Compressor, Steam Turbine, and Motor Symbols*

the multi-stage centrifugal compressor, the symbol clearly describes how the gas is compressed prior to being released. This is in sharp contrast to the steam turbine symbol which illustrates the opposite effect as the steam expands while passing over the rotor. Modern P&IDs show the motor symbol connected to the driven equipment. This equipment may be a pump, compressor, mixer, or generator. Figure 7.2–8 illustrates the standardized symbols for compressors, steam turbines, and motors.

Heat exchangers and cooling towers are two types of industrial equipment that share a unique relationship. A heat exchanger is a device used to transfer heat energy between two process flows. The cooling tower performs a similar function, however cooling towers and heat exchangers use different scientific principles to operate. Heat exchangers transfer heat energy through conductive and convective heat transfer while cooling towers transfer heat energy to the outside

Chapter 7 • *Process Instrumentation*

SHELL SIDE
TUBE SIDE

Plate and Frame Heat Exchanger

Hairpin Exchanger

Air Cooled Exchanger (Louvers Optional)

U-Tube Heat Exchanger

Double-Pipe Heat Exchanger

Single Pass Heat Exchanger

Spiral Heat Exchanger

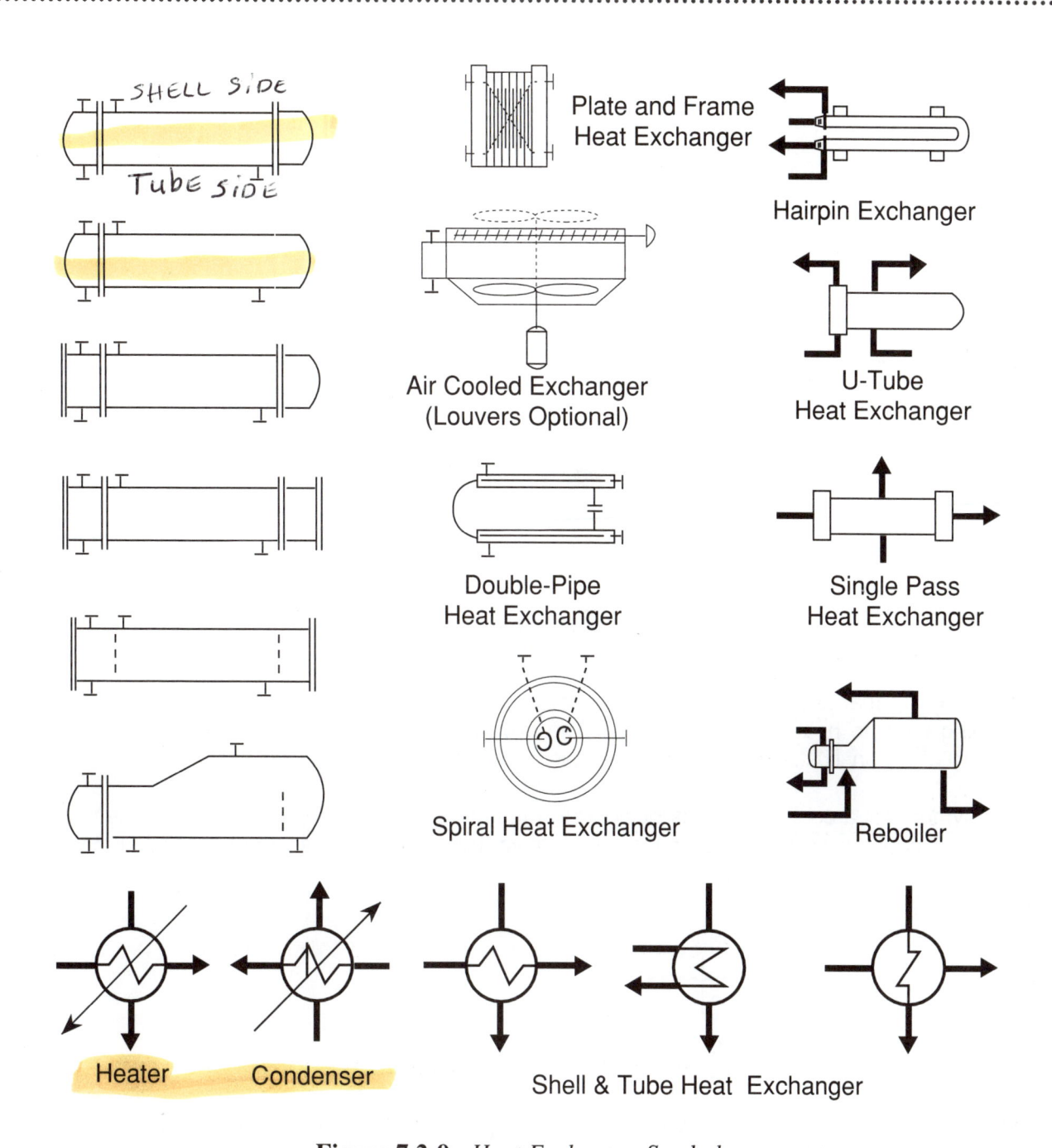

Reboiler

Heater   Condenser   Shell & Tube Heat Exchanger

Figure 7.2-9 *Heat Exchanger Symbols*

air through the principle of evaporation. Figures 7.2–9 and 7.2–10 illustrate the standard symbols used for heat exchangers and cooling towers.

The symbol for a heat exchanger clearly illustrates the flows through the device. It is important for a process technician to be able to see the shell inlet and outlet and the tube inlet and outlet

135

Process Instrumentation • Chapter 7

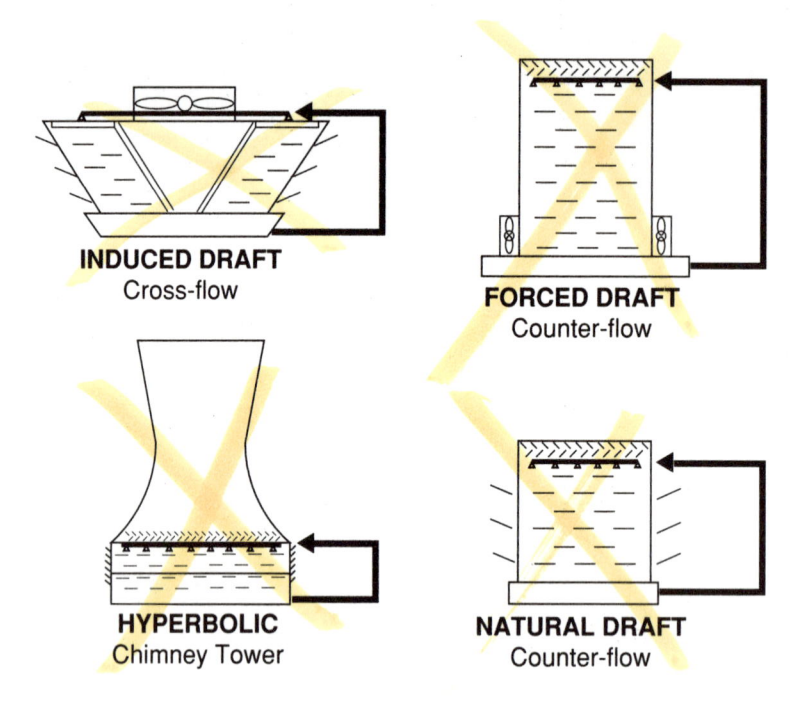

INDUCED DRAFT
Cross-flow

FORCED DRAFT
Counter-flow

HYPERBOLIC
Chimney Tower

NATURAL DRAFT
Counter-flow

Figure 7.2-10 *Cooling Tower*

flow paths. A heat exchanger with an arrow drawn through the body illustrates whether the device is being used to heat or cool a product. The downward direction indicates heating, while the upward direction illustrates cooling.

The symbol for a cooling tower is designed to resemble the actual device in the process unit. Cooled product flows out of the bottom of the tower and to the processing units, while hot water returns to a point located above the fill.

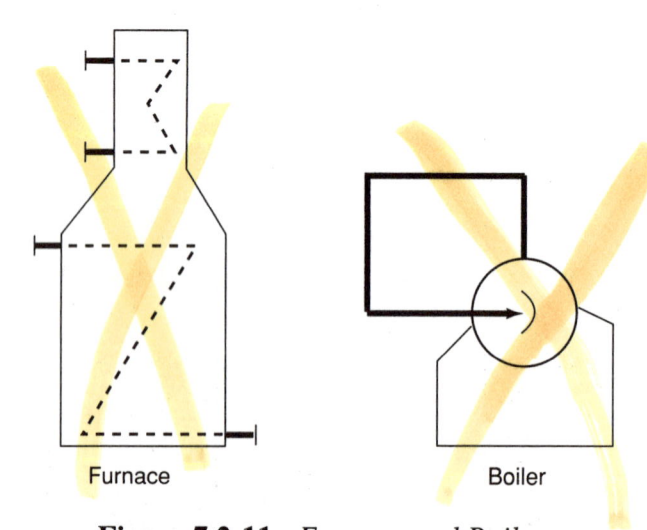

Furnace

Boiler

Figure 7.2-11 *Furnace and Boiler*

Chapter 7 • *Process Instrumentation*

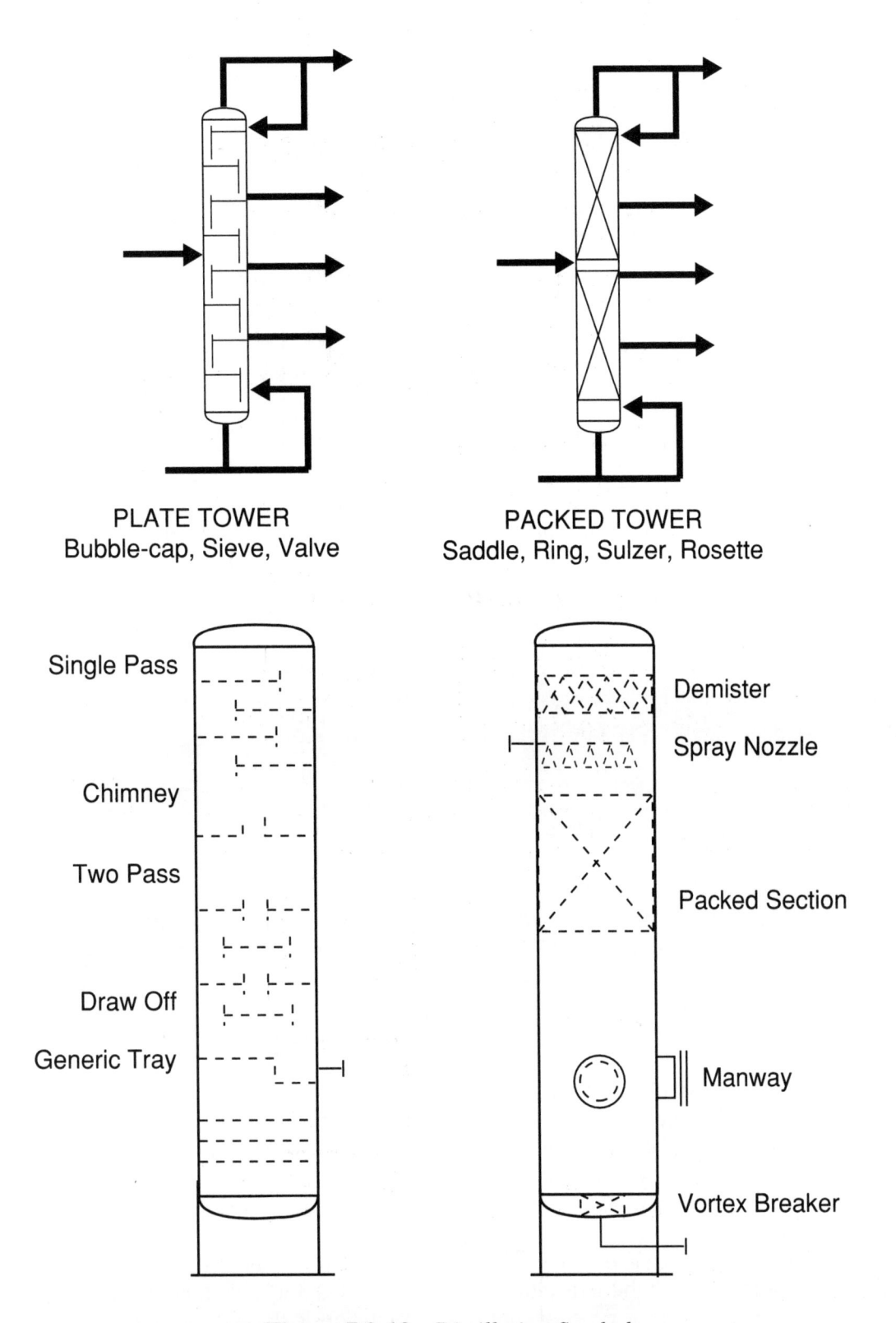

PLATE TOWER
Bubble-cap, Sieve, Valve

PACKED TOWER
Saddle, Ring, Sulzer, Rosette

Single Pass

Chimney

Two Pass

Draw Off

Generic Tray

Demister

Spray Nozzle

Packed Section

Manway

Vortex Breaker

Figure 7.2-12 *Distillation Symbols*

Process Instrumentation • Chapter 7

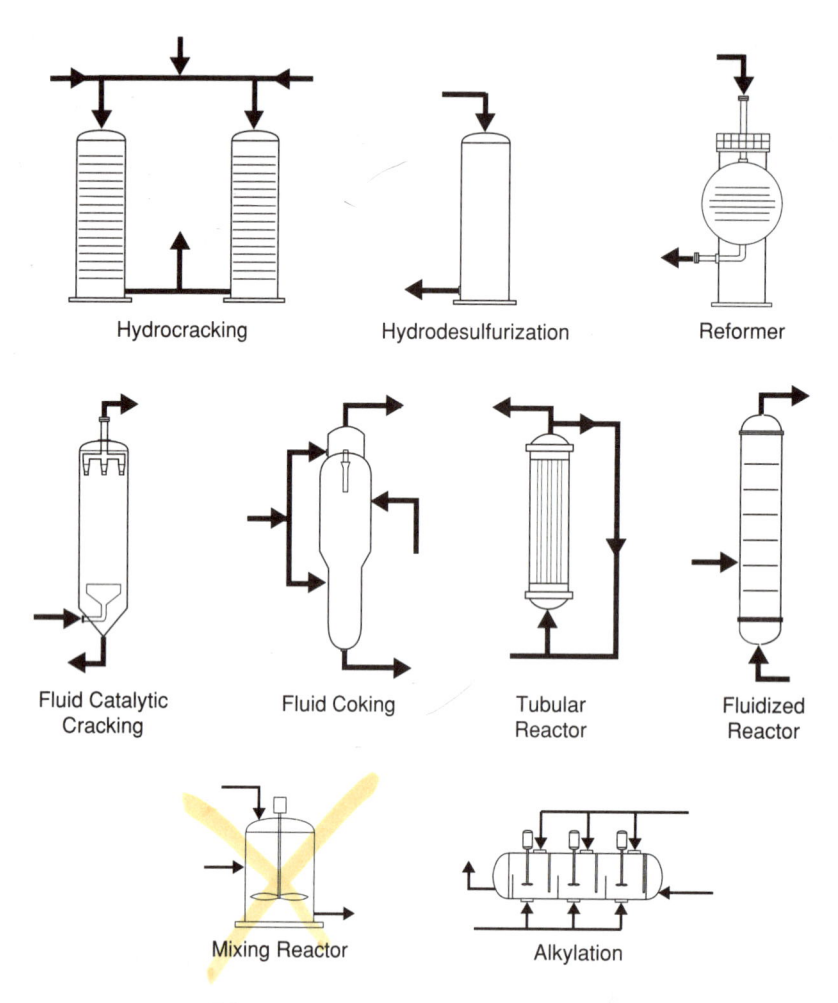

Figure 7.2-13 *Reactor Symbols*

On a typical P&ID, distillation columns, reactors, boilers, and furnaces are drawn as they visually appear in the plant. (Refer to Figure 7.2–11 thru 7.2–13 for the standard symbols file for these devices.) If a proprietary process includes several types of equipment not typically found on a standard symbol file, the designer draws the device as it visually appears in the unit. Figure 7.2–11 illustrates standard symbols used for boilers and furnaces.

Distillation columns come in two basic designs, plate and packed. Flow arrangements vary from process to process. The symbols allow the technician to identify primary and secondary flow paths. The two standard symbols for distillation columns can be found in Figure 7.2–12. The symbols for a plate column may resemble the top and lower left graphics in Figure 7.2–12. The graphics on the right are typical packed column symbols found on unit P&IDs.

Reactors are stationary vessels and can be classified as batch, semi-batch, or continuous. Some reactors use mixers to blend the individual components. Reactor design is dependent upon the type of service. Some of these processes include alkylation, catcracking, hydrodesulfurization, hydrocracking, fluid coking, reforming, polyethylene, and mixed xylenes.

Chapter 7 • *Process Instrumentation*

7.3 Process Instruments

Basic process instrumentation includes computers, gauges, recorders, transmitters, controllers, transducers, primary elements and sensors, switches, and control valves, Figure 7.3–1. Process technicians use instruments to control complex industrial processes. Thirty years ago, most operators controlled the processes in their plant manually. This type of process was "valve intensive," requiring the technician to open and close line-ups manually. Basic process instruments improved in the automation era. A single process technician can monitor and control a much larger process from a single control center.

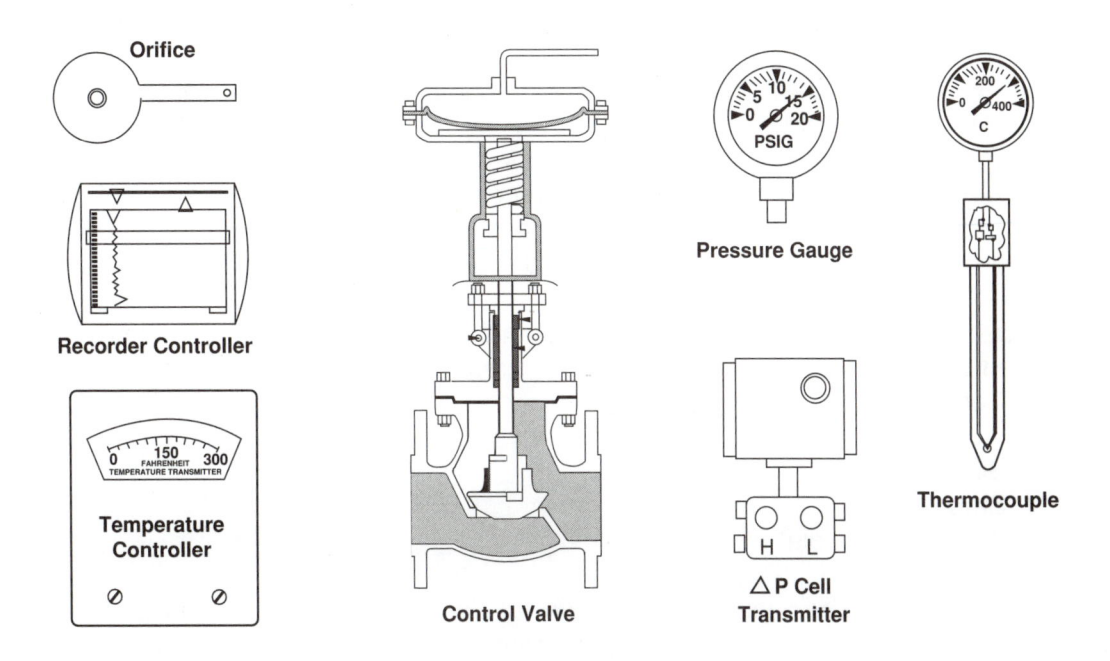

Figure 7.3-1 *Basic Instruments*

7.4 Basic Elements of a Control Loop

Process technicians use instrumentation to control a variety of automated processes. The key component of automatic control is the control loop, Figure 7.4–1. A control loop is a group of instruments that work together to control a process. These instruments typically include a transmitter coupled with a sensing device or primary element, a controller, a transducer, and a control valve. Process plants are composed of hundreds of control loops. These control loops are used to maintain pressure, temperature, flow, and level. The basic elements of a control loop are:

Measurement Device—Primary elements and sensors
- Flow—orifice plate, flow nozzle
- Level—float, displacer
- Pressure—helix, spiral, bellows
- Temperature—thermocouple, thermal and resistance bulb

RTD

139

Process Instrumentation • Chapter 7

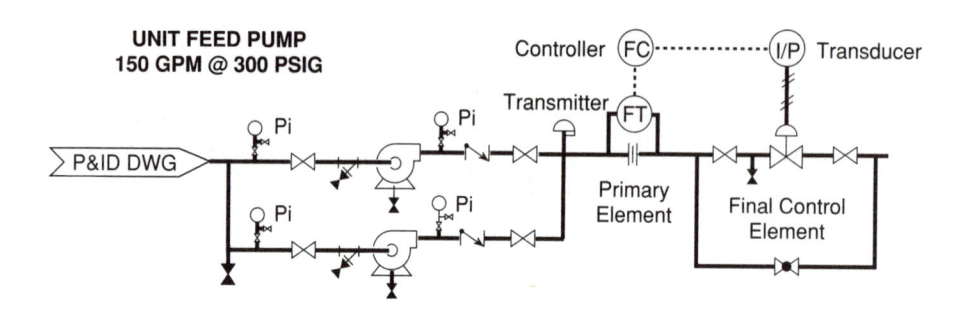

Figure 7.4-1 *Typical Control Loop*

Transmitter—a device designed to convert a measurement into a signal. This signal will be transmitted to another instrument.
- Pressure transmitter—tubing to process
- Temperature transmitter—tubing to process
- Flow transmitter—DP cell, high/low pressure taps
- Level transmitter—hooked to float or displacer

Controller—a device designed to compare a signal to a setpoint and transmit a signal to a final control element. *Brains of process most important part of loop*
- Recording
- Indicating
- Blind
- Strip chart
- Vertical and scale

Transducer—a device designed to convert an air signal to an electric signal or an electric signal to a pneumatic signal. Sometimes referred to as an I to P or as a converter.
- Air signal to an electric signal
- Electric signal to a pneumatic signal

Final control element—the part of a control loop that actually makes the change to control the process.
- Control valve
- Motor on a pump or compressor

7.5 Process Variables and Control Loops

Process variables typically fall into five different groups: pressure, temperature, flow, level, and analytical variables. Control loops are specifically designed to work with a selected variable. Process technicians monitor control process variables.

Figure 7.5–1 includes several examples of a flow control loop. Flow loops are typically designed so a measurement of the flow rate is taken first and then the flow is interrupted or controlled downstream. Flow control loops start at the primary element. Flow control primary elements could include orifice plates, venturi tubes, flow nozzles, nutating disks, oval gears, or turbine meters. The most common primary element is the orifice plate. Orifice plates artificially create a

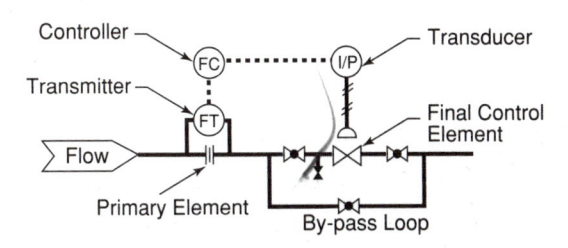

Figure 7.5-1 *Flow Control Loop*

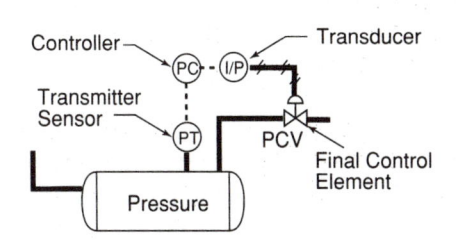

Figure 7.5-2 *Pressure Control Loop*

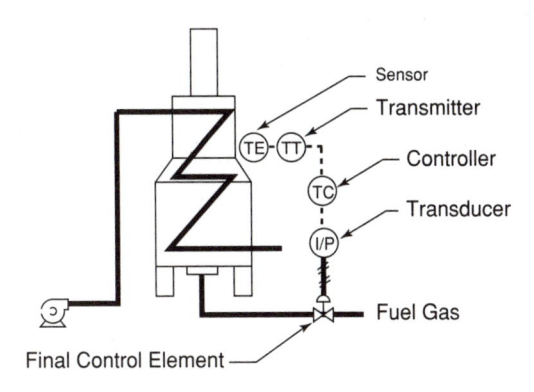

Figure 7.5-3 *Temperature Control Loop*

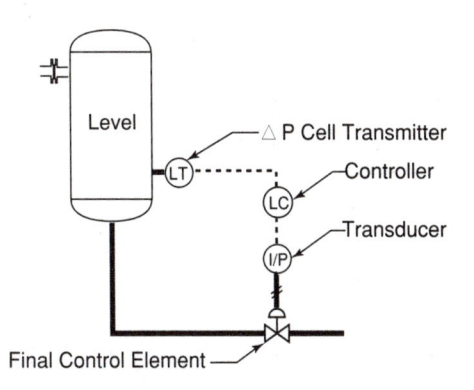

Figure 7.5-4 *Level Control Loop*

high pressure/low pressure situation that can be measured by the transmitter. Primary elements are typically used in conjunction with a transmitter. Although it appears that the primary element is interrupting the flow, this is not the case. Increased velocity across the orifice plate compensates for the restriction. The transmitted signal is sent to a controller that compares the incoming signal with the desired setpoint. If a change is required, the controller sends a signal to a final control element.

Control loop design uses the five elements of the control loop. The one area that changes consistently is the first: primary elements and sensors. Pressure control loops use devices to detect pressure changes. These primary elements are typically expansion type devices. Primary pressure elements include bourdon, helical, spiral, bellows, pressure capsule, or diaphragm. Figure 7.5–2 includes a pressure transmitter, controller, transducer, and control valve.

Figure 7.5–3 is a simple layout of a temperature control loop. In large fired furnaces a temperature measurement is taken at the furnace or from the exiting charge. The primary sensors used to detect temperature are thermocouples or RTDs, often called temperature elements. Temperature elements are linked to transmitters. A 4 to 20 milliamps (mA) signal is sent to a controller which compares it to a setpoint. Controllers may be located in the field near the equipment or in a remote location. The controller sends an electric signal to a transducer that is typically located near the valve to eliminate process lag. The transducer converts the electric signal to a pneumatic signal of 3 to 15 psi. The control valve in Figure 7.5–3 opens and closes depending upon the signal. Temperature is controlled by reducing or increasing fuel flow to the burners.

Level control loops use floats, displacers, or differential pressure transmitters. Figure 7.5–4 uses a differential pressure "ΔP cell" to detect level changes. The primary element or sensor is inside the transmitter. These two devices couple-up to detect and send a signal to a level controller. A transducer converts the signal and opens or closes the control valve.

Process Instrumentation • Chapter 7

7.6 Primary Elements and Sensors

	Primary Element	Sensor
Flow	orifice plate, flow nozzle, ΔP cell (diaphragm)	ΔP cell
Level	float, displacer ΔP cell (diaphragm)	ΔP cell
Pressure	helix, spiral, bellows bourdon tube, ΔP cell	ΔP cell
Temperature	capillary tubing thermal and resistance bulb	thermocouple, RTD

7.7 Transmitters and Control Loops

The ΔP cell transmitters, Figure 7.7–1, can be found in two basic designs, pneumatic and electronic. Controllers are typically mounted between 400 feet (closed loop) and 1000 feet (open loop) from the transmitter. The signal from an electronic transmitter is proportional to the difference in the high and low pressure legs. Standard output signals are 4 to 20 mA, 10 to 50 mA, and 1 to 5 V. The 10 to 50 mA transmitter is becoming very popular because it has a higher tolerance to outside interference. Pneumatic transmitters require a 20 psig air supply in order to run the standard 3 to 15 psig output.

The ΔP cells function by running a high and low pressure tap to each side of an internal twin diaphragm capsule. Pressure charges cause the diaphragms to move. This process increases or decreases the signal to the controller.

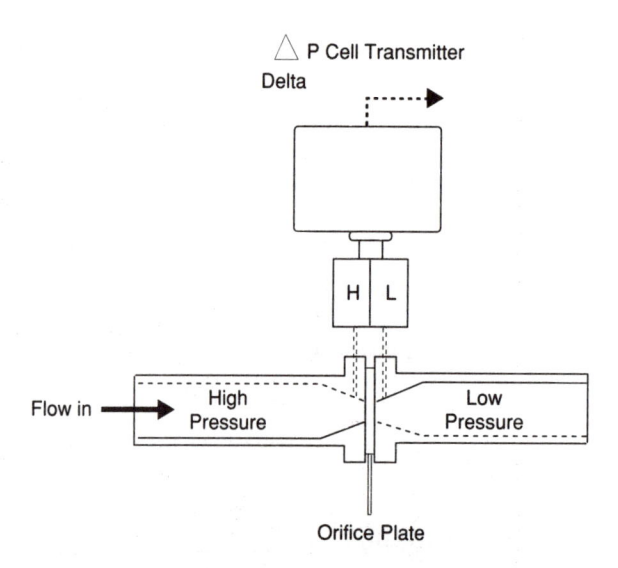

Figure 7.7-1 *ΔP Cell Transmitter*

Chapter 7 • *Process Instrumentation*

Smart transmitters are another type of transmitter frequently found in the chemical processing industry. This type of transmitter is very reliable and does not need constant attention. Smart transmitters have an internal diagnostic system that warns the operator if a problem is about to occur. This type of transmitter can be used with liquid or gas service, pressure, viscosity, temperature, flow, and level. Several features of the smart transmitter include speed, reliability, internal diagnostics, strong digital signal, and remote calibration capabilities.

7.8 Controllers and Control Modes

The primary purpose of a controller, Figure 7.8–1, is to receive a signal from a transmitter, compare this signal to a setpoint, and adjust the (final control element) process to stay within the range of the setpoint. Controllers come in three basic designs: pneumatic, electronic, and electric. Electronic controllers were first introduced in the early 1960s. Before this time period only pneumatic controllers were used. Pneumatic controllers require a clean air supply pressure of 20 psig. Several of the more attractive features in electronic controllers are the reduction of lag time in process changes, low installation expense, and ease of installation.

With the widespread use of the PC a number of applications were found for controller use. Distributed control systems (DCS) began to replace the older pneumatic and electronic controllers. The primary reason was the easy installation and the relatively few wires required to do it.

Most modern plants are still a combination of pneumatic, electronic, and electric systems. When looking at the control loop function it is almost impossible to identify what type of controller (pneumatic, electronic, or DCS) is being used. Controllers can be operated in manual, automatic, or cascade control. During plant start-up the controller is typically placed in the manual position and left there until the process has lined out. This process initiates a setpoint on the final control element; however it does not utilize a controlling function, it only opens the valve 50% or 25% and keeps it there until the technician changes the mode.

Figure 7.8-1 *Controller*

This keeps the process from swinging up and down during start-up. After the process is stable the operator places the controller in automatic and allows the controller to supervise the control loop function. At this point the controller attempts to open and close the control valve to maintain the setpoint. Cascade control is a term used to describe how one control loop controls or overrides the instructions of another control loop in order to achieve a desired setpoint. In this case, the primary control loop controller will override the secondary controller.

Proportional Band

The proportional band on a controller describes the scaling factor used to take a controller from 0% to 100% output. If the proportional band is set at 50% and the amount of lift the final control element (globe valve) has off the seat is four inches the control valve will open two inches. Range is defined as the portion of the process controlled by the controller. For example, the temperature range for a controller may be limited to 80 degrees F to 140 degrees F. Span is the difference Δ between the upper and lower range limits. This value is always recorded as a single number. For example the difference between 80 and 140 is 60.

Controller Modes

Controller modes include proportional (P), proportional plus integral (PI), proportional plus derivative (PR), and proportional-integral-derivative (PID). Proportional control is primarily used to provide gain where little or no load change typically occurs in the process. Proportional plus integral is used to eliminate offset between setpoint and process variables. PI works best where large changes occur slowly. Proportional plus derivative is designed to correct fast changing errors and reduce overshooting the setpoint. PD works best when frequent small changes are required. Proportional integral derivative is applied where massive rapid load changes occur. PID reduces swinging between the process variable and setpoint.

Rate Mode

The rate or derivative mode enhances controller output by increasing the output in relation to the changing process variable. As the process variable approaches the setpoint the rate or derivative mode relaxes, providing a braking action that prevents overshooting the setpoint. The rate responds aggressively to rapid changes and passively to smaller changes in the process variable.

Reset Mode

The reset or integral mode is designed to reduce the difference between the setpoint and process variable by continuously adjusting the controller's output until the offset is eliminated. The reset mode responds proportionally to the size of the error, the length of time that it lasts, and the integral gain setting.

Tuning Controllers

Turning controllers have several functions:
- Turn rate action off
- Set integral (reset) action to minimum
- Establish arbitrary gain
- Set controller to AUTO mode:
  - Reduce gain if process swings
  - Increase gain if process response is too slow
  - The process should draw a straight line when it is in control

7.9 Final Control Elements and Control Loops

Automatic Valves

Final control elements are typically classified as automated valves, however motors or other electrical devices can be used. The final control element is the last link in the modern control loop and is the device that actually makes the change in the process. Automatic valves will open or close to regulate the process. Control loops usually have (1) a sensing device, (2) a transmitter, (3) a controller, (4) a transducer, and (5) an automatic valve. Automatic valves can be controlled from remote locations making them invaluable in modern processing. To automate a valve, a device known as an actuator is installed. The actuator controls the position of the flow control element by moving and controlling the position of the valve stem.

Actuators come in three basic designs: pneumatic, electric, and hydraulic.

- Pneumatically (AIR) operated—this is the most common type of actuator. Pneumatic actuators convert air pressure to mechanical energy. They can be found in three designs: (1) diaphragm, (2) piston, and (3) vane.
  Diaphragm—The diaphragm actuator is a dome-shaped device that has a flexible diaphragm running through the center. It is typically mounted on the top of the valve. The center of the diaphragm in the dome is attached to the stem. The valve position (on or off) is held in place by a powerful spring. When air enters the dome on one side of the flexible diaphragm it opens, closes, or throttles the valve depending on design.
  Piston—The piston actuator uses an airtight cylinder and piston to move or position the stem. Commonly found in use with automated gate valves or slide valves. Used where a lot of stem travel is needed.
  Vane—Vane actuator's direct air against paddles or vanes.
- Electrically operated—Converts electricity to mechanical energy. Examples: solenoid valve and motor-driven actuator.
  Solenoid–Solenoid valves are designed for on-off service. The internal structure of a solenoid resembles a globe valve. The disc rests in the seat, stopping flow. The stem is attached to a metal core or armature that is held in place by a spring. A wire coil surrounds the upper spring and stem. When the wire coil is energized, a magnetic field is set up causing the armature to lift, compressing the spring. The armature is held in place until the current stops.
  Motor—A motor-driven actuator is attached to the stem of a valve by a set of gears. Gear movement controls the position of the stem.
- Hydraulically operated—Converts liquid pressure to mechanical energy.
  The hydraulic actuator uses a liquid-tight cylinder and piston to move or position the stem. Commonly found in use with automated gate valves or slide valves. Used where a lot of stem travel is needed.

Common terminology for actuators:
- Air to open/Spring to close. Fails in the closed position if air system goes down. Air line is typically located on the bottom of the dome.
- Air to close/Spring to open. Fails in the open position if air system goes down. Air line is typically located on the top of the dome.
- Double-acting/No spring. Air lines located on both sides of the dome.

145

The most common type of automated valve is a globe valve because of its versatile on-off or throttling feature. Control loops use on-off or throttling type valves to regulate the flow of fluid in and out of a system. Automatic valves can be used to control pressure, temperature, flow, or level.

Automatic valves fall into the following categories:
- Control valve—air operated, electrically operated, hydraulically operated.
- Spring or weight operated valve—spring operated valves hold the flow control element in place until pressure from under the disk grows strong enough to lift the element from the seat. Example: check valve.

7.10 Interlocks and Permissives

An interlock is a device designed to prevent damage to equipment and personnel. This is accomplished by stopping or preventing the start of certain equipment functions unless a preset condition has been met. There are two types of interlocks: softwire and hardwire. Softwire interlocks are contained within the logic of the control computer software. Hardwire interlocks are a physical arrangement. The hardwire interlock usually involves electrical relays that operate independent of the control computer. In many cases they run side by side with the computer interlocks. However, hardwire interlocks cannot be bypassed. They must be satisfied before the process they are part of can take place.

A permissive is a special type of interlock that controls a set of conditions that must be satisfied before a piece of equipment can be started. Permissives deal with start-up items whereas hardwire interlocks deal with shutdown items. A permissive is an interlock controlled by the DCS. This type of interlock will not necessarily shut down the equipment if one or more of its conditions are not met. It will, however, keep the equipment from starting up.

7.11 P&ID Components

The basic components of a piping and instrument drawing are the process legend, foundation, elevation, electrical, equipment location drawings, simple flow diagram, and the P&ID.

Simple Flow Diagram
A simple flow diagram provides a quick snapshot of the operating unit. Flow diagrams include all primary equipment, flows, and numbers. A technician can use this document to trace the primary flow of chemicals through the unit. Secondary or minor flows are not included. Complex control loops and instrumentation are not included. The flow diagram is used for visitor information and new employee training.

Process Legends
The process legend, Figure 7.11–1, provides the information needed to interpret and read the P&ID. Process legends are found at the front of the P&ID. The legend includes information about piping, instrument and equipment symbols, abbreviations, title block, drawing number, revision number, approvals, and company prefixes. At the present time symbol and diagram standardization is not complete. Many companies use their own symbols file to display unit drawings. Unique and unusual equipment also require a modified symbols file.

Chapter 7 • *Process Instrumentation*

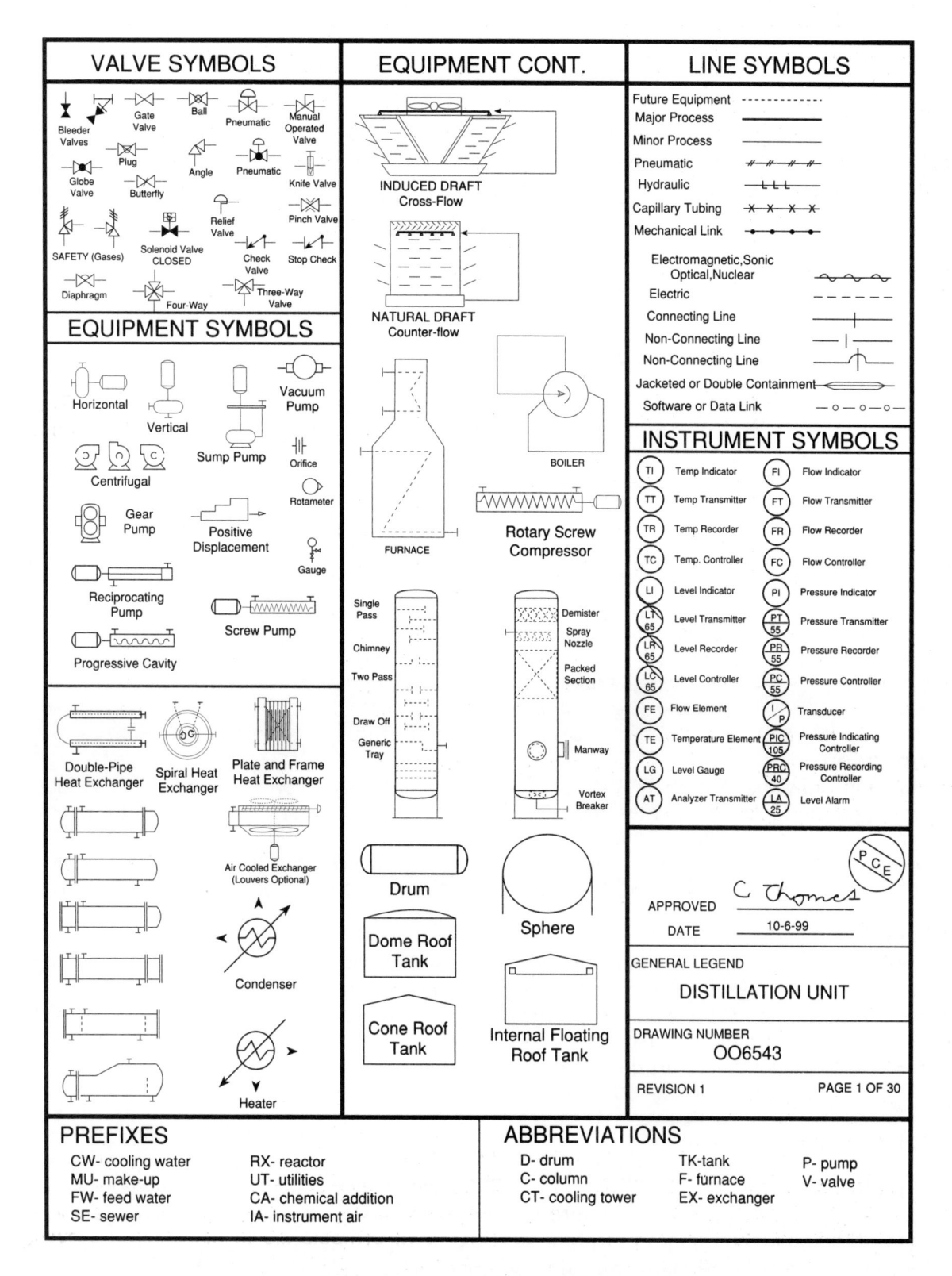

Figure 7.11-1 *Process Legend*

Process Instrumentation • Chapter 7

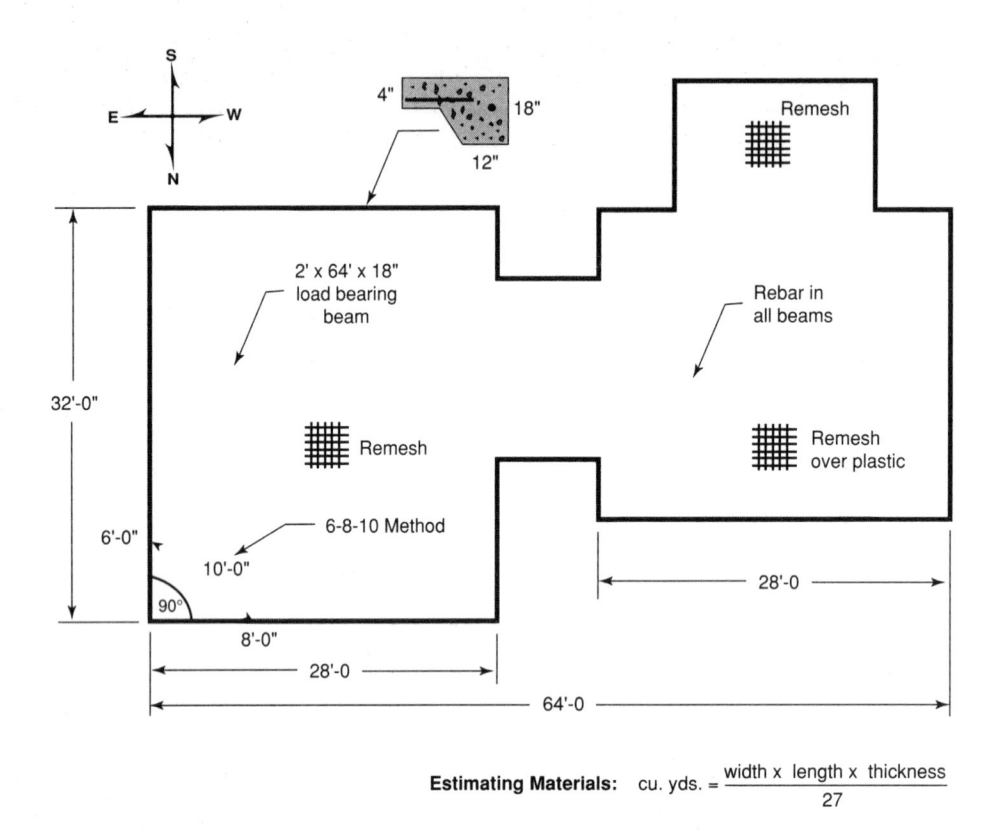

Figure 7.11-2 *Foundation*

Estimating Materials: cu. yds. = $\frac{\text{width} \times \text{length} \times \text{thickness}}{27}$

Foundation
Foundation drawings, Figure 7.11–2, are used by the construction crew pouring the footers, beams, and foundation. Concrete and steel specifications are designed to support equipment, integrate underground piping, and provide support for exterior and interior walls. Foundation drawings are typically not used by process technicians, however they are useful when questions arise about piping that disappears under the ground and when new equipment is being added.

Elevation
Elevation drawings convey a graphical representation that shows the location of process equipment in relation to existing structures and ground level. In a multistory structure the elevation drawing provides the technician with information about equipment operation and location. This information is important for making rounds, equipment checks, checklist development, catching samples, and performing start-ups and shutdowns. The elevation plan on Figure 7.11–3 illustrates equipment and structure locations.

Electrical
Electrical drawings include symbols and diagrams that depict an electrical process system. Electrical drawings show unit electricians where power transmission lines run and show the places where it is stepped down or up for operational purposes. A complex P&ID is designed to be used by a variety of crafts. The primary users of the document after plant start-up include process technicians, instrument and electrical, mechanical, safety, and engineering.

A process technician typically traces power to the unit from a motor control center (MCC). The primary components of an electrical system include the MCC, motors, transformers, breakers,

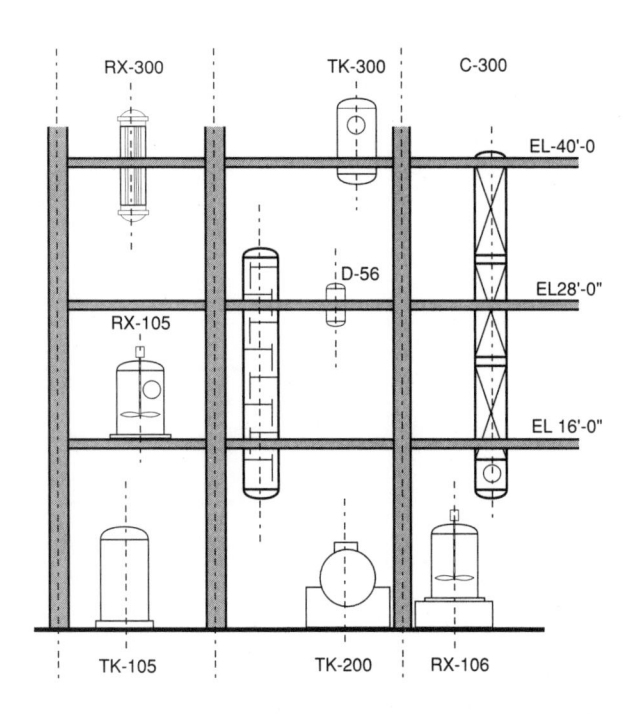

Figure 7.11-3 *Elevation*

fuses, switch gears, starters, and switches. The primary safety system is the isolation of hazardous energy "lock-out, tag-out." Process technicians are required to have training in this area. Figure 7.11–4 shows the basic symbols and flow path associated with an electrical drawing. Electrical lines are typically run in cable trays to switches, motors, ammeters, substations, and control rooms.

A transformer is a device used by industry to convert high voltage to low voltage. Problems with transformers are always handled by the electrical department. Electrical breakers are designed to interrupt current flow if design conditions are exceeded. Breakers are not switches and should not be turned on or off. If a tripping problem occurs the technician should call for an electrician. Fuses are devices designed to protect equipment from excess current. A thin strip of metal will melt if design specifications are exceeded. During operational rounds technicians check the ammeters inside the MCC for current flow to their electrical systems. Voltmeters (electrical devices used to monitor voltage in an electrical system) are also checked during routine rounds.

Equipment Location Drawings

Equipment location drawings show the exact floor plan location of equipment in relation to the plant's physical boundaries. Figure 7.11–5 illustrates this layout. Location drawings provide benefits similar to elevation drawings. The entire P&ID provides a three-dimensional look at the unit.

Refer to Figure 7.2–4 for an illustration of a P&ID.

Summary

Process flow diagrams (PFDs) and process instrumentation drawings (P&IDs) are used to outline or explain the complex flows, equipment, instrumentation, electronics, elevations, footings,

Process Instrumentation • Chapter 7

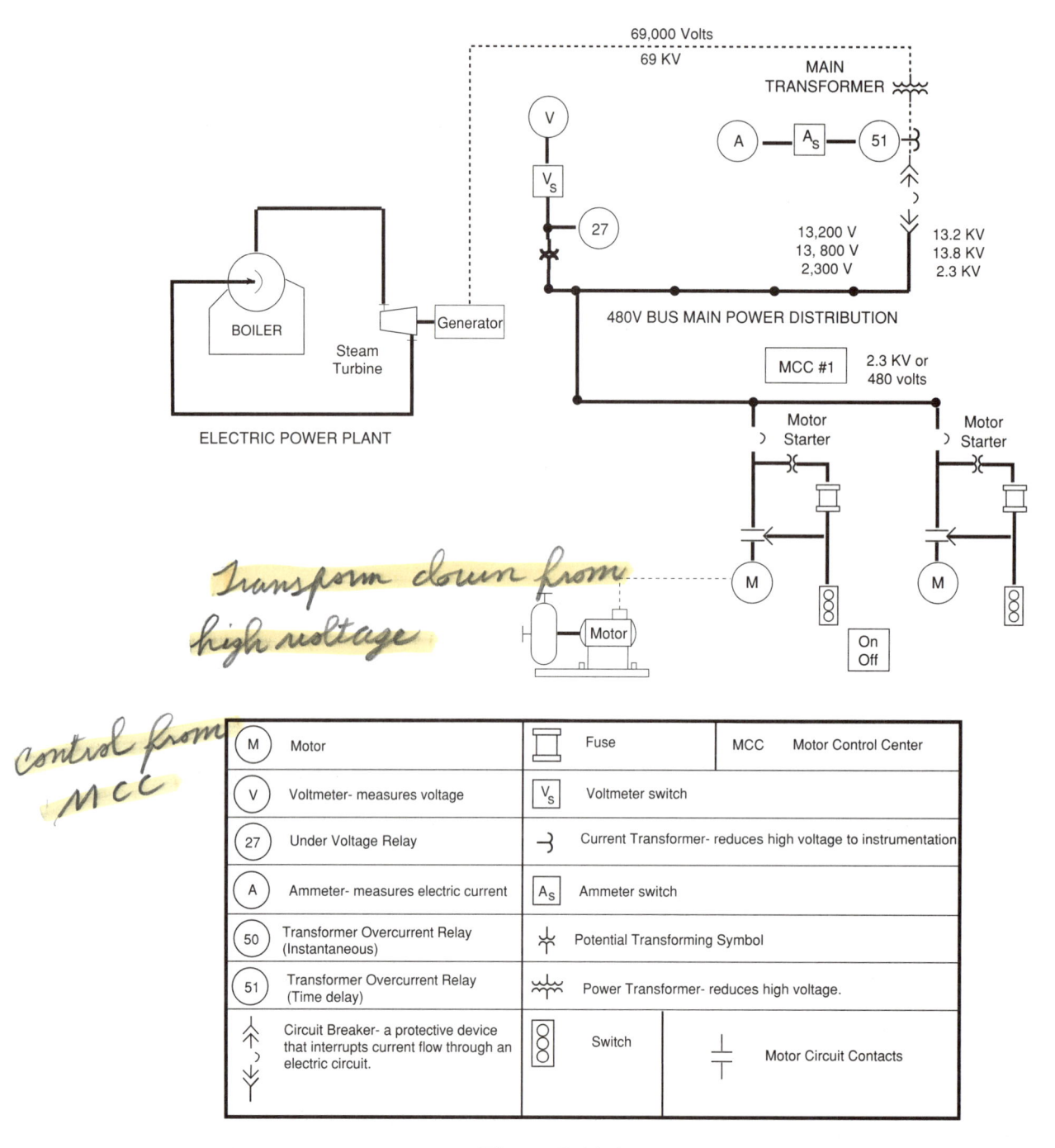

Figure 7.11-4 *Electrical*

(Diagram shows an electrical one-line diagram with a 69,000 Volts / 69 KV line to the Main Transformer, stepping down to 13,200 V / 13,800 V / 2,300 V (13.2 KV / 13.8 KV / 2.3 KV) on the 480V Bus Main Power Distribution, then to MCC #1 at 2.3 KV or 480 volts, feeding Motor Starters and Motors. An Electric Power Plant is shown with Boiler, Steam Turbine, and Generator. Instrumentation includes V (voltmeter), Vs (voltmeter switch), 27 (under voltage relay), A (ammeter), As (ammeter switch), and 51 (transformer overcurrent relay - time delay).)

Handwritten notes: *"Transform down from high voltage"* and *"Control from MCC"*

Legend:

Symbol	Description	Symbol	Description
M	Motor	Fuse	
V	Voltmeter- measures voltage	MCC	Motor Control Center
Vs		Vs	Voltmeter switch
27	Under Voltage Relay		Current Transformer- reduces high voltage to instrumentation
A	Ammeter- measures electric current	As	Ammeter switch
50	Transformer Overcurrent Relay (Instantaneous)		Potential Transforming Symbol
51	Transformer Overcurrent Relay (Time delay)		Power Transformer- reduces high voltage.
	Circuit Breaker- a protective device that interrupts current flow through an electric circuit.		Switch / Motor Circuit Contacts

and foundation that exist in a process unit. A P&ID is a complex representation of the various units found in a plant. A simple PFD is typically used to describe the primary flow path through a unit. Process diagrams can be broken down into two major categories: flow diagrams and P&IDs. A flow diagram is a simplified illustration that uses process symbols to describe the primary flow path through a unit. A P&ID is a complex diagram that uses process symbols to

Chapter 7 • *Process Instrumentation*

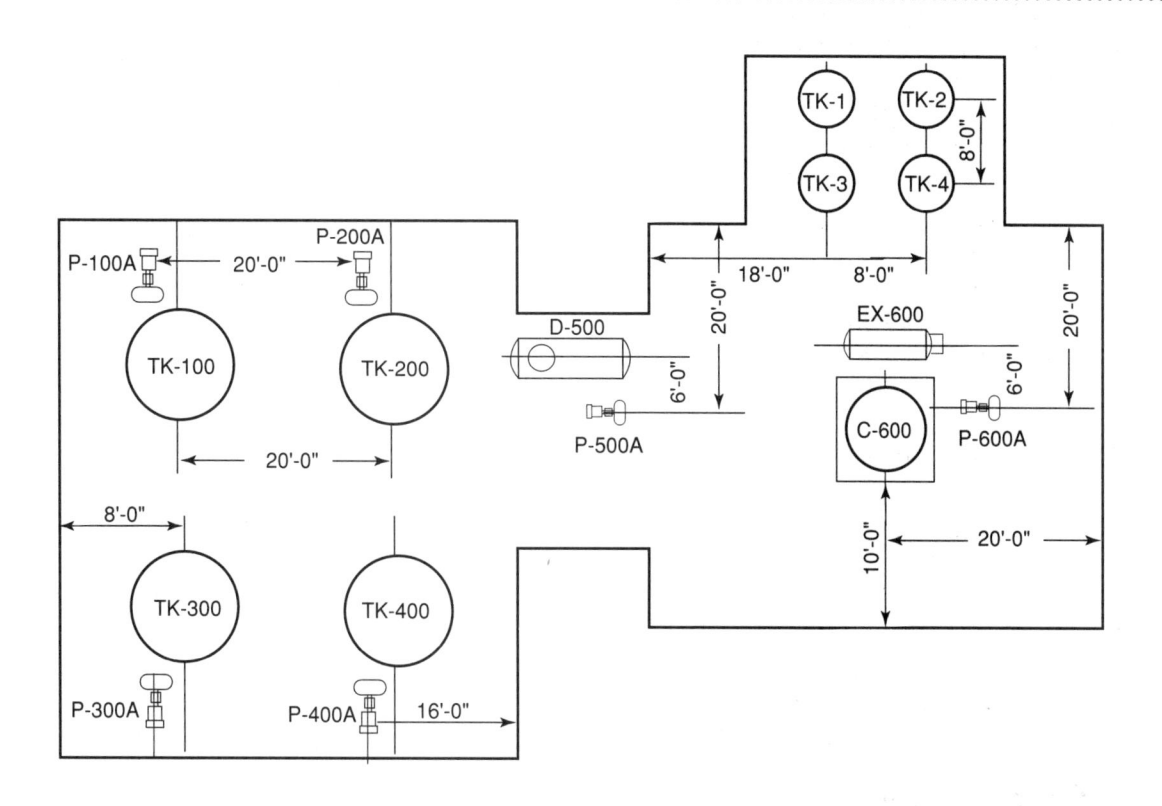

Figure 7.11-5 *Equipment Location*

process unit. Symbols and diagrams have been developed for most pieces of industrial equipment, process flows, and instrumentation. The symbols covered in this chapter include those typically used with valves, pumps, compressors, steam turbines, heat exchangers, cooling towers, furnaces, boilers, distillation columns, instrumentation, and process flows.

A control loop is defined as a group of instruments that work together to control a process. These instruments typically include a transmitter coupled with a sensing device or primary element, a controller, a transducer, and a control valve. Process variables typically fall into five different groups; pressure, temperature, flow, level, and analytical variables. Control loops are specifically designed to work with a selected variable. The primary purpose of a controller is to receive a signal from a transmitter, compare this signal to a setpoint, and adjust the (final control element) process to stay within the range of the setpoint. Controllers can be pneumatic, electronic, or electric. Final control elements are typically classified as automated valves, however motors or other electrical devices can be used. The final control element is the last link in the modern control loop and is the device that actually makes the change in the process.

An interlock is a device designed to prevent damage to equipment and personnel. This is accomplished by stopping or preventing the start of certain equipment functions unless a preset condition has been met. A permissive is a special type of interlock that controls a set of conditions that must be satisfied before a piece of equipment can be started. Permissives deal with start-up items whereas hardwire interlocks deal with shutdown items. This type of interlock will not necessarily shut down the equipment if one or more of its conditions are not met, however it will keep the equipment from starting up.

Process Instrumentation • Chapter 7

Chapter 7

Review Questions

1. Describe a process flow diagram (PFD).

2. Describe a piping and instrumentation drawing (P&ID).

3. How are instrumentation symbols shown on a P&ID?

4. Draw the symbols for a gate and globe valve.

5. Draw the symbols for a centrifugal pump and a PD pump.

6. Draw the symbols for a blower and a reciprocating compressor.

7. Draw the symbol for a steam turbine.

8. Draw the symbol for a heat exchanger.

9. Draw the symbol for a cooling tower.

10. Draw the symbol for a packed column.

11. Draw the symbol for a plate column.

12. Draw the symbol for a furnace.

13. Draw the symbol for a boiler.

14. Draw the symbol for an automatic valve.

15. Draw a pressure control loop.

16. Draw a flow control loop.

17. Draw a level control loop.

18. Draw a temperature control loop.

19. Draw a simple flow diagram. Include piping, pumps, two tanks, and six different valves. Provide a way to circulate and blend the material using the pumps-piping-valves relationship.

20. Draw a simple P&ID. (Do not copy the example from the book. Be original.)

Chapter 8

Systems

OBJECTIVES

After studying this chapter, the student will be able to:

- *Describe the key terms and definitions associated with process systems.*
- *Describe a simple pump-around system.*
- *Identify the key components of a compressor system.*
- *Draw a heat exchanger, cooling tower system.*
- *Sketch a simple lubrication system.*
- *Identify the elements of a process electrical system.*
- *Describe a furnace system.*
- *Describe a plastics system.*
- *Draw a typical reactor system.*
- *List the key elements of a steam generation system.*
- *Draw a distillation system.*

KEY TERMS

Compressor system—key elements of this system include a dynamic or positive displacement compressor, valves, piping, storage tank, dryers, and instrument air header.

Cooling tower, heat exchanger system—system includes a central cooling tower connected to a number of heat exchangers in series or parallel flow arrangements, piping, valves, pumps, and instruments.

Distillation system—system including tanks, piping, valves, pumps, heat exchangers, furnace, reactor, cooling tower, instruments, distillation column, reboiler, boiler, and compressor system.

Electrical system—system composed of motor, start/stop switch, substation breaker, and main substation breaker.

Equipment system—system composed of a set of equipment that shares a common or unique relationship.

Furnace system—typical system components include a furnace, piping, valves, pumps, fans, distillation columns, reactors, heat exchangers, and instrumentation.

Hydraulic system—system including fluid reservoir, pump, piping, pressure control valve, flow control valve, directional control valve, and actuator (cylinder, piston).

Industrial process—an industrial process could be one of three hundred different refinery or chemical plant processes. These processes use specific equipment systems to produce their products.

Lubrication system—system includes piping, valves, pumps, bottle oilers, sight glasses, dipsticks, pressure gauges, lubricant, seals, bearings, reservoirs, and rotating equipment.

Plastics system—system includes granules blender, rotary feeder, compressor and air system, valves, feed tanks, piping, feeders, ribbon blenders, homogenizers, extruder, pelletizer, water tank, classifier, dryer, and product tanks.

Pump-around system—system consists of a series of pumps, piping, valves, and tanks, connected together to form a system.

Reactor system—system includes piping, valves, instrumentation, a reactor, distillation column, and furnace.

Steam generation system—system consists of a boiler, piping, valves, tanks, steam header, steam turbine, reboiler, and condensate return header.

Chapter 8 • *Systems*

8.2 Flow Diagrams and Equipment Relationships

New technicians have difficulty determining which pieces of industrial equipment go together when asked to develop a simple flow diagram. There are a variety of relationships that are common between industrial facilities. The typical relationships that will be covered in this module include key terms listed above. Figure 8.2–1 illustrates the basic equipment found in a simple pump-around system. Using a simple pump-around system, technicians can learn how to perform equipment line-ups, start-ups, operational checks, and shutdowns. The key elements of a pump-around system include process piping, storage tank(s), valves, gauges, and a pump.

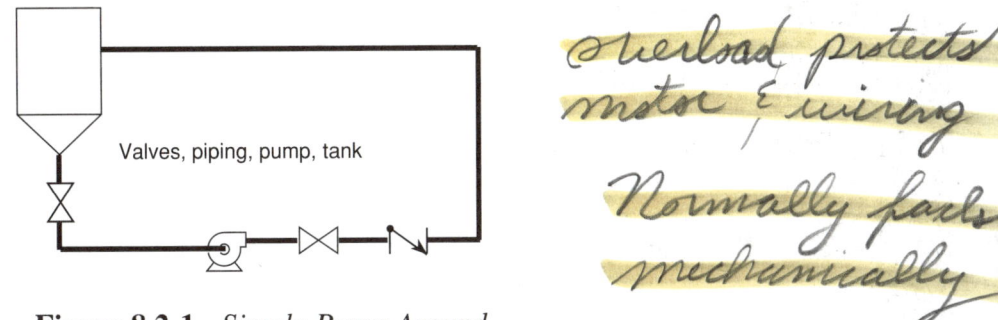

Figure 8.2-1 *Simple Pump Around*

8.3 Compressor System

A compressor system is a simple arrangement of equipment designed to produce clean, dry, compressed air or gas for industrial applications. A compressor system typically includes process piping, valves, a compressor, a receiver, heat exchangers, dryers, back pressure regulators, gauges, and moisture removal equipment. The sequence and equipment arrangement is illustrated in Figure 8.3–1. Additional information about compressor systems can be found in the equipment and systems courses.

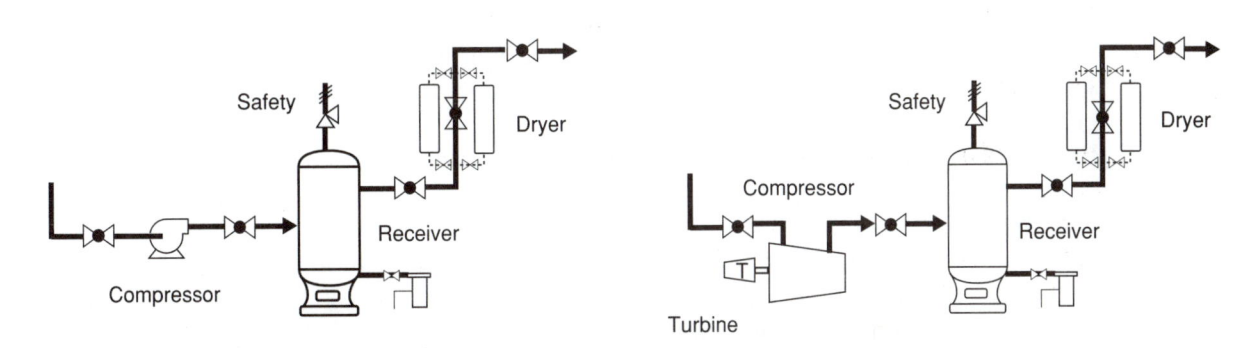

Figure 8.3-1 *Compressor Systems*

8.4 Heat Exchanger and Cooling Tower System

Cooling towers and heat exchangers share a unique relationship although each device responds to different scientific principles. Heat exchangers transfer heat energy between two process flows without allowing the two streams to come into physical contact with each other. This process takes place primarily through conduction and convective heat transfer. A cooling tower transfers heat energy from hot water primarily through evaporation. Evaporative heat transfer accounts for 80 to 90% of the heat transfer process inside a cooling tower. A cooling tower provides cool water to a heat exchanger so the hot process can be cooled down. The warm water is returned to the cooling tower for heat removal. A simple heat exchanger, cooling tower system includes multiple heat exchangers, pumps, valves, piping, instrumentation, and a cooling tower, Figure 8.4–1.

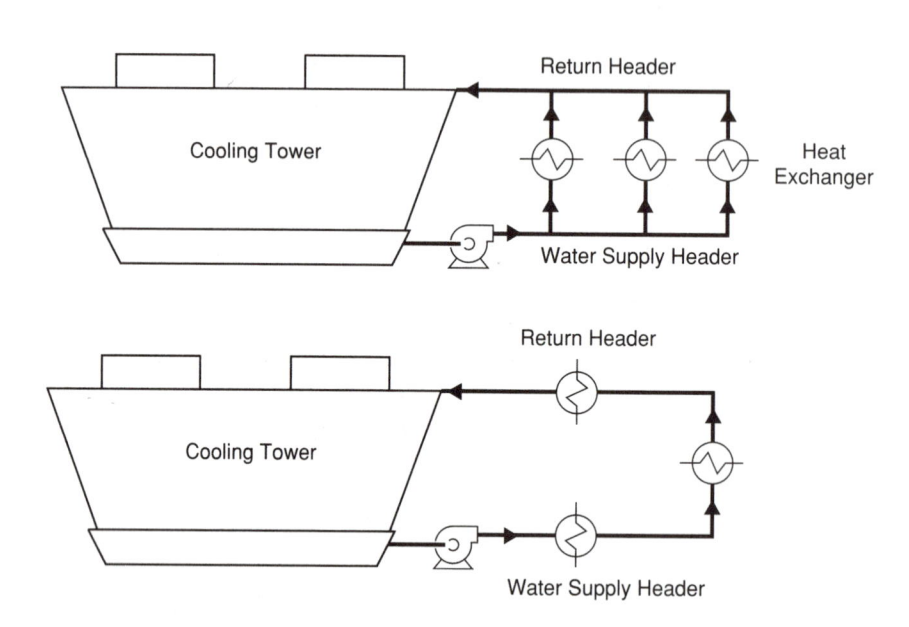

Figure 8.4-1 *Cooling Tower/Heat Exchanger*

8.5 Lubrication System

Lubrication systems provide a constant source of clean oil to pump and compressor bearings, gearboxes, steam turbines, and rotating or moving equipment. A typical lubrication system includes a lubricant reservoir, pump, valves, heat exchanger, and piping. Figure 8.5–1 illustrates how an industrial lubrication system operates.

8.6 Electrical System

Electrical systems, Figure 8.6–1, are a collection of complex processes that include a boiler to generate steam, a steam turbine driven electric generator to produce electricity, a main substation with transformers to reduce the electrical output, a motor control center (MCC) to centralize local power distribution, and electrically powered equipment that is run by the electrical system.

8.7 Furnace System

Chapter 8 • **Systems**

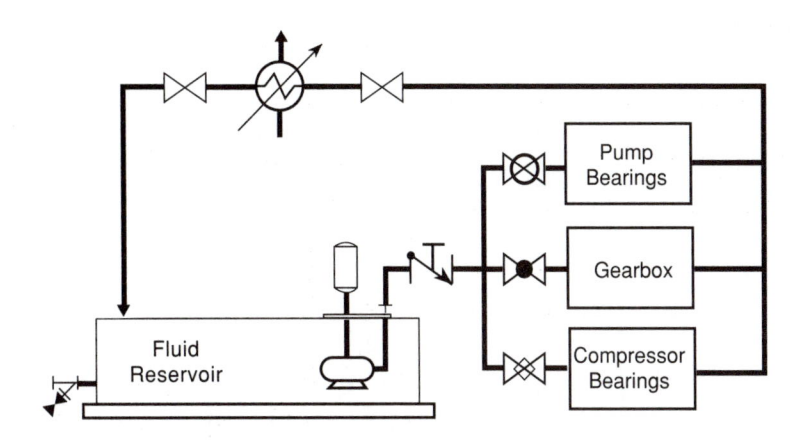

Figure 8.5-1 *Lubrication System*

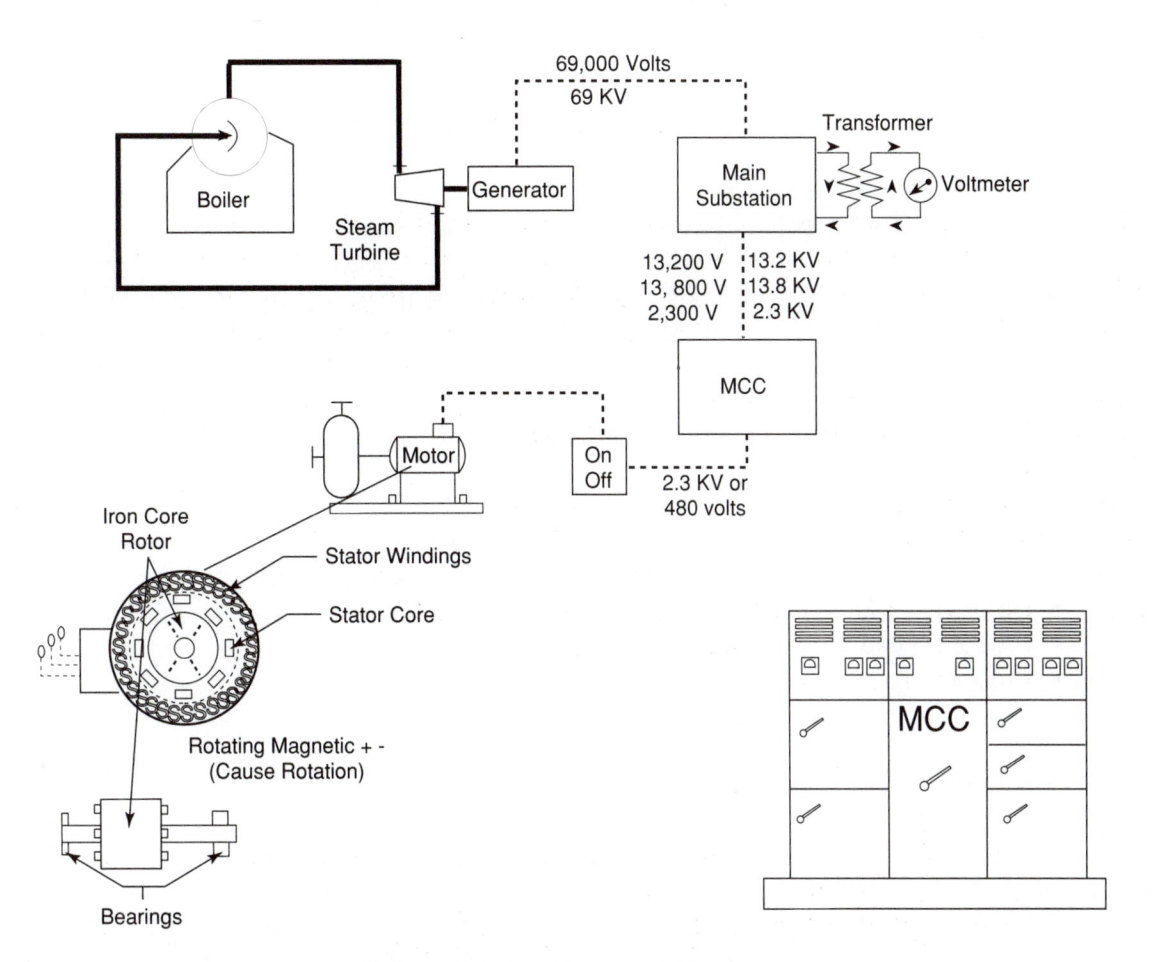

Figure 8.6-1 *Electrical System*

157

Systems • Chapter 8

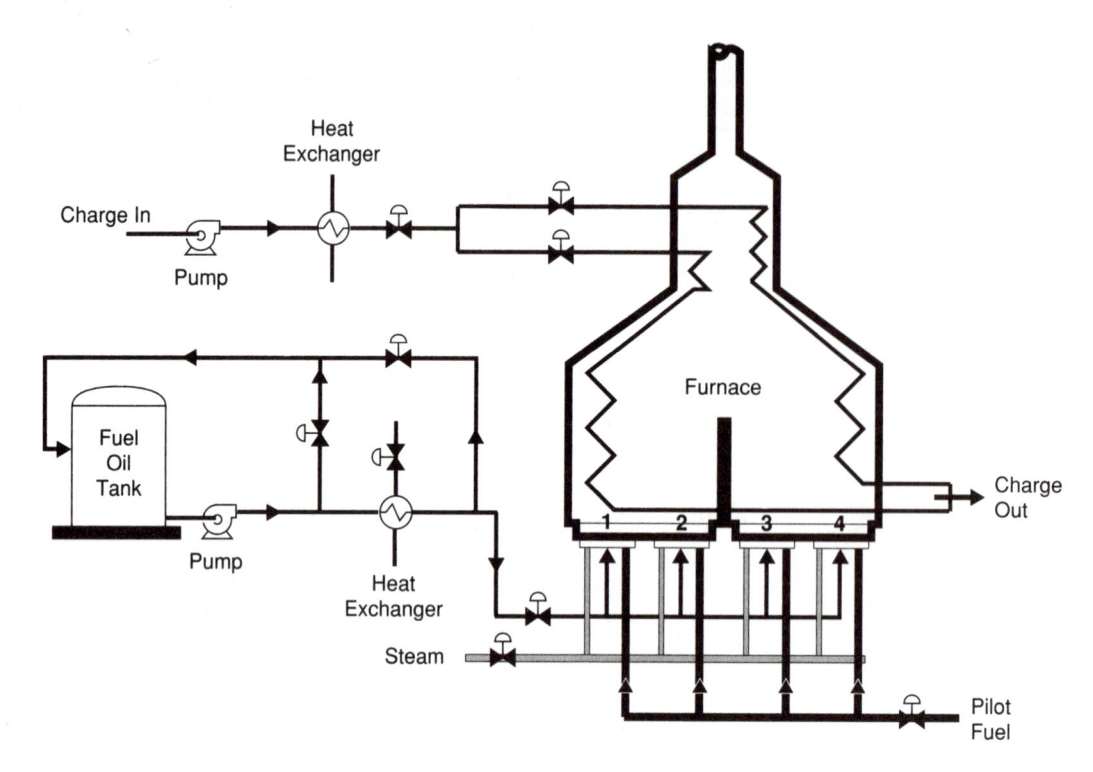

Figure 8.7-1 *Furnace System*

The chemical processing industry uses fired heaters to heat large quantities of crude oil or other hydrocarbon feedstocks to operating temperatures for processing. A furnace is composed of a firebox, outer shell, lower radiant section, upper convection section, insulation, refractory, convection tubes, radiant tubes, stack, damper, and burners. The furnace system, Figure 8.7–1, is a collection of other systems linked together to produce a specific result.

Typical systems found in a furnace system include:
- Feed section—tanks, piping, valves, and pumps
- Pre-heating section—heat exchangers, boilers, valves, and piping
- Heating section—the furnace
- Process—example, distillation column or reactor
- Products

8.8 Plastics System

A simple plastics system includes equipment that could be used for molding, extrusion, laminating, casting, or calendering.

Chapter 8 • **Systems**

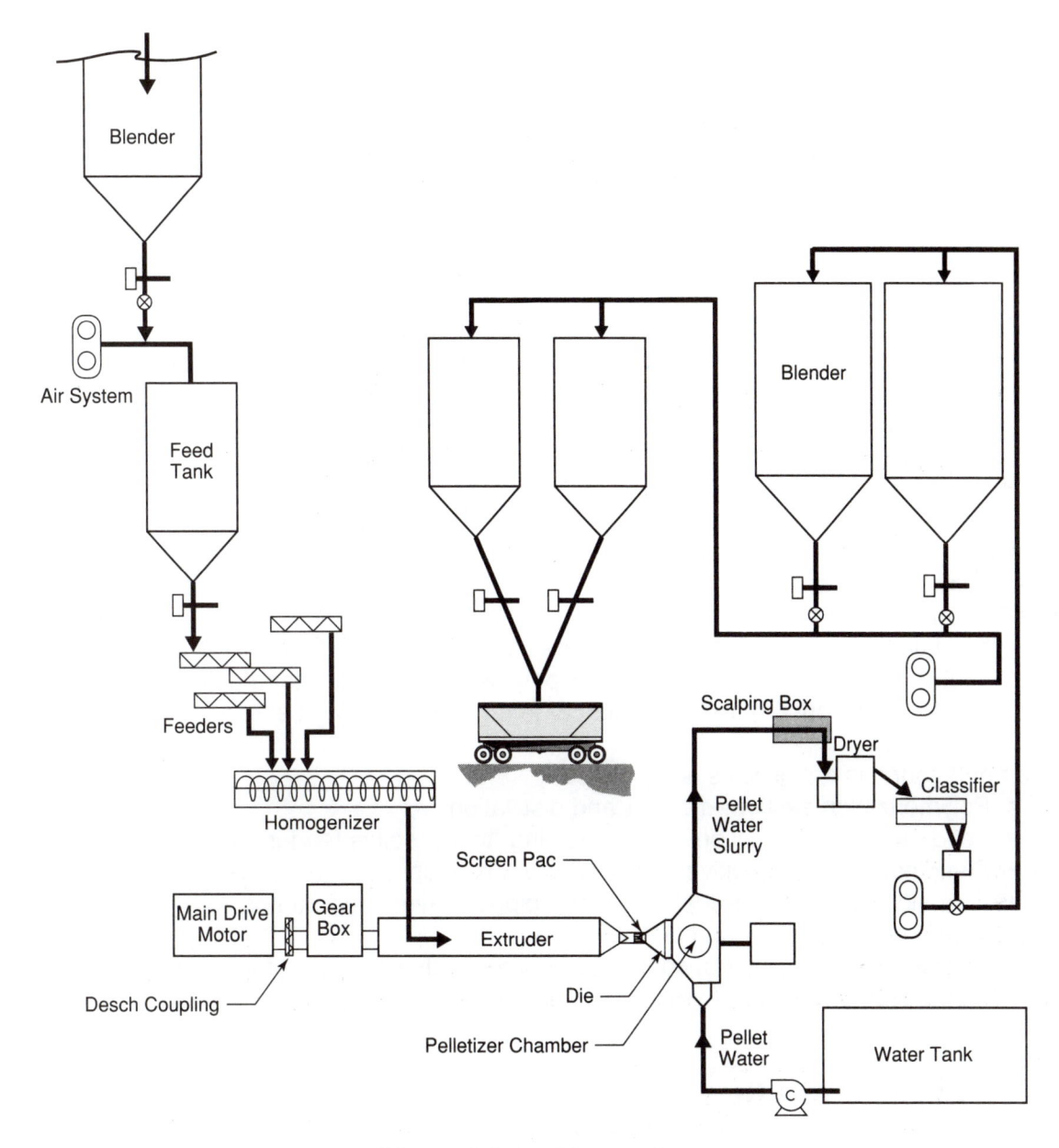

Figure 8.8-1 *Plastics System*

Molding
Injection molding, compression molding, and blow molding are processes that resemble the way we make waffles for breakfast. Molten polymer is squeezed into a mold to produce a product. Examples of products include tableware, plastic toys, or baby bottles.

Extrusion
Extrusion is a process that takes molten polymer and squeezes it down a barrel. This process is comparable to squeezing shampoo out of a plastic bottle. Examples of products made from the plastic pellets produced by extruders are baby diapers, washing machine parts, and car parts.

Laminating
Laminating is a process that takes aluminum foil, paper, or cloth and coats it with melted resin. This process is similar to building a sandwich. Examples of laminating include motherboards and electronic circuits.

Casting
Casting is a process that most closely reflects baking a cake. Just as the batter is poured into a pan, molten polymer is poured into a mold. This process produces items like eyeglass lenses.

Calendering
The last process, calendering, closely resembles spreading butter over hot pancakes. During the calendering process rollers spread molten resins over thin sheets of paper or cloth. Calendering is used to make face cards.

Initial Process
The initial process for creating plastic is very complex. Plastic is made from synthetic resins. The atoms that make up the molecules of synthetic resins are composed of carbon, hydrogen, oxygen, and nitrogen. Process technicians make synthetic resins by combining chemical compounds like ammonia, benzene, hexamethylenetetramine, or a variety of other chemicals. The reaction that takes place creates synthetic links between molecules called monomers. This process creates long chain molecules called polymers. Industrial manufacturers refer to this process as polymerization. Figure 8.8–1 illustrates a typical finishing section in a plastics plant.

Key elements found in a plastics system include:
- Polymerization section—reactor and distillation
- Feed and transfer section—valves, piping, tanks, solids feeders, and compressor
- Blending section—additive blenders and homogenizer
- Extrusion section—extruder, pelletizer, pipes, valves, and pumps
- Drying section—dryer and classifier
- Products—solids feeders, compressor, pipes, valves, storage tanks
- Transportation—railroad, truck, bags, and boxes

8.9 Reactor System

Reaction technology can be categorized as batch, semi-batch or continuous operation. Reactors are vessels designed to allow a reaction to occur as two or more flows are exposed to each other under a variety of conditions; heat, cold, pressure, time or a catalyst. During the reaction process the two or more flows react to form a new product. A typical mixing reactor will include a vessel, a mixer, valves, piping, two or more inlet ports and a single outlet port, Figure 8.9–1.

8.10 Steam Generation System

The production of steam is very important to the operation of an industrial facility. Steam is used in a variety of operations including steam turbines, steam tracing, heat exchangers, reboilers, stripping, distillation, etc. Steam is produced in a device called a boiler. Boilers are classified as

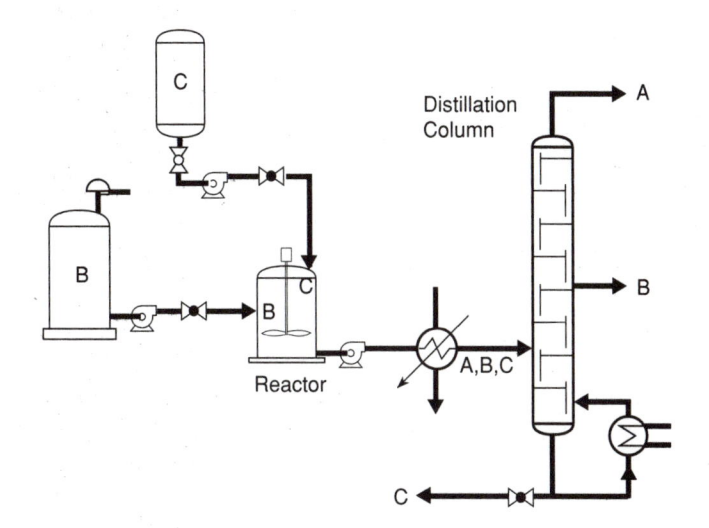

Figure 8.9-1 *Typical Reactor System*

water or fire tube. Both types are designed to do one thing—boil water. A typical boiler is composed of:
- Water pump, valves, and piping
- Large furnace
- Large upper steam generating drum
- Lower mud drum
- Downcomer tubes, riser tubes, steam generating tubes, superheated tubes, and desuperheated tubes

Figure 8.10–1 (page 162) shows a steam generation system.

8.11 Distillation System

Distillation is a process that separates the various components in a process stream by their individual boiling points. In this type of system a distillation column is the central piece of equipment, Figure 8.11–1. Distillation columns can be classified as either plate or packed. Plate columns have sieve trays, valve trays, or bubble-cap trays. Packed columns are filled with packing material like sulzer packing, rasching ring, pall ring, saddles, metal intalox, teller rosette, or mini-ring.

During the distillation process, hydrocarbon feed is stored in a feed tank. Prior to sending the feed to the column it is tested to ensure consistency. Some blending may occur at this point to ensure feed uniformity. Before the charge can be sent to the column it needs to be heated to operating temperatures. This part of the process involves sending the feed through a series of heat exchangers and a fired furnace. Feedstock temperatures are gradually stepped up as the flow moves through the system.

As the heated charge leaves the furnace and enters the distillation column, a fraction of the feed vaporizes and rises up the column; the heavier components in liquid state drop down the col-

161

Systems • Chapter 8

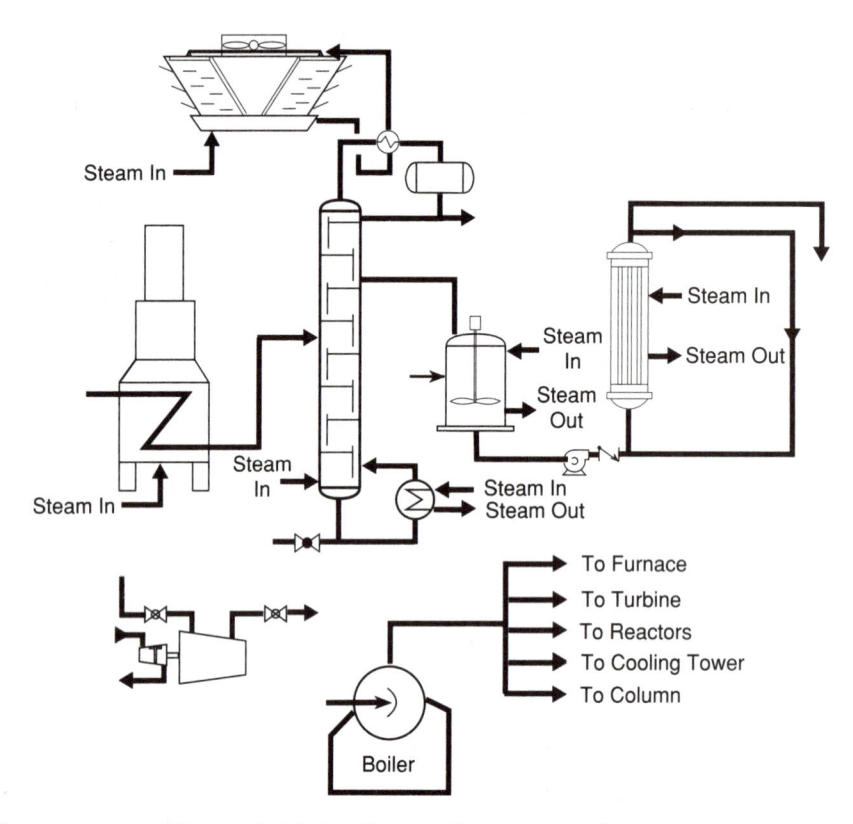

Figure 8.10-1 *Steam Generation System*

umn. This initiates the separation by boiling point process. Since the energy in the process stream begins to dissipate immediately, a reboiler or heating source is attached to the column. This allows the separation process to continue. Some distillation columns are steam traced to ensure even temperature control. The distillation process is represented by four distinct systems and one "super system":

- Pre-heating & heating system—heat exchangers and furnace
- Process—example, distillation column or reactor
- Products—tanks, piping, valves, and pumps
- Utilities super system—boiler system, compressor systems, cooling tower system, electrical system, and water system
- Feed system—tanks, piping, valves, and pumps

8.12 Refrigeration System

Heating and cooling are two important aspects of modern process control. Refrigeration systems, Figure 8.12–1, are used to provide cooling to industrial applications like air conditioning. Refrigeration units are composed of:

- Compressor—high pressure refrigeration gas
- Heat exchanger—cooling tower
- Receiver

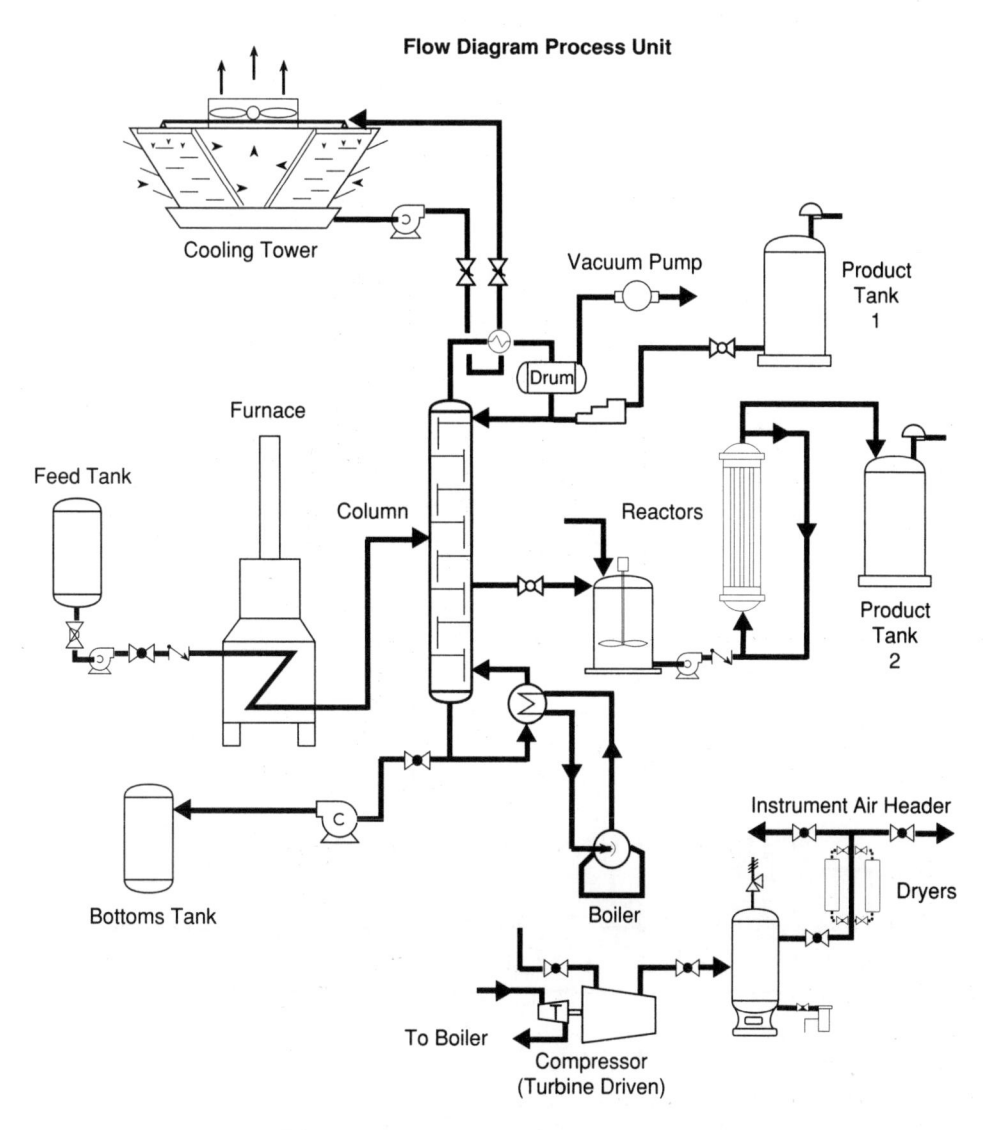

Figure 8.11-1 *Distillation System*

- Expansion valve—low pressure refrigeration liquid
- Heat exchanger (evaporator)—low pressure refrigerant gas

In the refrigeration process, low pressure refrigerant gas is drawn into a compressor, converted into high pressure refrigeration gas, and pushed into a shell and tube heat exchanger. During the compression process a tremendous amount of heat is generated and must be removed by the exchanger. During the cooling process the gas condenses into liquid phase and is collected in a receiver. From the receiver the high pressure liquid refrigerant is pushed through a small opening in an expansion valve. As the liquid expands it changes phase. Since the boiling point of the refrigerant is so low a cooling effect occurs in the evaporator. As the low pressure refrigerant leaves the evaporator it enters into the suction side of the compressor and the process begins again.

Systems • Chapter 8

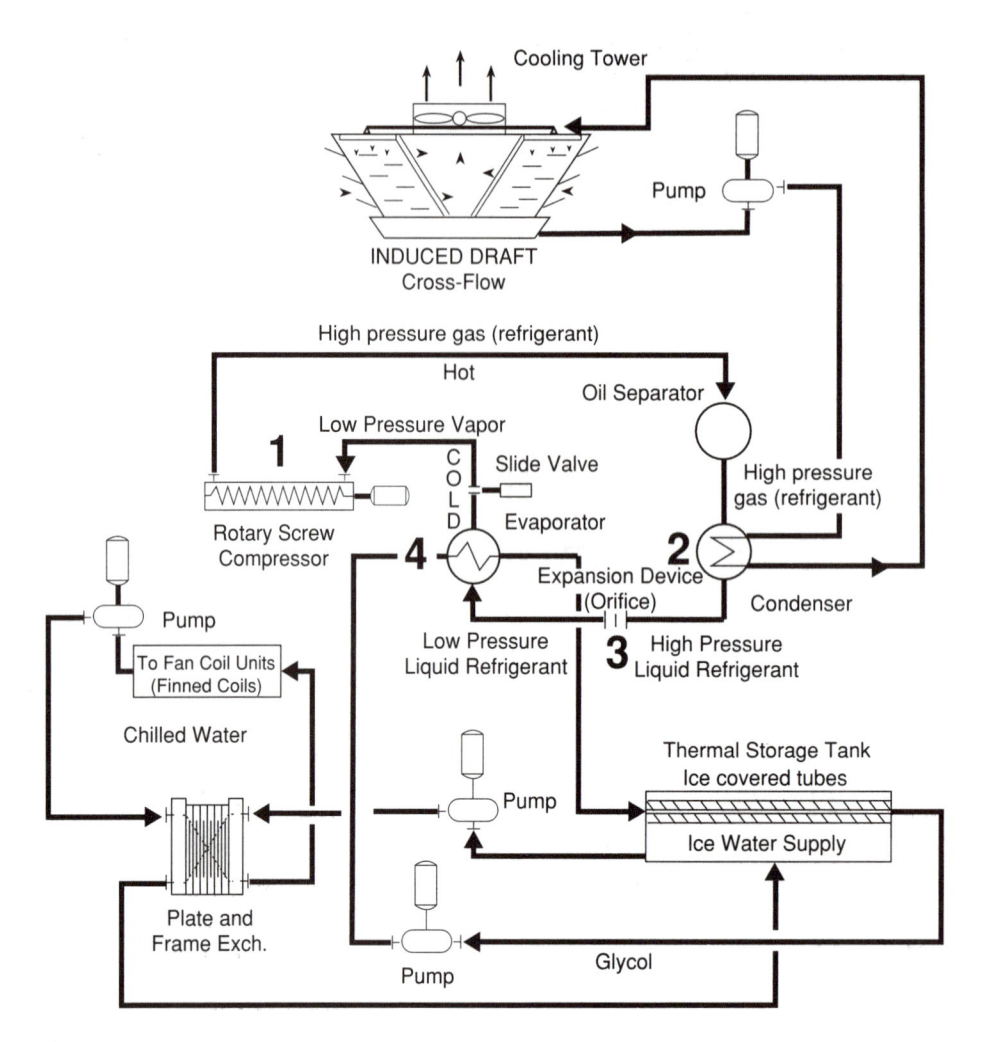

Figure 8.12-1a *Two Step Refrigeration System*

8.13 Water Treatment System

Twenty years ago the chemical processing industry pumped a tremendous amount of water out of the ground for industrial applications. This process was stopped after it was discovered that as the water table dropped so did the surrounding countryside. The CPI uses surface water for most industrial applications. Surface water is defined as water that is drawn in for industrial applications from lakes, rivers, and oceans.

As water enters the plant it is stored in large holding basins and allowed to settle out. A series of large pumps take suction off the basin and send water to a series of filters for additional purification. Some filtered water is sent to demineralizers for additional treatment to remove dissolved impurities. Figure 8.13–1 illustrates a water treatment system.

Chapter 8 • *Systems*

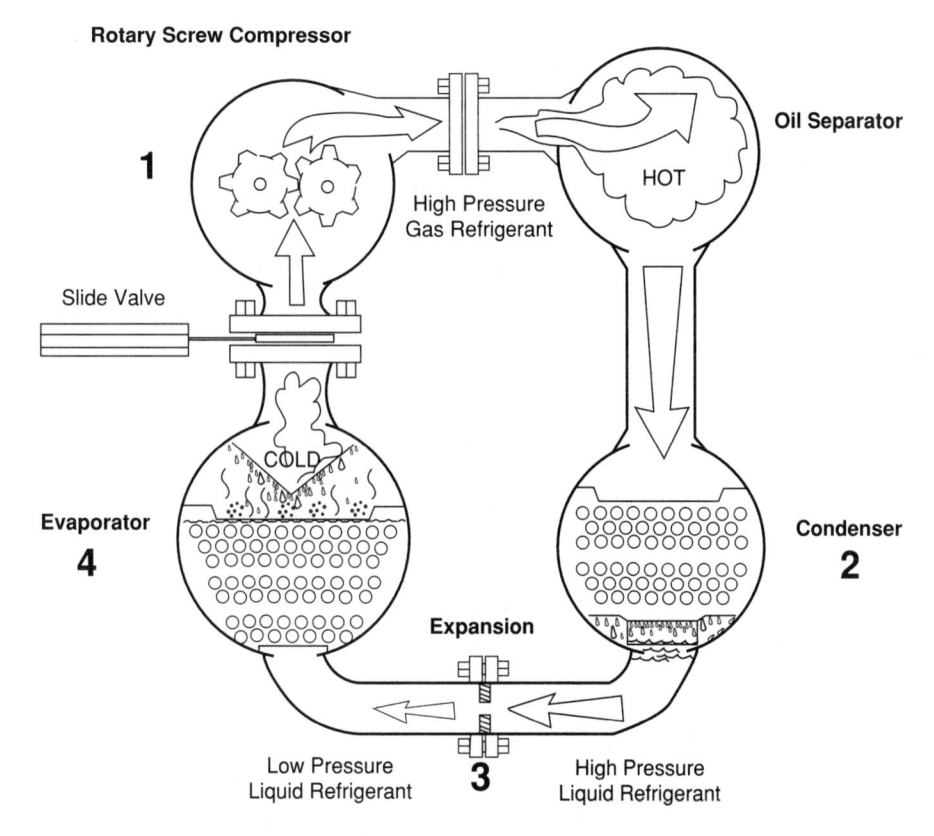

Figure 8.12-1b *Refrigeration System*

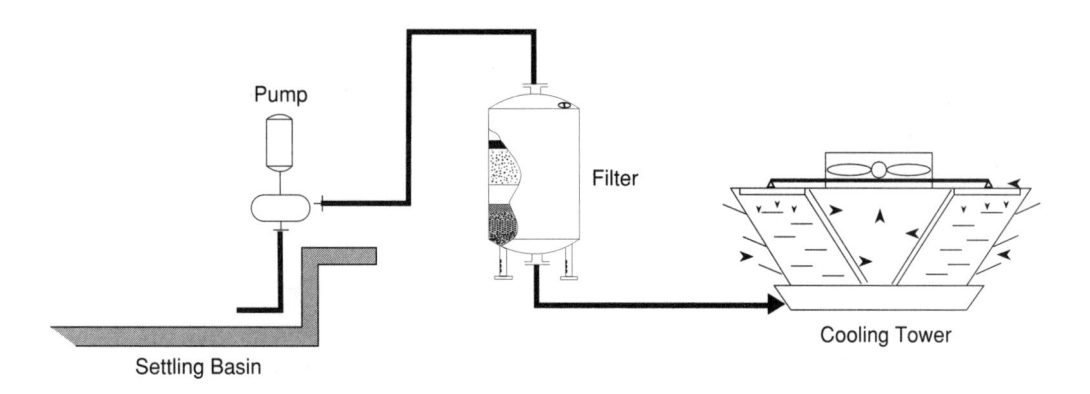

Figure 8.13-1 *Water Treatment System*

8.14 Hydraulics

Process technicians use hydraulic systems, Figure 8.14–1, to open or close valves, lift heavy objects, run hydraulic motors, and stop the rotation of a rotary or reciprocating device. A hydraulic

165

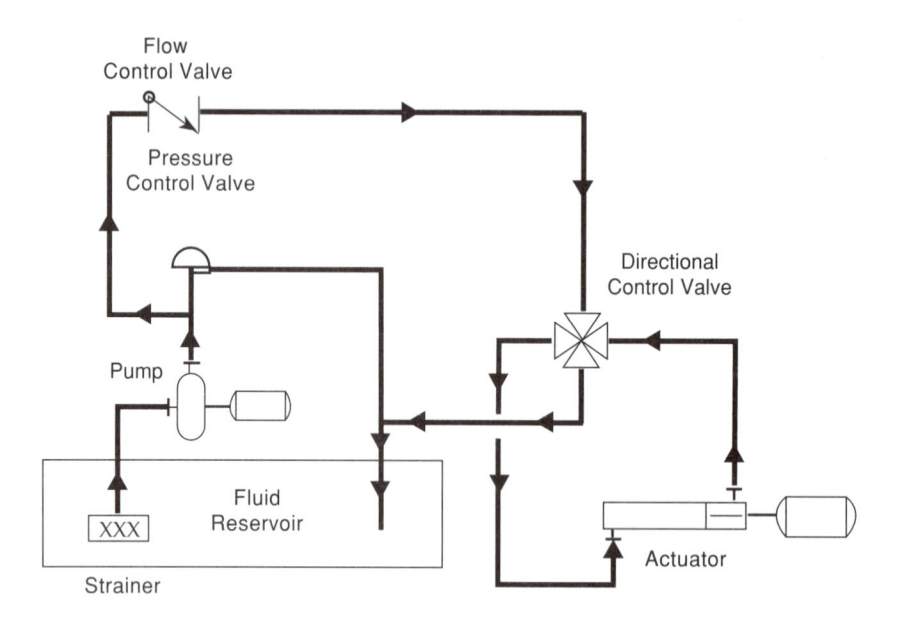

Figure 8.14-1 *Hydraulic System*

system is a collection of equipment designed to apply pressure on a confined liquid in order to perform work. A similar process is used in the brake systems of most cars and trucks. A hydraulic system is composed of a fluid reservoir, strainer, pump, piping, flow control valve, pressure control valve, four-way directional control valve, and actuator.

8.15 Utilities

Process plants have utility sections that specialize in water treatment, steam generation, cooling tower systems, and compressed gases. Each of these systems has been discussed in this chapter. Process utilities are typically defined as water and compressed gases. Plant water can be classified as boiler feed water, drinking water, fire water, cooling tower water, potable water, and waste water. Compressed gases include air, nitrogen, hydrogen, chlorine, etc.

Summary

Equipment systems are composed of equipment that shares a common or unique relationship. Industrial processes can be classified as refinery or petrochemical. Industrial processes produce a specific product while an equipment system layout does not. System layouts are designed to show how the equipment works together. Common equipment systems include pump around, compressor, cooling tower/heat exchanger system, lubrication, hydraulic, electrical, furnace, plastics, reactor, steam generation, and distillation system.

Chapter 8 — Review Questions

1. Identify the equipment used in a pump-around system.
2. Identify the equipment used in a compressor system.
3. Identify the equipment used in a heat exchanger/cooling tower system.
4. Identify the equipment used in a lubrication system.
5. Identify the equipment used in an electrical system.
6. Identify the equipment used in a furnace system.
7. Identify the equipment used in a plastics system.
8. Identify the equipment used in a reactor system.
9. Identify the equipment used in a steam generation system.
10. Identify the equipment used in a distillation system.
11. Identify the equipment used in a refrigeration system.
12. Identify the equipment used in a water treatment system.
13. Identify the equipment used in a hydraulics system.

Chapter 9

Industrial Processes

OBJECTIVES

After studying this chapter, the student will be able to:

- *Define terms and definitions.*
- *Explain petrochemical processes.*
- *Describe the benzene, BTX aromatics, and ethylbenzene processes.*
- *Describe the ethylene glycols, mixed xylenes, and olefins processes.*
- *Describe the paraxylene, polyethylene, and xylene isomerization processes.*
- *Review refinery processes.*
- *Describe the alkylation, catalytic cracking, and catalytic reforming processes.*
- *Describe the coking, crude distillation, and deasphalting processes.*
- *Describe the fluid catalytic cracking, hydrocracking, and isomerization processes.*
- *Contrast petrochemical and refinery processes.*

Industrial Processes • Chapter 9

KEY TERMS

Alkylation—uses a reactor to make one large molecule out of two small molecules.

Alkylation unit—The central piece of equipment in an alkylation unit is a reactor filled catalyst. The catalyst causes a chemical reaction to occur which produces the desired product.

Catcracker—uses a fixed bed catalyst to separate smaller hydrocarbons from larger ones.

Distillation Tower—a series of stills arranged so the vapor and liquid products from each tray flow counter-currently to each other.

Fixed bed reactor—the fixed medium in a reactor remains in place as raw materials pass over it.

Fluid Catalytic Cracking—a process that uses a reactor to split large gas oil molecules into smaller, more useful ones.

Fluidized bed reactor—suspend solids in a reactor by counter-current flow of gas. Particle segregation occurs over time as heavier components fall to the bottom and lighter ones move to the top.

Fluid coking—a process that uses a reactor to scrape the bottom of the barrel and squeeze light products out of the residue.

Hydrocracking—uses a multi-stage reactor system to boost yields of gasoline from crude oil.

Hydrodesulfurization reactor—sweetens products by removing sulfur.

Reactor—a device used to combine raw materials, heat, pressure, and catalysts in the right proportions to initiate reactions and form products.

Reboiler—a heat exchanger used to maintain the heat balance on a distillation tower.

Reformer—a reactor filled with a catalyst designed to break large molecules into smaller ones through chemical reactions that remove hydrogen atoms.

Regenerator—used to recycle or regenerate contaminated catalyst.

9.2 Common Industrial Processes

During World War I (1914 to 1918) oil production became as important as ammunition production. Oil was used to operate ships, airplanes, tanks, automobiles, motor cycles, and other motorized equipment. As technology improved so did farming techniques around the world. Tractor technology and other motorized farming implements increased productivity. The increased productivity of gasoline led to a new tax-generating source for the government. Another by-product of gas production was asphalt. This new material enabled the federal government, and state and local authorities to upgrade existing road systems and launch new road building ventures.

During World War II (1939 to 1945) technology took a few steps forward. New process equipment was tested on naval vessels, submarines, aircraft, land vehicles, and communication technology. American oil companies demonstrated the ability to adapt quickly to wartime needs by producing over 80% of the aviation fuel used by the Allies. Huge quantities of oil and new specialty chemicals (such as butadiene and toluene) were needed during the war. Butadiene was used to make synthetic rubber and toluene is a major ingredient in medicinal oils and TNT. World War II saw significant improvements in the industrial processes of alkylation and catalytic cracking. These two processes greatly enhanced the production of high octane aviation gasoline.

Post-war years saw a tremendous increase in oil consumption. Process technicians could easily find lifelong jobs at many of the large refineries. In the early 50s a HUMBLE Oil company (EXXON) employee could secure a car loan in the Baytown, Texas area simply by showing the salesperson an employee badge. This experience was common in cities where oil refining, gas processing, and petrochemicals pumped so much money into the local economy.

From 1950 to 1972 the government continued to draft large numbers of process technicians into the military. Most companies worked with employees who were drafted and allowed them to return to their jobs after their tour of duty. Some companies counted an employee's service time in the military as uninterrupted company service time. This group of employees greatly influenced the military type environment found in the chemical processing industry.

Work Force Trends 1960–1980
As the complexity of the industrial processes increased, a significant change was encountered in the make-up of the technical work force. Starting in the 1960s and building into the 70s the chemical processing industry began to employ large numbers of engineers. As this fact became known, engineering programs around the United States began to fill up. Colleges started turning out record numbers of:
- Electrical engineers
- Chemical engineers
- Mechanical engineers
- Petroleum engineers
- Industrial engineers
- Nuclear engineers

Industrial Processes • Chapter 9

Engineers were employed in the chemical processing industry as technical support to the operations groups. This relationship was not new since engineers, chemists, and technicians had worked together as a team for many years, however the increased numbers were new.

Industrial processes can be broken into three basic groups: refinery, gases, and petrochemical. The most common industrial processes include:

Petrochemical Processes
- Alkylbenzene, linear
- Amines, methyl
- Ammonia (5)
- Benzene
- Bisphenol- A (2)
- BTX aromatics (3)
- Butadiene
- Butanediol, 1,4-
- Butyraldehyde, n and i
- Caprolactam
- Cumene (3)
- Dimethyl terephthalate (2)
- EDC via oxychlorination
- Ethanolamines
- Ethylbenzene (3)
- Ethylene (6)
- Ethylene lycols
- Ethylene oxide
- Formaldehyde
- Maleic anhydride
- Methanol (4)
- Mixed xylenes
- Olefins (3)
- Paraxylene (2)
- Phenol (2)
- Phthalic anhydride
- Polycaproamide
- Polyethylene (5)
- Polyethylene terephthalate (PET)
- Polyethylene, LDPE-EVA
- Polypropylene (3)
- Propylene (3)
- PVC (suspension) (2)
- Styrene (2)
- Terephthalic acid
- Urea
- Vinyl chloride
- Vinyl chloride monomer
- Xylene isomerization
- Xylene isomers

Refining Processes
- Alkylation (4)
- Benzene reduction
- Benzene saturation
- Catalytic cracking
- Catalytic reforming (3)
- Coking (4)
- Crude distillation
- Deasphalting (3)
- Deep catalytic cracking
- Electrical desalting
- Ethers (7)
- Fluid catalytic cracking (6)
- Hydrocracking (6)
- Hydrogenation
- Hydrotreating (7)
- Isomerization (6)
- Resid catalytic cracking
- Treating
- Visbreaking (3)

The numbers in parenthesis indicate the number of registered processes or ways to manufacture the chemical.

9.3 Petrochemical Processes

Since 1960, rapid developments in the petrochemical area have developed. Many of these processes have more than one way they can be produced. At the present time there are hundreds of petrochemical processes. Petrochemical processes are far more numerous than the basic core refinery processes. This chapter lists many of the primary processes found throughout the country. For example, a process technician working in ethylene production could have over five different operational arrangements. In addition to these new processes the first ten years of the new millennium will see a tremendous surge in the development of small specialty chemical companies.

9.4 Benzene

The benzene process is designed to produce high-purity benzene and heavy aromatics from a mixture of toluene and heavier aromatics, Figure 9.4–1. Heated hydrogen and feedstock are passed over a special catalyst bed (1) which reacts to form a mixture of benzene, unreacted toluene, xylene and heavy aromatics. This mixture is condensed (2) in a drum and stabilized (3). Stabilized bottoms are sent to a fixed bed clay treater for acid wash color specifications and then distilled (4) to separate benzene, xylenes, toluene, and C9+ aromatics.

Yields: 99 mol% aromatic yield of fresh toluene. Typical production yields for xylenes and benzene are:

Wt% Feedstock	Benzene	Xylene
Nonaromatics	3.2	2.3
Benzene	0	11.3
Toluene	47.3	0.7
C8 aromatics	49.5	0.3
C9+ aromatics	0	85.4
	100%	100%
Wt% Products		
Benzene	75.7	36.9
C8 aromatics	0	37.7

9.5 BTX Aromatics

BTX aromatics, Figure 9.5–1, is a process based upon extractive distillation. The process is designed to produce yields of benzene, toluene hydrogen, xylenes, and C5+. The BTX process starts with a feedstock composed of paraffins 57%, naphthenes 37% and aromatics 6% being fed into a series of moving bed reactors 1 through 4. The feed flows downward over the special catalyst bed and out the lower section of the number 1 reactor where the process is repeated

Industrial Processes • Chapter 9

Benzene Process

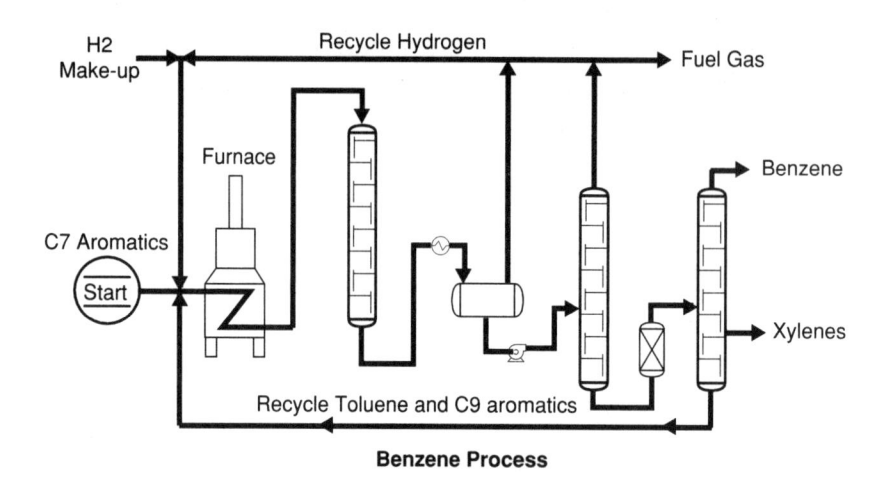

Figure 9.4-1 *Benzene*

with reactors 2, 3, and 4. The catalyst and feed mixtures are designed to flow from reactors 1 through 4. Since solids and liquids have different flow characteristics the solid catalyst requires a unique gas lift transfer process to be moved from one reactor to the next. The gas lifted catalyst is "series fed" into each reactor's feed hopper until it reaches reactor 4. As the catalyst moves between reactors it accumulates coke deposits. To eliminate the coke deposits the fourth reactor transfers the catalyst to a regenerator where the coke is removed. The regenerated catalyst is gas lifted back to the feed hopper section on reactor 1 where the process starts over.

BTX Aromatics Process

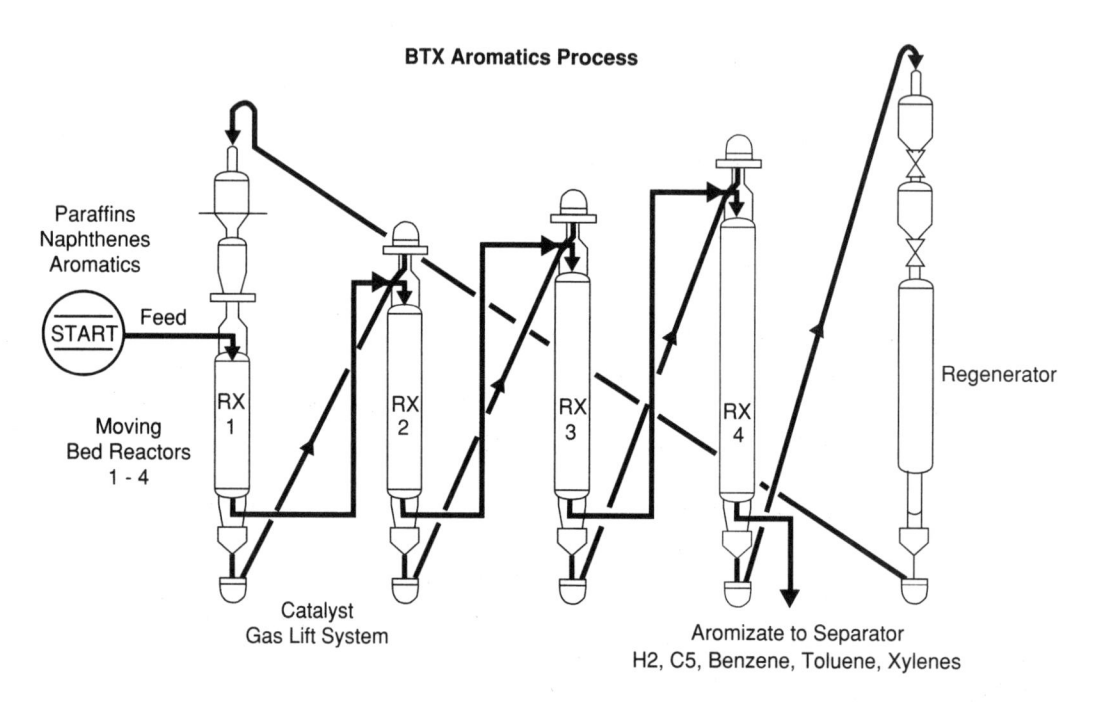

Figure 9.5-1 *BTX*

Chapter 9 • *Industrial Processes*

Ethylbenzene Process

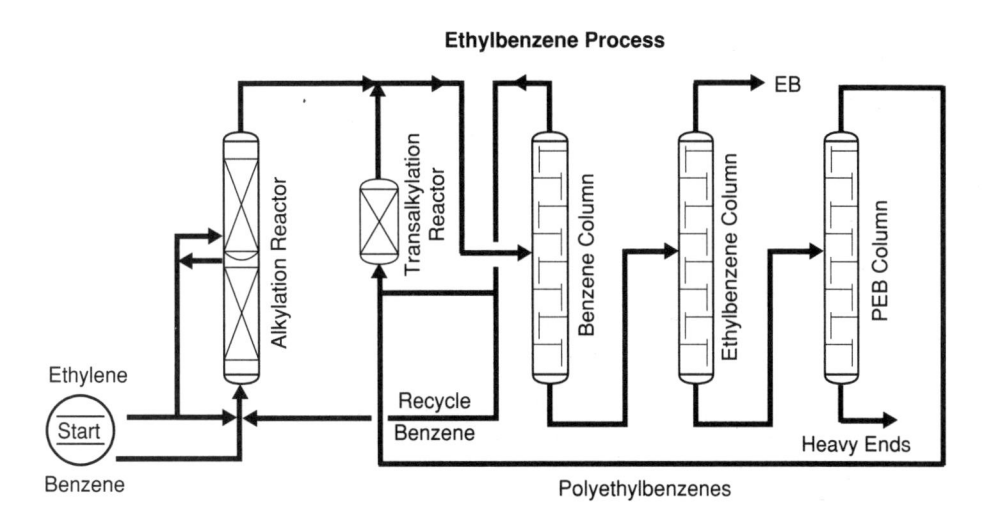

Figure 9.6-1 *Ethylbenzene*

9.6 Ethylbenzene

One of the more common ways to manufacture ethylbenzene, Figure 9.6–1, is to use a fixed bed reactor filled with a special catalyst, a series of distillation columns, and a special alkylation of benzene/ethylene process.

9.7 Ethylene Glycols

The raw feedstock for an ethylene glycols unit includes refined ethylene oxide and pure water. A mixture of ethylene oxide, and recycled and pure water is pumped to a feed tank where it is blended and heated prior to being sent to the (2) glycol reactor. Residence times in the reactor are long enough to allow all of the ethylene oxide to react. After the reaction is complete the water glycol mixture is pumped to a multistage evaporator. A thermosyphon reboiler is used to maintain temperature on the column. A total of six glycol columns are utilized to purify and separate the various components of the process streams. As the glycol water mixture flows from one column to the next, successively lower pressures are encountered. The last four columns are operating under a vacuum. The plant process is designed to produce purified monoethylene glycol EB, however, a number of secondary products (triethylene glycol TEG and diethylene glycol DEG) are formed. Figure 9.7–1 is an illustration of the ethylene glycol process.

9.8 Mixed Xylenes

The mixed xylene process, Figure 9.8–1, selectively converts toluene to high purity benzene, mixed xylenes, C9+ aromatics, and C5-. The feedstock is composed of dry toluene, C9 aromatics, and hydrogen-rich recycle gas. The raw feedstock is introduced to the unit by being passed through a heat exchanger, fired heater, and into a reactor. Mixed xylenes and benzene are produced during toluene disproportionation in vapor phase. Products from the reactor are pumped

Industrial Processes • Chapter 9

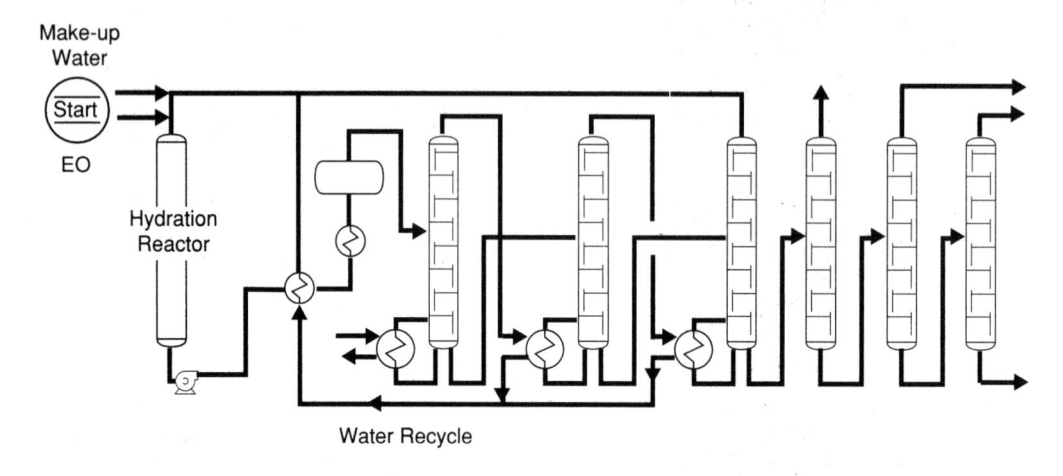

Figure 9.7-1 *Ethylene Glycol*

to a separator where hydrogen-rich gas is recycled to the reactor and primary bottoms products are pumped into a series of fractionation columns for product separation.

Reactor Wt% Yields	Feed	Product
C5 and lighter		1.3
C9+ aromatics		1.8
Benzene		19.8
Toluene	100	52.0
Ethylbenzene		0.6
m-Xylene		12.8
o-Xylene		5.4
p-Xylene		6.3

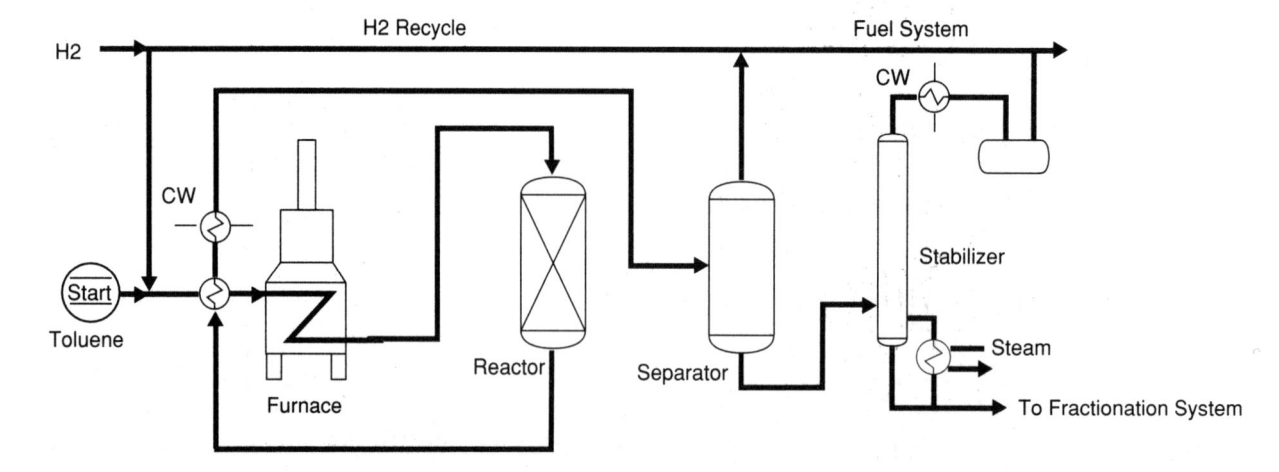

Figure 9.8-1 *Mixed Xylenes*

176

9.9 Olefins

There are three processes associated with olefins production. One process is designed to convert natural gas or raw methanol to ethylene, propylene, and butane. The second process is designed to produce isobutylene and isoamylene feedstocks from hydrocarbon feedstock. This material is used in ethers production, polymerization, and linear olefins for alkylation. The third process selectively converts gas oil feedstocks into high octane gasoline and distillate, and C2-C5 olefins.

9.10 Paraxylene

The paraxylene process, Figure 9.10–1, takes mixed xylenes from reformers or steam crackers to produce high purity paraxylene. Feedstock is pumped to a feed rerun column that removes C9 and heavier out the bottom and mixed xylenes out the top. The overhead product is sent to a set of adsorption columns where paraxylene is removed and purified to 99.9%. A series of distillation columns is used to separate or recycle the rest of the products.

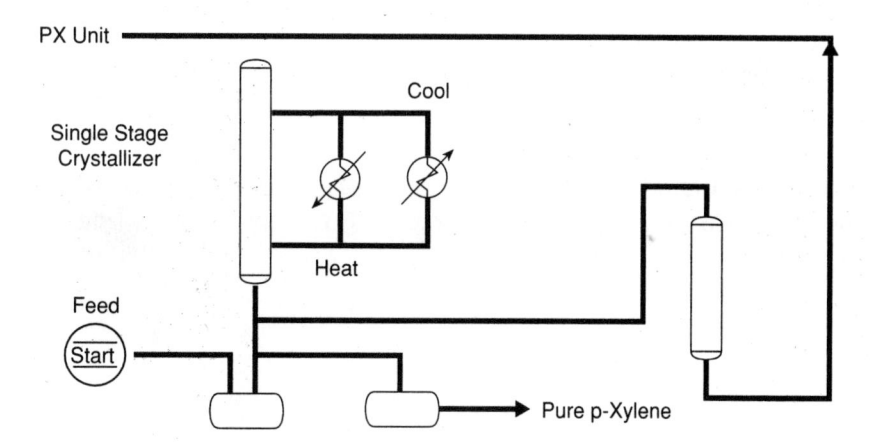

Figure 9.10-1 *Paraxylene*

9.11 Polyethylene

A variety of polyethylene processes exist and are popular with industrial manufacturers, Figure 9.11–1. These applications can be used to produce high or low density polyethylene, linear polyethylene, or linear low density polyethylene.

9.12 Xylene Isomerization

Xylene isomerization takes depleted paraxylene and orthoxylene streams from the paraxylene unit and passes them over a dual, fixed bed catalyst. As the process flow moves through the reactor and over the catalyst it is combined with hydrogen-rich recycle gas. The upper section of the reactor is utilized for EB dealkylation and the lower section is optimized for xylene isomer-

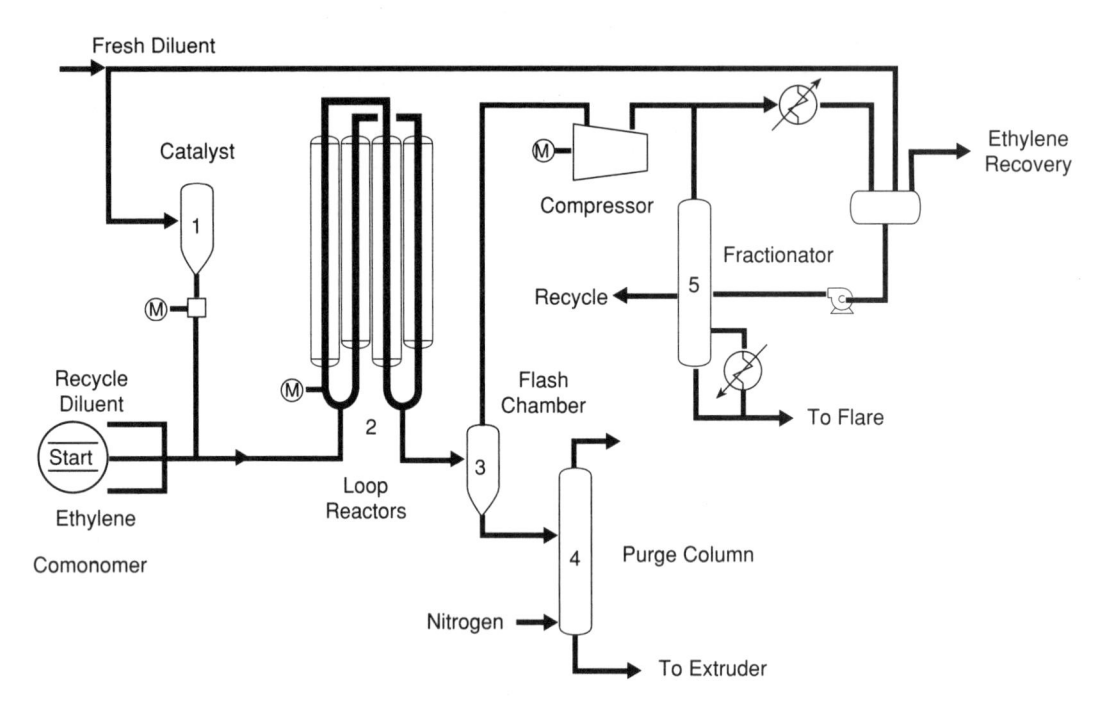

Figure 9.11-1 *Polyethylene*

the reactor is utilized for EB dealkylation and the lower section is optimized for xylene isomerization. EB conversion rates are typically in excess of 65% while paraxylene concentrations are typically 102% greater than equilibrium.

9.13 Ethylene

There are six popular methods used by industrial manufacturers to produce ethylene. The Lummus method is used to produce over 45% of the ethylene sold in the world. This process produces 99.95 vol.% polymer grade ethylene. Some of the by-products created during this process are propylene, butadiene rich C4's, aromatic rich C6-C8 pyrolysis gasoline, and pure grade hydrogen.

9.14 Refining Processes

There are 19 common refining processes used by industrial manufacturers. Refining processes are typically linked to the large branches of the crude oil tree. The global economy has allowed the chemical processing industry to diversify into a variety of business ventures. During the 60s and 70s a large number of petrochemical processes were developed. Each of these processes had its roots in the refinery operation. This formula can also be applied the opposite way. The refining group has the oldest set of processes. Products from the refinery are typically used as feedstocks for modern petrochemical processes.

Chapter 9 • *Industrial Processes*

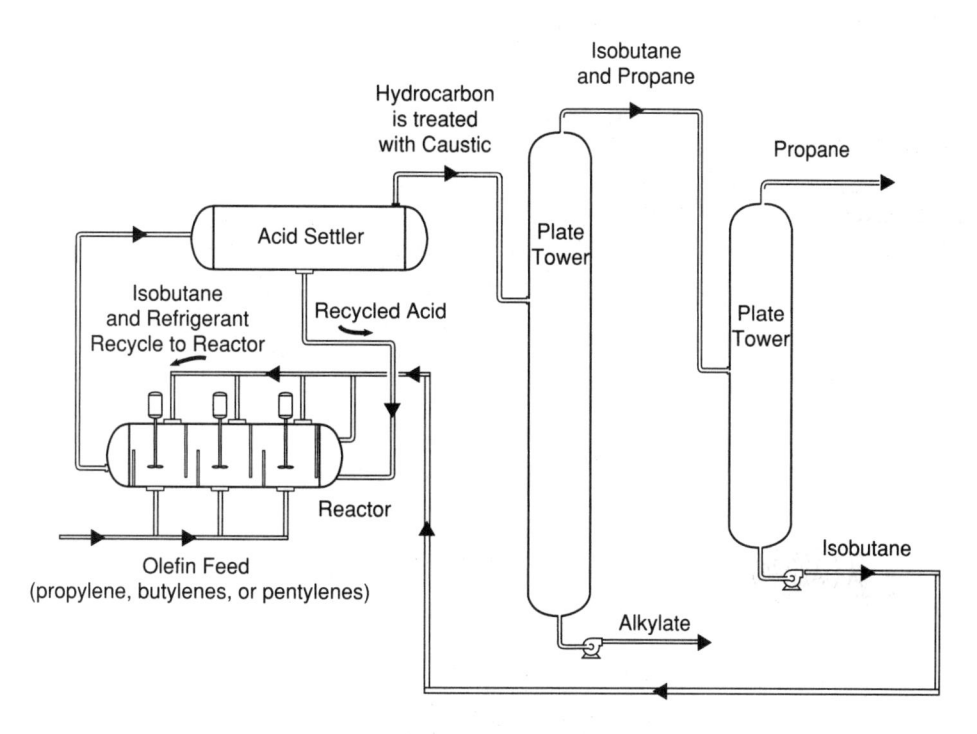

Figure 9.15-1 *Alkylation*

9.15 Alkylation

Alkylation units take two small molecules of isobutane and olefin (propylene, butylenes, or pentylenes) and combine them into one large molecule of high-octane liquid called alkylate, Figure 9.15–1. This combining process takes place inside a reactor filled with an acid catalyst. Alkylate is a superior anti-knock product that is used in blending unleaded gasoline.

After the reaction a number of products are formed that require further processing to separate and clean the desired chemical streams. A separator and an alkaline substance are used to remove (strip) the acid. The stripped acid is sent back to the reactor while the remaining reactor products are sent to a distillation tower. Alkylate, isobutane, and propane gas are fractionally separated in the tower. Isobutane is returned to the alkylation reactor for further processing. Alkylate is sent on to the gasoline blending unit.

9.16 Fluid Catalytic Cracking

Crude oil comes into a refinery and is processed in an atmospheric pipe still. The side-stream of the pipe still is rich with light gas oil. Fluid catalytic cracking units split this gas oil into smaller, more useful molecules. Fluid catalytic cracking units use the following equipment during operation:
- Catalyst regenerator
- Reactor
- Fractionating tower

Industrial Processes • Chapter 9

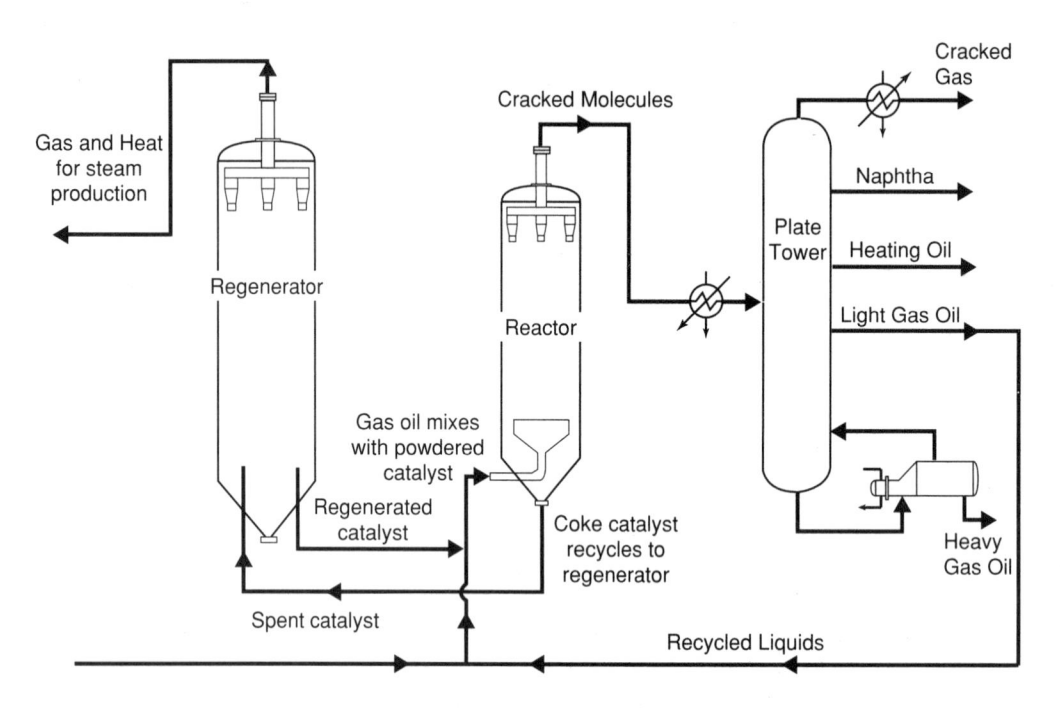

Figure 9.16-1 *Catcracking*

During operation gas oil enters the reactor and is mixed with a superheated powdered catalyst. The term cracking is applied to the process because during vaporization the molecules literally split and are sent to a fractionation tower for further processing. The chemical reaction between the catalyst and light gas oil produces a solid carbon deposit. This deposit forms on the powdered catalyst and deactivates it. The spent catalyst is drawn off and sent to the regenerator where the coke is burned off. Catalyst regeneration is a continuous process during operation. In the fractionation tower the light gas oil is separated into five different cuts:

1. Cat cracked gas
2. Cat cracked naphtha
3. Cat cracked heating oil
4. Light gas oil
5. Resid

9.17 Hydrodesulfurization

Crude oil is a mixture of hydrocarbons, clay, water, and sulfur. Some crude mixtures have higher concentrations of sulfur than others. These mixtures are referred to as "sour feed." Hydrodesulfurization, Figure 9.17–1, is a process used by industrial manufacturers to "sweeten" or remove the sulfur. Hydrodesulfurization units use the following equipment during operation:

- Fired heater
- Separator
- Reactor

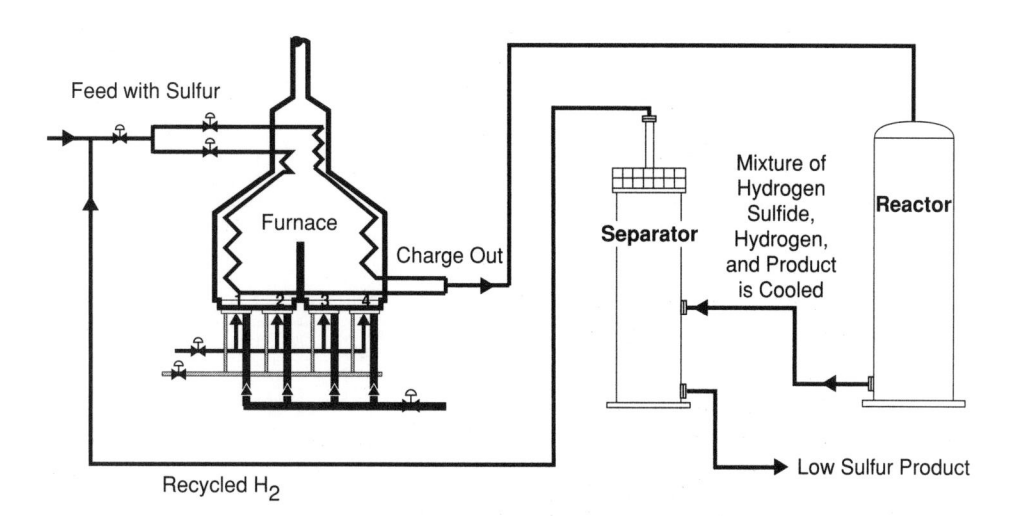

Figure 9.17-1 *Hydrodesulfurization*

During operation sour feed is mixed with hydrogen and heated up in a fired furnace. The heated mixture is sent to a reactor where the hydrogen combines with the sulfur to form hydrogen sulfide. By lowering the temperature slightly the sweet crude condenses leaving the hydrogen sulfide in a vapor state. This vapor and liquid mixture is sent to a separator where the low sulfur sweet feed is removed. The hydrogen sulfide and hydrogen are sent for further processing where the hydrogen is separated and returned to the original system.

9.18 Hydrocracking

Hydrocracking is a process that industrial manufacturers use to boost gasoline yields. The process splits heavy gas oil molecules into smaller, lighter molecules called hydrocrackate. Hydrocracking units use the following equipment during operation:
- First- and second-stage reactor
- Separator drum
- Fractionating tower

The hydrocracking process mixes heavy gas oil feed with hydrogen before sending it to the first-stage reactor. The reactor is filled with a fixed bed of catalyst. As process flow moves from the top of the reactor to the bottom the cracking reaction takes place. First-stage hydrocrackate is sent to a separator drum where the hydrogen is reclaimed and the hydrocrackate is moved to a fractionation tower. In the fractionation tower the hydrocrackate is separated into five different cuts:

- Butane
- Light hydrocrackate
- Heavy hydrocrackate
- Heating oil
- Heavy bottoms

Industrial Processes • Chapter 9

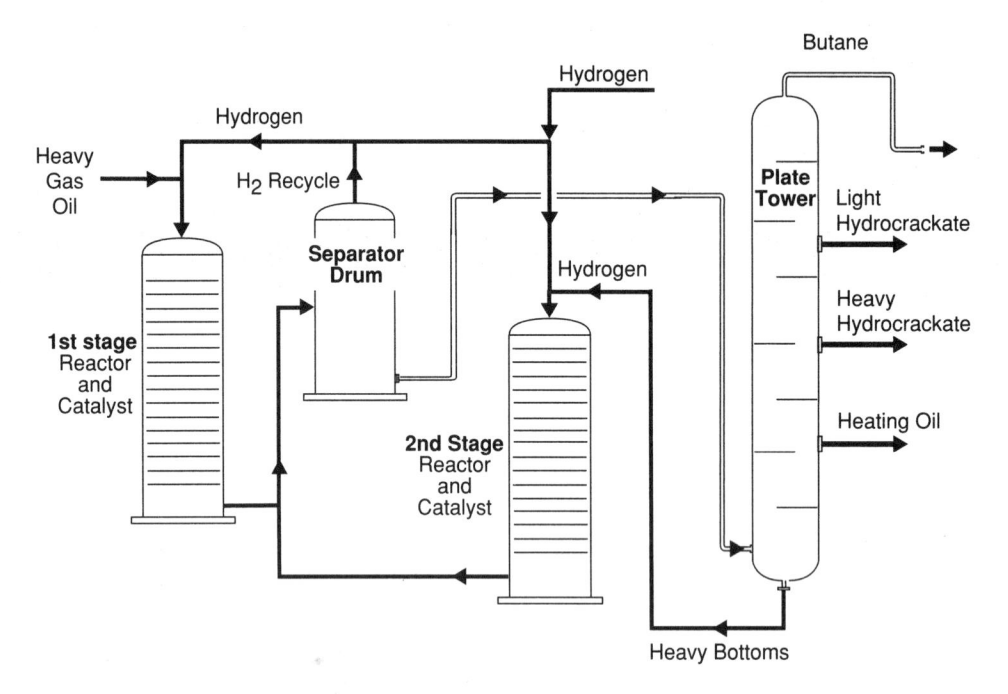

Figure 9.18-1 *Hydrocracking*

The heavy bottoms is mixed with hydrogen and sent to the second-stage reactor for further processing. The second-stage reactor reclaims as much of the hydrocrackate as possible before sending it to the separator and tower.

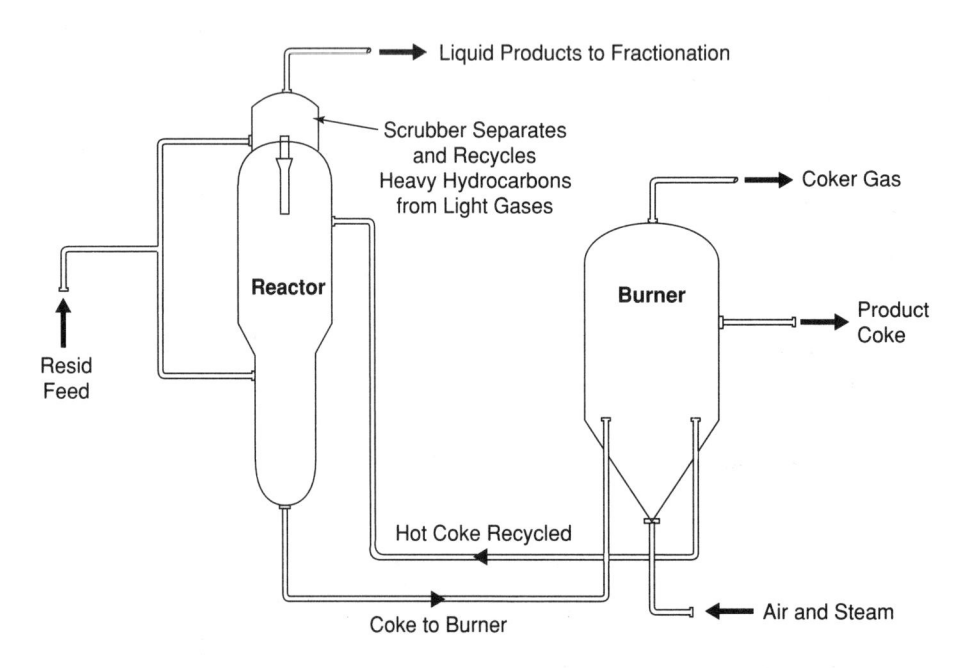

Figure 9.19-1 *Fluid Coking*

Chapter 9 • *Industrial Processes*

9.19 Fluid Coking

Fluid coking, Figure 9.19–1, is a process used by industrial manufacturers to squeeze every last useful molecule out of heavy residues. Residue from other processes flows into a specially designed, high temperature reactor. Light products vaporize and flow to a fractionation column. The remaining material is sent to a burner where further processing takes place. The burner produces three separate products:
- Coker gas for use in plant
- Product coke for sale
- Recycled coke for reactor

9.20 Catalytic Reforming

Catalytic reforming, Figure 9.20–1, is a process that utilizes refinery naphtha to produce high octane reformate. The advanced design utilizes a set of four moving bed reactors and one regenerator. The process is similar to that incorporated by BTX aromatics. The design utilizes continuous catalyst regeneration, continuous liquid, and solid flow movements between reactors.

9.21 Crude Distillation

Crude distillation, Figure 9.21–1, is a process that separates the various fractions of crude oil by their boiling points. During the distillation process crude oil goes through a number of phases. The initial charge is heated and desalted. This heating process is gradual as the charge moves through a series of heat exchangers before it enters the fired furnace. In the furnace the charge splits into a number of passes that are combined when the feed stock exits the furnace. A typical inlet temperature for a fired furnace is 550 degrees F while the outlet temperature varies between 675 and 725 degrees F. The heated charge is pumped to a distillation column where a

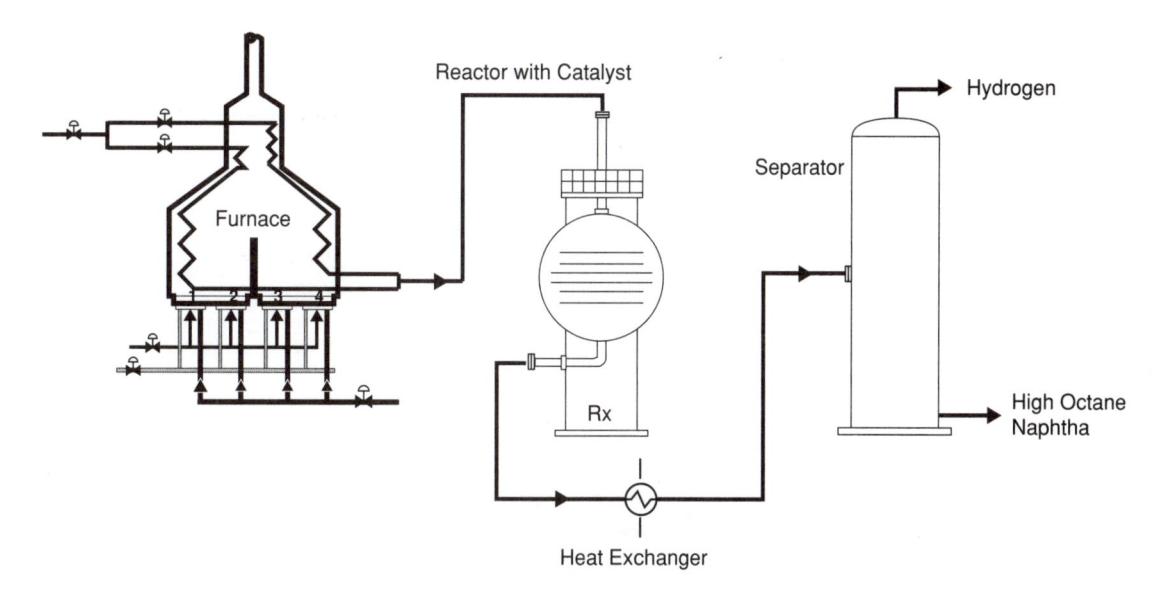

Figure 9.20-1 *Reforming*

183

Industrial Processes • Chapter 9

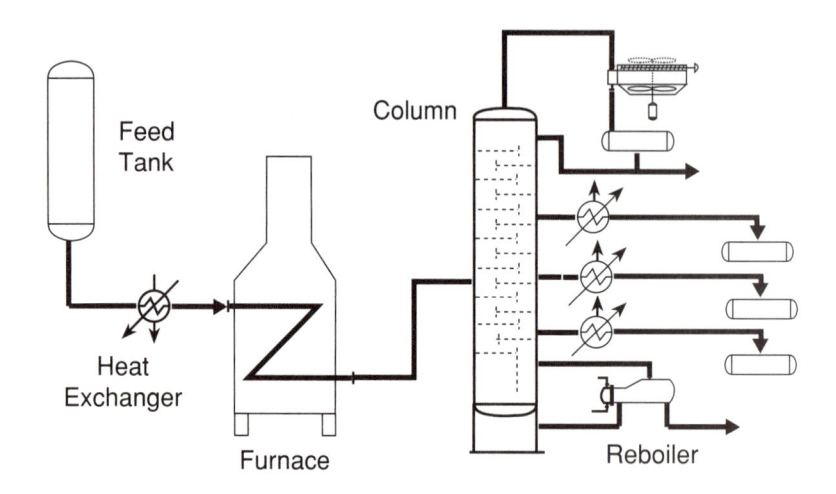

Figure 9.21-1 *Crude Distillation*

fraction of the feed vaporizes and moves up the column. The distillation column is a device that incorporates a still upon a still design. Vapors rise up the column while liquids drop down. Molecular distribution on each tray in the distillation column is different. A typical distillation column has one feed line, one overhead line, one reflux line, four side streams, and one bottoms line. Different products exit at each point on the column. Crude distillation columns produce flash gas, light and heavy naphtha, kerosene, diesel, cracker feed, gas oils, and asphalt.

Summary

Industrial processes can be broken into three distinct groups: refining (19 processes), gas processing (15 processes), and petrochemical processing (40 processes). The oldest and best established group is refining processing. The refining processes discussed in this chapter included alkylation, fluid catalytic cracking, hydrodesulfurization, hydrocracking, fluid coking, catalytic reforming, and crude distillation. Alkylation takes two small molecules of isobutane and olefin and combines them into one large molecule called alkylate. Fluid catalytic cracking uses a heated catalyst to break large gas oil molecules into smaller ones.

Hydrodesulfurization is a refining process that removes sulfur from a process stream. Hydrocracking uses a multistage reactor system to boost yields of gasoline from crude oil. Fluid coking is a refinery process applied to heavy resid to remove or break loose usable products. Catalytic reforming uses a reactor, catalyst approach to break hydrogen loose from high octane naphtha. The final refinery process covered in this chapter is crude distillation. Crude distillation separates the various components in crude oil by their boiling points. Refinery processing has been developing for many years.

Gas processing springs directly from the refining process. During the past thirty years the gas processing and the petrochemical area experienced rapid technological advances. The petrochemical processes covered in this chapter include benzene, BTX aromatics, ethylbenzene, ethylene glycol, mixed xylenes, olefins, paraxylene, polyethylene, xylene isomerization and ethylene.

Chapter 9

Review Questions

1. What is the primary difference between petrochemical and refinery processes?
2. List the significant events that occurred between 1914 and 1960.
3. Describe the evolution of the refinery process to petrochemical processes.
4. Describe the benzene process.
5. Describe the BTX aromatics process.
6. Sketch the ethylbenzene process.
7. Describe ethylene glycol operations.
8. Describe mixed xylenes operations.
9. Describe olefins operations.
10. Describe the paraxylene process.
11. Draw a simple sketch of the polyethylene process.
12. Describe xylene isomerization operations.
13. Describe the ethylene process.
14. Explain the basic refining process.
15. Explain alkylation.
16. Describe fluid catalytic cracking, catalytic reforming and hydrocracking.
17. Describe the hydrodesulfurization process.
18. Explain the principles of crude distillation.
19. Explain fluid coking.

Chapter 10
Operations

OBJECTIVES

After studying this chapter, the student will be able to:

- *Describe key terms and definitions used in operations.*
- *Combine systems into operating processes.*
- *Describe a process technician's role during plant operations.*
- *Write operating procedures.*
- *Demonstrate application of operating procedures.*
- *Operate equipment under various conditions.*

Operations • Chapter 10

KEY TERMS

Lower control limit—control limit for sample plot points below x-bar or the center line.

Mean—the average of all samples obtained for developing a statistical process control chart.

Quality—can be defined as understanding the customer's expectations and needs and providing products and services that meet those needs.

Quality tools—used by process technicians, these include statistical process control, flowcharts, cause and effect diagrams, fishbone, pareto charts, run charts, control charts, planned experimentation, histograms or frequency plots, forms for collecting data (checklists), and scatter plots.

R (range)—the delta (Δ) between two or more numbers. Example: 2.5, 3.5, 1.5. The range of these numbers is between 1.5 and 3.5 or 2.

Team skills—interpersonal skills such as interviewing, communications, listening skills, working in self-directed work teams, diversity skills, and applying process technical skills learned in college training programs.

Upper control limit—control limit for sample plot points above x-bar or the center line.

X-bar—the center line on an SPC chart.

10.2 Team Skills

There are several college courses that will help a process technician with fundamental team skills. Some of these skills include interpersonal skills, interviewing, communications, listening skills, working in self-directed work teams, diversity skills, and applying process technical skills learned in college training programs.

College course selection beyond the core process curriculum should include classes that provide role-playing, case studies and analysis, group learning experiences and learning objectives that deal with improving a person's ability to cope with communication and confidence problems. Students participating in advanced process classes are given opportunities to apply knowledge and skills they have acquired during the program. Some students are very good at theory type tests inside a classroom while others are very strong within a laboratory or work environment. Merging the two skills is a major objective of advanced process courses.

During benchtop, DCS, and pilot plant operations, process instructors look for:
(Supervisors Evaluation; See operations section of this chapter)
- Students backing away from operating the unit
- Students who have difficulty understanding what is going on
- Student-to-student conflicts
- Diversity awareness and sexual harassment
- Students who are late or absent frequently
- Participation in group activities
- Hands-on skills and mechanical aptitude
- Ability to complete assignments
- Ability to operate the processes
- Ability to work with minimal supervision

Working in self-directed work teams is a fundamental skill that can be assessed during the laboratory portion of the program.

10.3 Quality Tools

Quality can be defined as understanding your customer's expectations and needs and providing products and services that meet these needs. The chemical processing industry has been using quality tools to meet these needs for many years. The leading gurus on quality control and their specific ideas on quality are:
- Dr. Walter Shewhart (1891-1967)—a physicist who worked for Bell Laboratories. He developed a quality system using statistics that could enhance product uniformity. Published findings in 1931 text The Economic Control of Manufactured Product.
- Dr. W. Edwards Deming—met Shewhart as he was developing his quality methodology. Deming understood Shewhart's concepts and began to enhance them. A critical element of the Deming approach is the need for "practical statistical methods which everyone could use to improve quality." Believed in on-line quality control and not end-line control. Quality improves as variability decreases.

- Leslie E. Simon—understood and embraced Shewhart's text. Convinced military to use these new quality techniques.
- Dr. Joseph M. Juran—quality improvement expert to Japanese. Advocated the adoption of a universal way of thinking about quality and quality training. Focused on management; wanted them to rank quality and financial returns on the same level. Quality linked to customer needs.
- Phillip B. Crosby—"Quality is conformance to requirements." Do it right the first time. Quality does not mean higher costs.
- Genichi Taguchi—poor products produce quality problems. "Quality is a virtue of design."

The list of quality tools used by process technicians include:
- Statistical process control (SPC)
- Flowchart
- Cause and effect diagrams (fishbone)
- Pareto chart
- Run chart
- Control chart
- Planned experimentation
- Histogram or frequency plot
- Forms for collecting data (checklists)
- Scatter plot

Figure 10.3–1 shows examples of four of these quality tools: fishbone, scatter plot, SPC, and pareto chart.

In modern manufacturing, raw materials are purchased from suppliers to create new products that are sold to their customers. In many cases the quality of a product is linked to the quality of the raw materials used to make the product. When the supplier is openly working with the customer, the lines of communication are open. Information and ideas move between the supplier and customer. The old saying "two heads are better than one" can be applied to the business goals of both the customer and supplier. The supplier should attempt to meet or exceed the customer's goals and help the customer be successful. It is possible to meet or exceed a customer's goals and still see them go out of business. When a supplier makes the success of their customers a priority, both entities succeed.

10.4 Statistical Process Control (SPC)

SPC is a term most process technicians are familiar with and use on-the-job. Statistical process control and control charts are used to determine if a process is in control or requires an adjustment. Most operations have normal variations of highs and lows for the equipment, Figure 10.4–1. For example, the thermostat in a home may be set at 75 degrees, however the temperature may range from 68 to 82 degrees as the air conditioning cycles on and off.

Before statistical process control, process changes were made based upon a desired setpoint. The normal cyclical variations of the process were not considered. This type of quality control

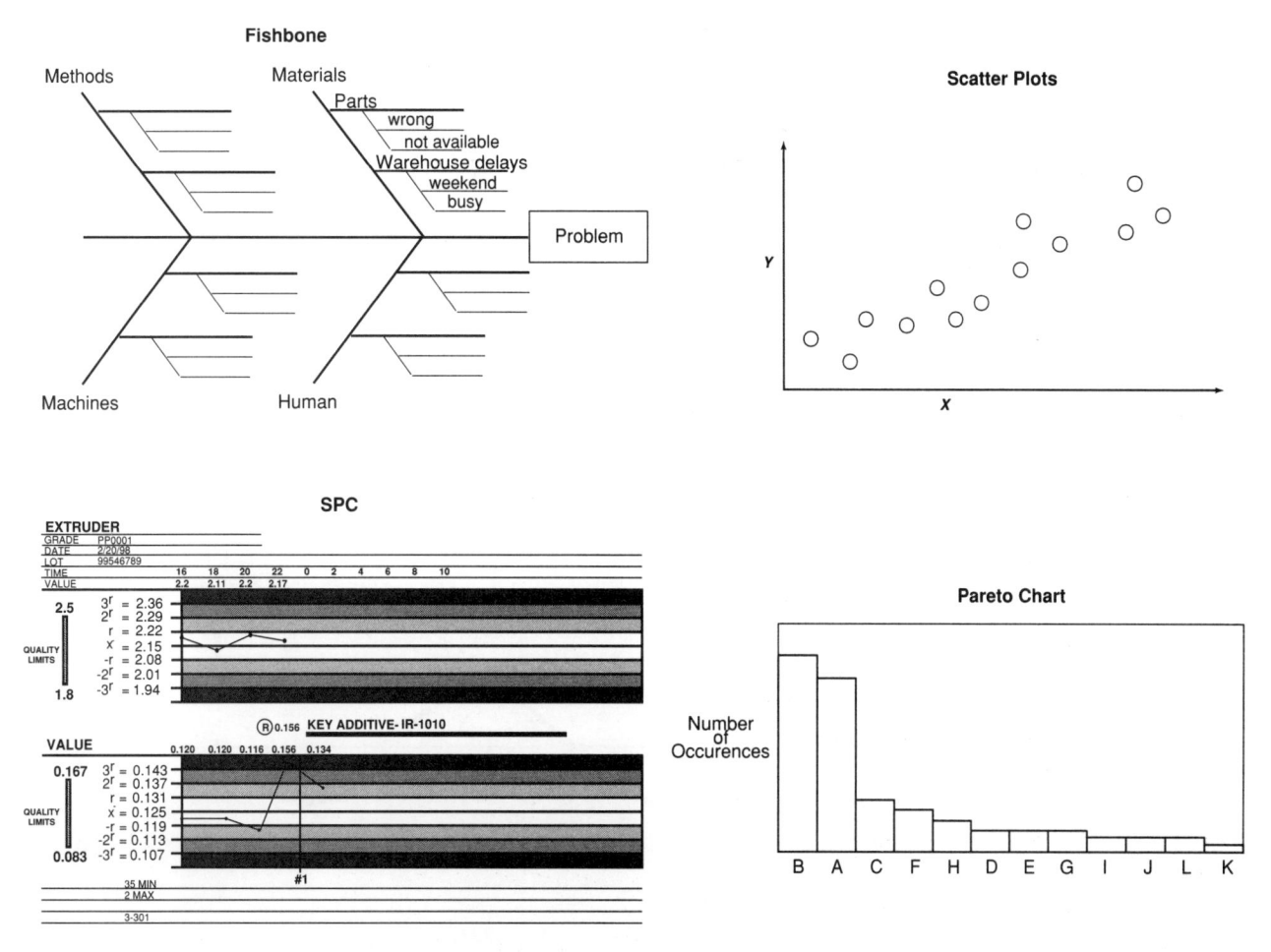

Figure 10.3-1 *Quality Tools*

made it difficult to control a process. If the technician compares the setpoint to the process variable as the temperature reaches the top of its cycle and is about to drop, an adjustment would be required to get it closer to the target. Since the process was about to drop in temperature naturally, a much lower temperature cycle will occur. This will initiate a swinging action as the process cycles out of control on each adjustment.

The following process is used to develop a SPC chart:
- Collect samples (typically twenty-five sampling periods are used)
- Add samples and divide the total by 25. This will identify the mean xbar.
- Calculate range samples, add up, and divide by 25. This will identify R.
- Calculate the upper control limit UCLxbar.
  UCLxbar = x + A2R
- Calculate the lower control limit LCLxbar.
  LCLxbar = x - A2R

Factors for control limit calculations can be found on a table. Selections are made based upon total number of samples collected and the control limit equation.

Operations • Chapter 10

NORMAL VARIATION

ADJUSTMENTS + NORMAL VARIATION

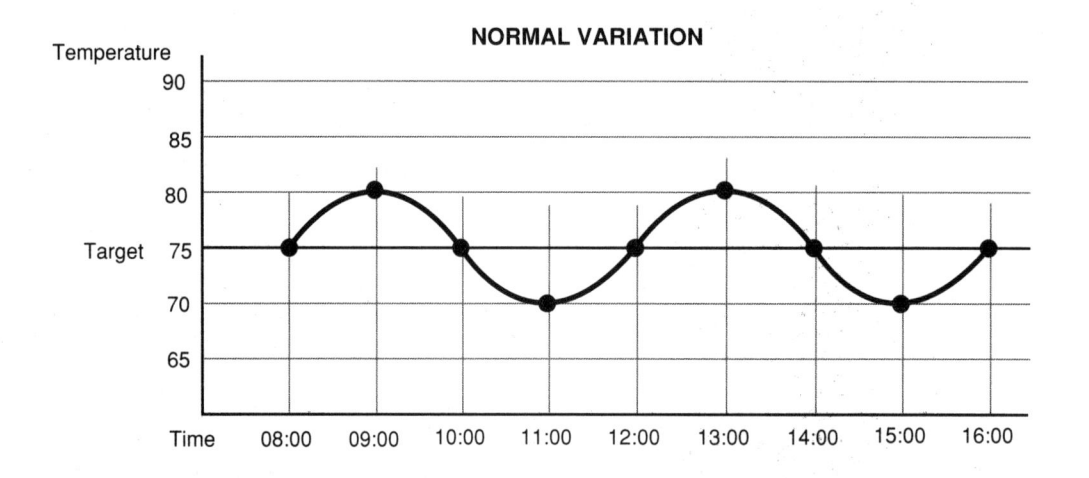

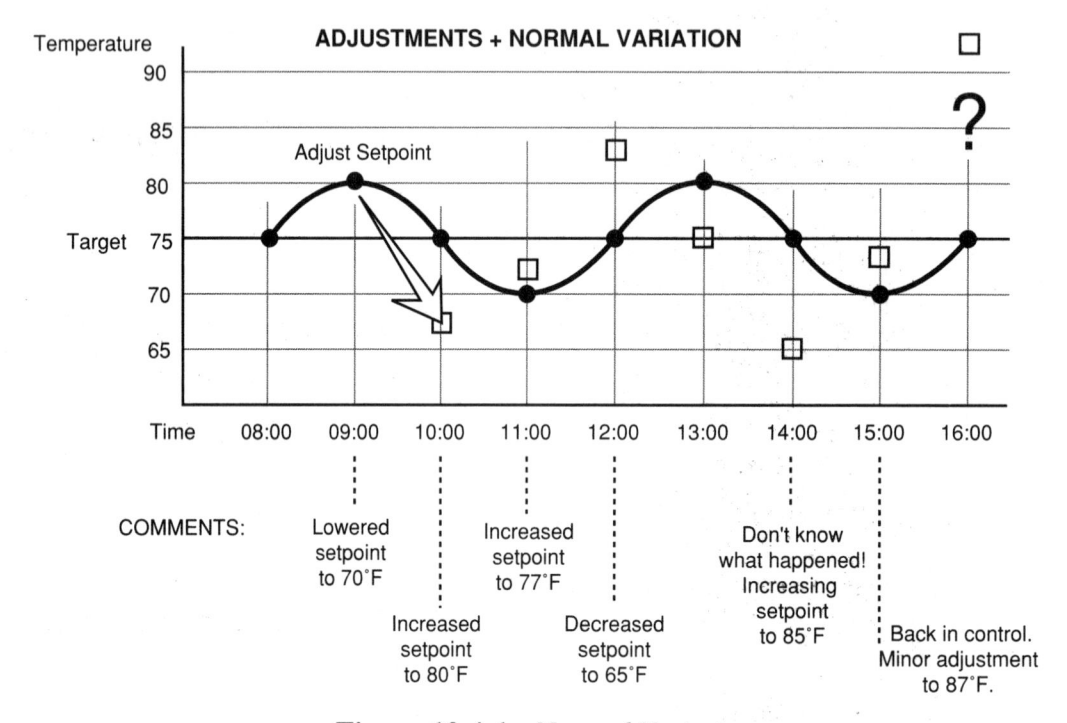

COMMENTS:
- Lowered setpoint to 70°F
- Increased setpoint to 80°F
- Increased setpoint to 77°F
- Decreased setpoint to 65°F
- Don't know what happened! Increasing setpoint to 85°F
- Back in control. Minor adjustment to 87°F.

Figure 10.4-1 *Normal Variation*

Factors for Control Limits

n	A2	D4	d2
2	1.880	3.268	1.128
3	1.023	2.574	1.693
4	0.729	2.282	2.059
5	0.577	2.114	2.326
6	0.483	2.004	2.534

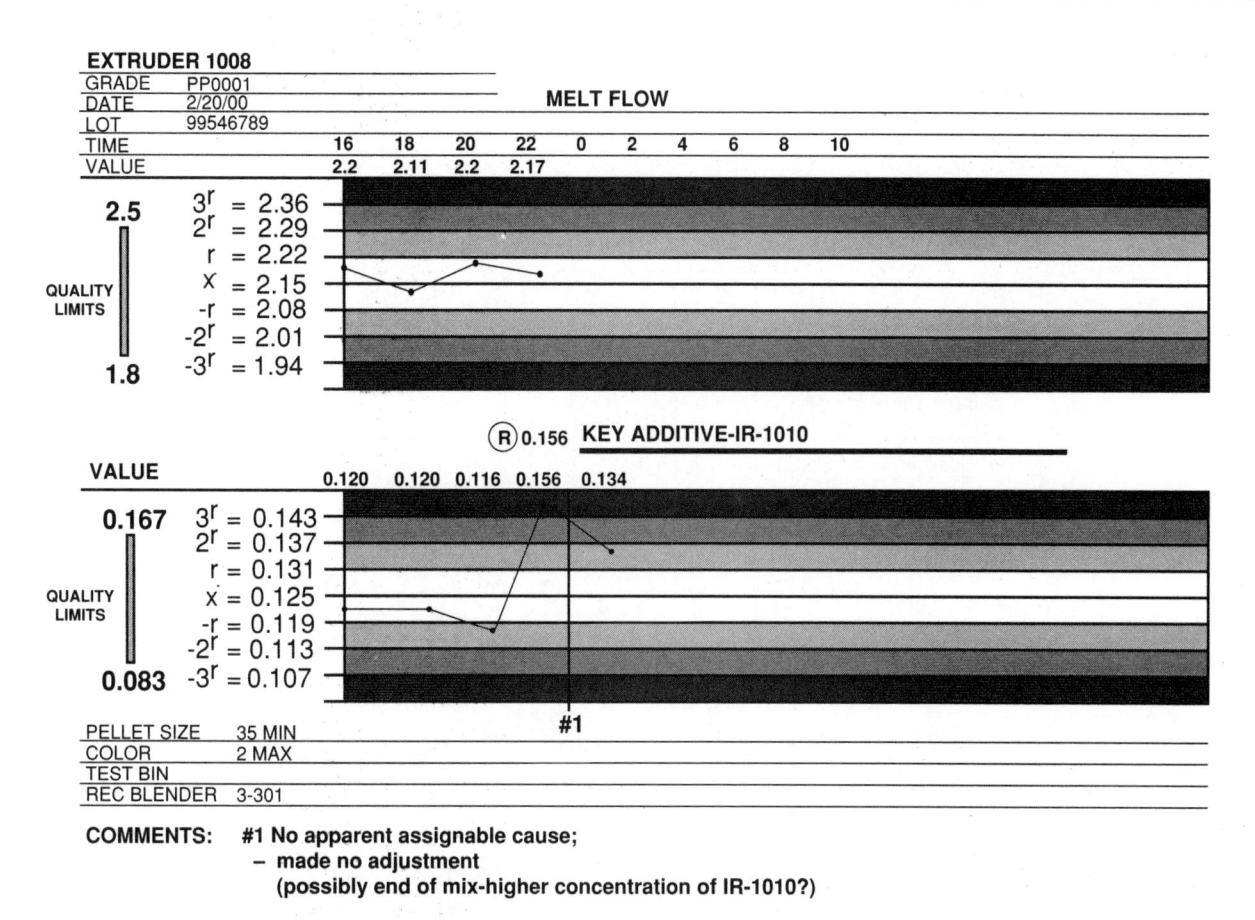

Figure 10.4-2 *Control Chart*

Control Limits for X bar (x) and Range (R)
　　Upper control limit UCLxbar = x bar + A2R (See factors for control limits)
　　Lower control limit LCLxbar = x bar - A2R (See factors for control limits)
　　UCLr = D4R

Figure 10.4–2 shows a sample control chart.

10.5 Troubleshooting

New technicians are assigned to complete unit checklists as they make their rounds. Levels, temperatures, pressures, and flows are carefully controlled inside a process unit. Data collection and analysis are part of the complex troubleshooting processes that take place inside the chemical processing industry. The data collection procedure is used to complete statistical process control charts, histograms, and other quality tools used to operate the unit. Checklists are valuable indicators that a problem is developing on a process unit.

Operations • Chapter 10

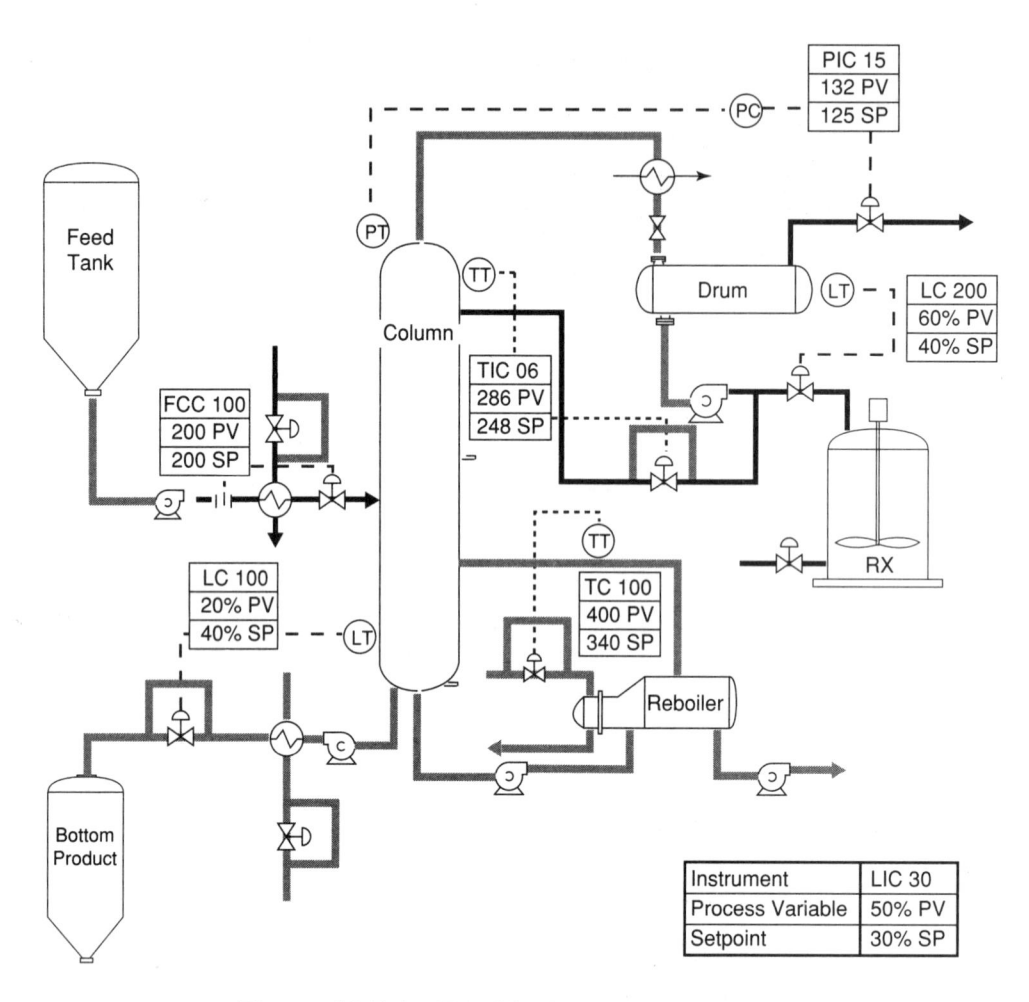

Figure 10.5-1 *Troubleshooting Program*

Troubleshooting is a complex technology that requires a thorough knowledge of the equipment and processes. Process technicians inspect and maintain equipment, place equipment in service and remove from service, make rounds, relief, complete checklists and control documentation, use statistical process control, respond to emergencies, and troubleshoot system problems. Preventive maintenance programs are designed to keep equipment in good condition.

Troubleshooting technology is an acquired skill. A process technician can learn a variety of troubleshooting methods.

Simple Troubleshooting Model
Model One—Cause and Effect. Cause and effect relationships can be identified when a primary problem initiates a series of secondary problems. For example, if the steam valve sticks in the Open position on the reboiler, higher temperatures than normal are experienced. These higher temperatures cause more product to vaporize and go over the top of the distillation column. This

creates high fluid levels in the overhead accumulator and low levels in the bottom of the column. During this time frame the feed rate to the column remains constant. The key control loops involved are as follows:

CONTROL LOOP	SETPOINT	PROCESS VARIABLE
FC-100	200 gpm	200 gpm
LC-100	40%	20%
LC-200	40%	60%
TC-100	350F	400F

The temperature control loop setpoint calls for 350 degrees while the process variable records 400 degrees. In the cause and effect troubleshooting model each control loop that is above or below the setpoint is identified and listed on a sheet of paper. At this point process technicians use their understanding of the process to review each control loop and see if the first control loop could cause the second loop to be out of control. This process of elimination allows technicians to immediately identify those control loops that are only secondary problems and not the primary problem. For example, could LC-100 column bottoms level cause LC-200 (the overhead accumulator) to be high? Could LC-100 cause TC-100 to be above temperature? Now turn the problem around and see if TC-100 could cause the level to be low in the column and high in the overhead accumulator. Figure 10.5–1 is an example of a troubleshooting problem.

10.6   Benchtop Operations

Most refinery and petrochemical processes can trace their roots back to a theory that was applied in the laboratory and operated on a small benchtop unit that can be as small as a desktop. Benchtop operations iron out many of the concerns found in the theory of the process. Process or research technicians work with chemists and engineers to produce and repeat laboratory data. Work on the benchtop unit leads to a larger process called pilot plant operation. This is the final step before the process goes commercial.

A simple and safe example of a benchtop operation is a red dye and water separation, Figure 10.6–1. The system includes a small distillation column, heater, 1/4 inch tubing, valves, and small tanks.

10.7   Pilot Plant Operation

During the first few weeks of training a process technician will be involved in the following activities:
- Review the safety hazards associated with pilot unit.
- Study operation procedures.
- Walk through unit with trainer and study process flow diagrams.
- Trace lines and study instrumentation.
- Fill out checklists and SPC charts.
- Watch trainer start up and shut down the unit.

Operations • Chapter 10

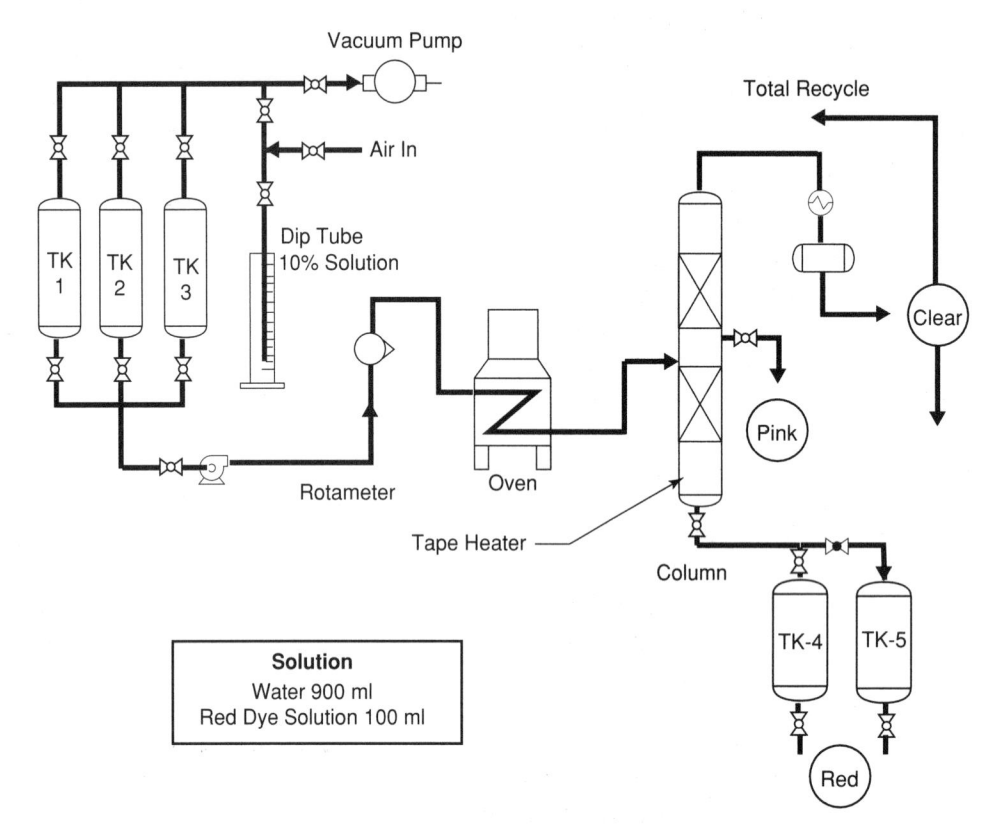

Figure 10.6-1 *Simple Benchtop Unit*

- Study process and instrument drawings.
- Start up and shut down the unit.
- Work in self-directed work teams.
- Troubleshoot problems.
- Catch samples and make rounds.
- Perform housekeeping and routine maintenance.

The operations course is the capstone experience for students in college process programs and should reflect an on-the-job training experience. The course combines systems into operational processes, while allowing students to focus on operations under various conditions. Typical roles and responsibilities of an operator are generally applied.

Pilot Unit Modules
1. Unit Overview
2. Safety
3. Basic Equipment
4. Systems
5. Quality Control
6. Operation
7. Technical Notebook
8. Supervisor Evaluation
9. Program Exit Exam

Unit Overview
Before a new technician is assigned to an operating unit a number of steps should be followed by the college program director. An initial orientation allows the instructor to describe what the student is about to experience. Initial information should include a description of the process, potential hazards and safety rules, and team assignments. Students will be required to wear minimal personal protective equipment (PPE) including hard hat and safety glasses. During scheduled times each team is required to trace lines and produce individual process and instrumentation drawings (P&IDs). A description of the process and a list of hazards should also be produced by the student.

Safety Overview
Applied safety training is a new experience for students in a process technology pilot plant program. Operating units provide a unique experience that allows students to:
- Wear PPE and isolate equipment for maintenance
- Wear respirators, identify hazards, and apply general safety rules
- Have safety meetings with team leader
- Conduct hazard analysis, and study Process Safety Management
- Analyze various aspects of HAZCOM
- Simulate fire prevention, and protection and control
- Develop and apply HAZWOPER
- Develop spill release guidelines
- Apply hearing conservation and industrial noise principles
- Use fall protection principles
- Fill out permits; demonstrate lock-out, tag-out, confined space entry, opening and blinding, and cold and hot work

Basic Equipment
At the end of the college training program students are given an opportunity to use equipment they have studied in class. During the operations class students produce a pencil sketch, component list, and basic description of each piece of equipment they will be operating. A typical equipment list includes:

- Valves—gate, globe, ball, diaphragm, automatic, regulators
- Piping—SS, carbon steel, fittings, tubing
- Pumps—external gear, mag drives, centrifugal, gear box
- Compressors—piston
- Stationary vessels
- Heat exchangers
- Cooling towers
- Boiler—firetube, steam traps
- Distillation—packed
- Basic instruments

Systems
After the P&ID is developed technicians should be encouraged to identify the various systems in the drawing. A separate section should be developed that includes a pencil sketch of each of

these systems and descriptions of each. Examples of basic systems include:
- Feed system
- Pre-heat system
- Distillation system
- Steam system
- Refrigeration system
- Air system
- Electric system
- Tank farm
- Pressure relief

Quality Control
A variety of quality principles can be applied during the operation of a pilot plant. Some of these topics include applied statistical process control and the development of checklists, SPC charts, sampling procedures, logbook entries, and the use of quality tools.

Unit Operations
A critical component of the unit operations course is operations. The primary objective of the class is to provide students with the opportunity to operate a process unit. Successful completion of the course depends on demonstrating this ability to the program director.

Operations students are given the opportunity to use small hand tools for routine maintenance, maintain housekeeping, make rounds, start up and shut down the unit, troubleshoot problems, and write procedures.

Technical Notebook
The technical notebook allows students to document their experience in the program and provide the instructor with the following information:
- P&ID and process description
- Safety section
- Unit equipment section
- Systems section
- Quality control
- Unit operations—start-up/shutdown procedures

Supervisor Evaluation
The supervisor evaluation allows the instructor to utilize additional information not found in the earlier sections. Prior to the evaluation of a trainee, attendance records, team leader evaluations, and personal impressions are collected.

Exit Exam
The exit exam is a comprehensive program final that allows the student to review and demonstrate competency on core objectives found in the process technology program.

Summary

Team skills include interpersonal skills, interviewing skills, communications, listening skills, ability to work in a self-directed work team, diversity training, and the ability to apply process technical skills learned in college training programs.

Quality can be defined as understanding your customer's expectations and needs and providing products and services that meet these needs. The leading experts on quality control are Dr. Walter Shewhart, Dr. W. Edwards Deming, Leslie E. Simon, Dr. Joseph M. Juran, Phillip B. Crosby, and Genichi Taguchi. The list of quality tools used by process technicians includes statistical process controls, flowcharts, cause and effect diagrams (fishbone), pareto charts, run charts, control charts, planned experimentation, histograms or frequency plots, forms for collecting data (checklists), and scatter plots.

Troubleshooting is a complex technology that requires a thorough knowledge of equipment and processes. Process technicians inspect and maintain equipment, place equipment in service and remove from service, make rounds, relief, complete checklists and control documentation, use statistical process control, respond to emergencies, and troubleshoot system problems.

Chapter 10

Review Questions

1. Sketch a simple pump around unit.

2. Describe how the system operates.

3. Identify physical hazards.

4. Write a procedure to lock-out and isolate the pumps for maintenance.

5. Design a simple chemical list. Include three chemicals.

6. Identify health hazards.

7. Select required personal protective equipment.

8. Secure a material safety data sheet.

9. Write a start-up procedure for your unit.

10. Write a shutdown procedure for your unit.

11. Design a checklist for your unit.

12. Design a statistical process control chart for your pump's output.

13. Draw the equipment found on your unit and label each component.

14. List three negative activities a process instructor looks for during lab time.

15. List three positive activities a process technician can be involved in during lab time.

16. Who is Dr. Walter Shewhart?

17. In your own words define quality.

18. What are quality tools?

19. Who is Dr. W. Edwards Deming?

20. List the traits required to work in a self-directed work team.

Chapter 11
Process Chemistry

OBJECTIVES

After studying this chapter, the student will be able to:
- *Define fundamental chemistry terms.*
- *Describe the fundamental principles of chemistry.*
- *Describe and use a chemical equation and periodic table.*
- *Describe these chemical reactions:*
    exothermic
    endothermic
    replacement
    neutralization
    combustion
    heat and pressure
    catalysts
- *Perform a material balance.*
- *Perform a percent-by-weight calculation.*
- *Describe pH measurements.*
- *Describe hydrocarbons.*
- *Review applied concepts of chemical processing.*

KEY TERMS

Absorption—removes one or more components from a gas mixture by exposing it to a gas or liquid.

LOW   HIGH → CAUSTIC OR Alky

Acid and base—an acid is a bitter-tasting chemical compound that has a pH value below 7.0, changes blue litmus to red, yields hydrogen ions in water, and has a high concentration of hydrogen ions. A base is a bitter-tasting chemical compound that has a soapy feel and a pH value above 7.0. It turns red litmus paper blue and yields hydroxyl ions.

Adsorption—uses a porous solid to remove gases or liquids from a mixture.

Anion—negatively charged ion.

Atom—the smallest particle of a chemical element that still retains the properties of an element. An atom is composed of protons and neutrons in a central nucleus surrounded by electrons. Nearly all of an atom's mass is located in the nucleus.

Atomic mass unit (AMU)—the sum of the masses in the nucleus of an atom.

Atomic number—identifies the position of the element on the periodic table and the total number of protons in the atom.

Balanced equation—the sum of the reactants (atoms) equals the sum of the products (atoms).

Catalyst—a chemical that can increase or decrease reaction rate without becoming part of the product.

Catcracking—a process designed to increase the yield of desirable products from a barrel of crude oil. Catcracking utilizes a catalyst to accelerate the separation process.

Cations—positively charged ions.

Chemical bond (covalent)—occurs when elements react with each other by sharing electrons. This forms an electrically neutral molecule.

Chemical bond (ionic)—occurs when positively charged elements react with negatively charged elements to form ionic bonds through the transferring of valence electrons. Ionic bonds have higher melting points and are held together by electrostatic attraction.

Chemical equation—numbers and symbols that represent a description of a chemical reaction.

Chemical reaction—a term used to describe the breaking of chemical bonds, forming of chemical bonds, or the breaking and forming of chemical bonds.

Chemistry—the science and laws that deal with the characteristics or structure of elements and the changes that take place when they combine to form other substances.

Combining processes—alkylation, polymerization, reforming. Rearranges or combines various chemical fractions.

Compound—a substance formed by the chemical combination of two or more substances in definite proportions by weight.

Crystallization—a process that separates wax and semi-solid substances from heavy fractions. This is accomplished by cooling the fractions and filtering out the solids (crystals).

Density—equals the mass of any substance in grams divided by the volume of the substance in milliliters.

Electron—a negatively charged particle that orbits the nucleus of an atom.

Element—composed of identical atoms.

Equilibrium—a condition of balance or where opposite forces equal or balance each other.

Fractional distillation—a process that separates the components in a mixture by their individual boiling points.

Gas chromatography—a process used to determine the concentration of chemicals in a mixture.

Hydrocarbons—a class of chemical compounds that contains hydrogen and carbon.

Hydrogen ion—positively charged hydrogen particle.

Hydroxyl ion—negatively charged OH particle.

Ion—electrically charged atom.

Inorganic chemistry—the branch of chemistry that studies compounds that do not contain carbon.

Process Chemistry ● Chapter 11

Limiting factor—the significant factor that limits production rates for equipment or raw materials.

Material balancing—a method for calculating reactant amounts vs. product target rates.

Material Safety Data Sheet (MSDS)—contains material safety, hazards, and handling information.

Matter—anything that occupies space and has mass.

Mixture—composed of two or more substances that are only physically mixed. Mixtures can be separated through physical means such as boiling or magnetic attraction.

Neutron—a neutral particle in the nucleus of an atom.

Organic chemistry—the scientific study of substances that contain carbon.

Percent-by-weight solution—the concentration of the solute is expressed as a percentage of the total weight of the solution.

Periodic table—provides information about all known elements, including atomic mass, symbol, atomic number, and boiling point.

pH—a measurement system used to determine the acidity or alkalinity of a solution.

Process reaction—produces new molecular arrangements.

Proton—a positively charged particle in the nucleus of an atom.

Reactants and products—raw materials or reactants are combined in specific proportions to form-finished products.

Reaction (combustion)—combustion reactions are exothermic reactions that require fuel, oxygen, and heat to occur. In this type of reaction, oxygen reacts with another material so rapidly that fire is created.

Reaction (endothermic)—a reaction that requires heat or energy.

Reaction (equilibrium)—a type of reaction that balances reactant to product formation with product to reactant formation.

Reaction (exothermic)—reactions that produce heat or energy.

Chapter 11 • *Process Chemistry*

Reaction (neutralization)—a reaction designed to remove hydrogen ions or hydroxyl ions from a liquid.

Reaction (replacement)—a reaction designed to break a bond and form a new bond by replacing one or more of the original compound's components.

Reaction rate—refers to the amount of time it takes a given amount of reactants to form a product.

Solute—the material dissolved in the solution.

Solution—homogenous mixture.

Solvent extraction—a process that uses a solvent (benzene, furfural, phenol) to dissolve certain fractions or separate out as solids. Examples include: kerosene, lubricating oils.

Specific gravity—equals the mass of a substance divided by the mass of an equal volume of water. When metric units are used, specific gravity has the same numerical value as density.

Thermal cracking—uses heat and pressure to improve yields from a barrel of oil.

11.2 Fundamental Principles of Chemistry

Chemistry is the study of the characteristics or structure of elements and the changes that take place when they combine to form other substances. Process operators play a major role in the production and manufacturing of raw materials that have been converted into finished products. Modern chemistry is an essential part of the process environment and for this reason a vital part in the initial training plans of most technicians.

Matter is anything that occupies space and has mass. The four physical states of matter are solid, liquid, gas, and plasma, which can be found in powerful magnetic fields.

Elements
The purest form of matter is called an element. Elements cannot be broken down or changed by chemical or physical means. The chemical structure of an element is composed of identical components called atoms. A list of all of the known elements, natural and synthetic, can be found on the chemical element chart (periodic table). The periodic table provides information about all known elements, such as atomic mass, symbol, atomic number, and boiling point.

Atoms
An atom is the smallest particle of an element that still retains the characteristics of an element. Atoms are composed of positively charged particles called protons, neutral particles called neu-

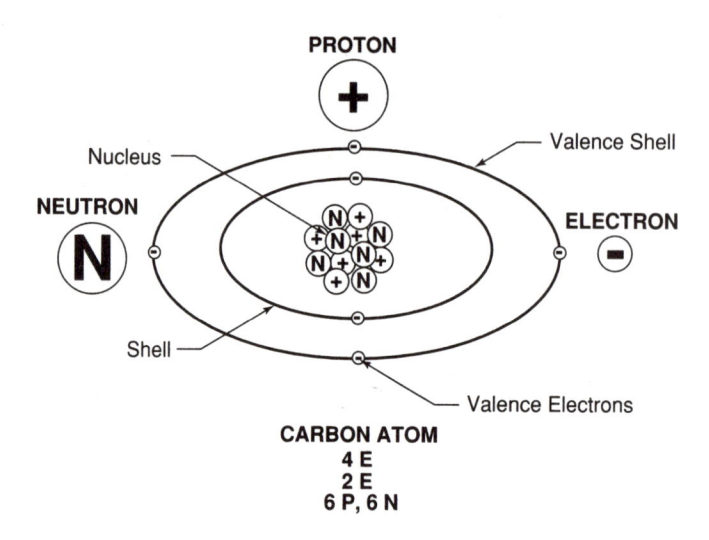

Figure 11.2-1 *Carbon Atom*

trons, and negatively charged particles called electrons, Figure 11.2–1. Protons and neutrons make up the majority of the mass in an atom and reside in an area referred to as the nucleus. The sum of the masses in the nucleus (protons and neutrons) is called the atomic mass unit (AMU).

Atomic Number
The atomic number of an element is determined by the number of protons in its nucleus. The atomic number is used to place the element in its proper place on the chemical element chart.

Electrons
Orbiting the nucleus are negatively charged particles known as electrons. Electrons and protons are equally balanced in an atom. This is important because it ensures that the atom is electrically neutral.

Valence Electrons
The electrons that reside in the outermost shell of an atom are referred to as valence electrons. Valence electrons are important to chemistry because they provide the links in virtually every chemical reaction. Atoms share their valence electrons to form chemical bonds.

Chemical Bonding—Covalent and Ionic
The two most common models for chemical bonding are covalent and ionic. Covalent bonds occur when elements react with each other by sharing electrons. This forms an electrically neutral molecule because the protons and electrons electrically balance each other.

Ions
If an atom has unequal numbers of protons or electrons, it is called an ion. Ions are electrically charged atoms (positive or negative). Ionic bonds occur when positively charged elements react

OXYGEN ATOM
6 E
2 E
8 P, 8 N

H

H

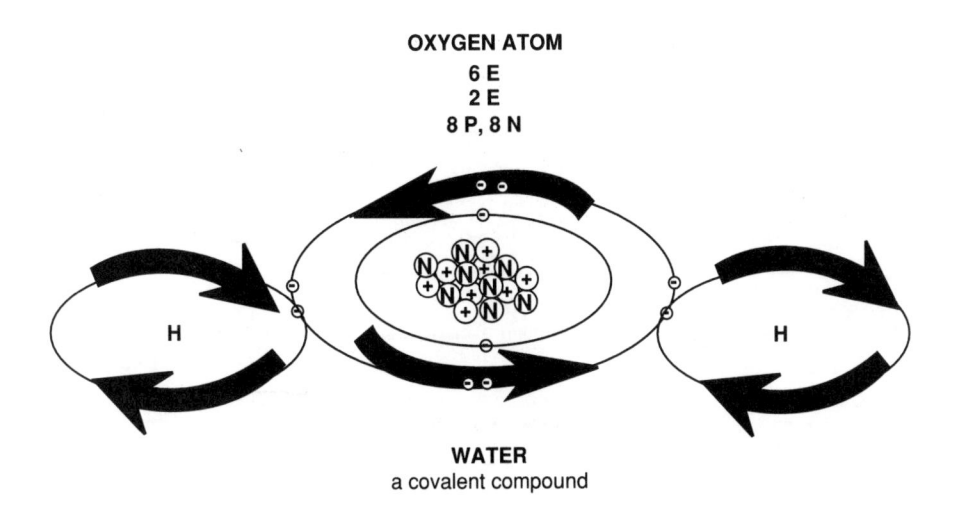

WATER
a covalent compound

Figure 11.2-2 *Compound*

with negatively charged elements to form ionic bonds through the sharing or lending of electrons. Ionic bonds have higher melting points and are held together by electrostatic attraction.

Molecules and Compounds
Compounds are the products of chemical reactions, Figure 11.2–2. A compound is a substance formed by the chemical combination of two or more substances in definite proportions by weight. A molecule is the smallest particle that retains the properties of the compound.

Solutions
Solutions are a type of homogenous mixture. The term *homogenous* refers to the evenly mixed composition of the solution. A common example of a homogenous solution is a drink mix. As the contents of the powdered drink mix are mixed with water, it is evenly dispersed throughout the solution.

Mixture
Mixtures do not have a definite composition. A mixture is composed of two or more substances that are only mixed physically. Because a mixture is not chemically combined, it can be separated through physical means, such as boiling or magnetic attraction. Crude oil is a simple example of a mixture, Figure 11.2–3. It is composed of hundreds of different hydrocarbons. Process operators separate the different components in the crude oil by heating to the boiling point in a distillation column.

11.3 Chemical Equations and the Periodic Table

The most common chemical substances are elements. Chemical elements are the building blocks for all substances. Each element is composed of atoms from only one kind of element. Chemists describe elements with letters from the alphabet. The letter symbol for hydrogen is *H*.

Process Chemistry • Chapter 11

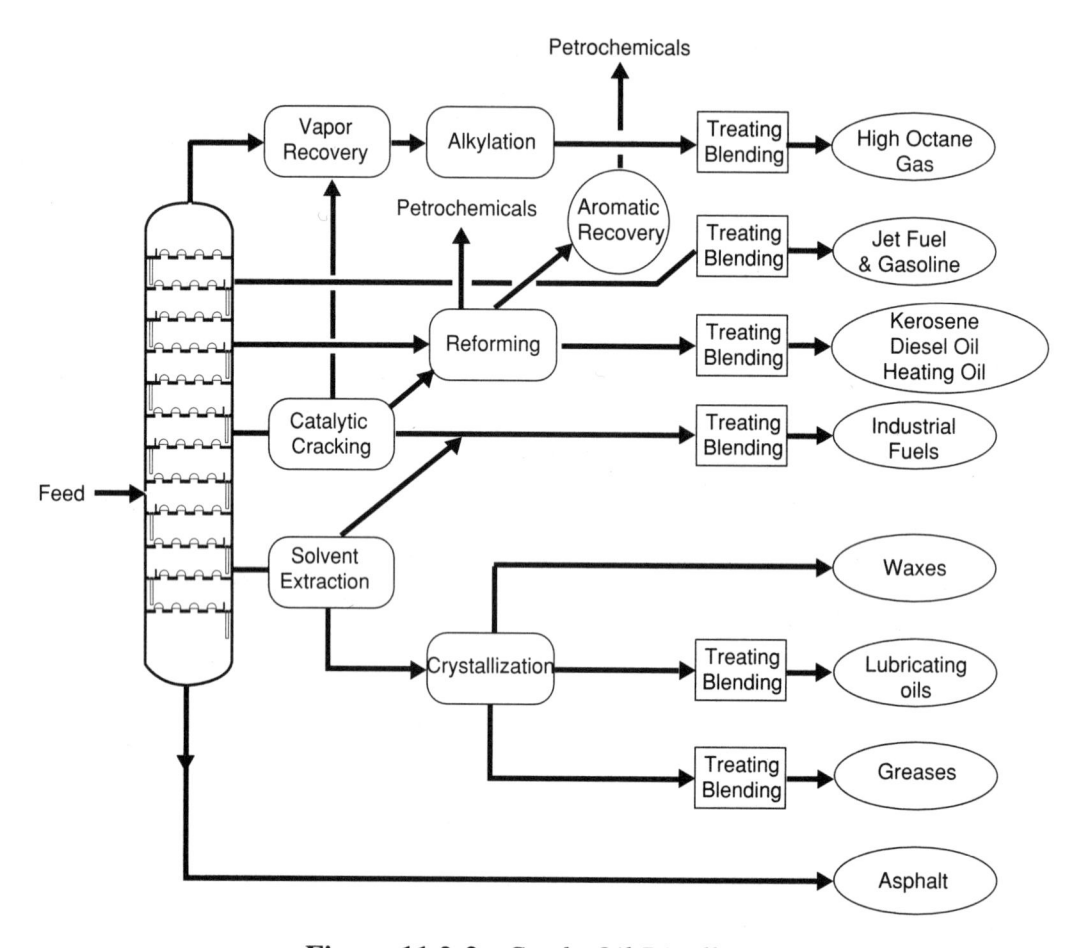

Figure 11.2-3 *Crude Oil Distillation*

The letter symbol for carbon is C, Figure 11.3–1. A list of all known chemical symbols can be found on a periodic table or chemical element chart, Figure 11.3–2. A good understanding of the chemical element chart helps a technician better understand chemical equations.

A chemical reaction can be described by associated numbers and symbols. The chemical number identifies how many protons are in an atom, and the atomic mass unit identifies how many units of an element are present.

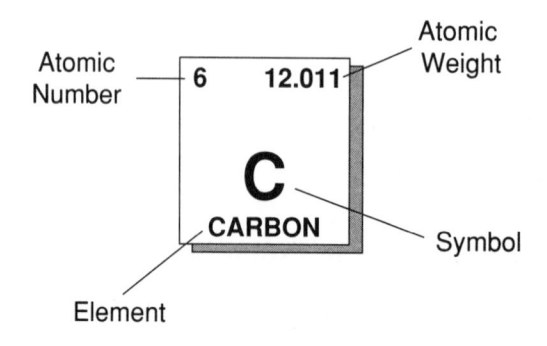

Figure 11.3-1 *Periodic Table Information Box*

GROUP

1A	11A	111A	1VA	VA	V1A	V11A	V111A	V111A	V111A	1B	11B	111B	1VB	VB	V1B	V11B	V111
1 1.0079 H HYDROGEN																	2 4.002 He HELIUM
3 6.941 Li LITHIUM	4 9.0126 Be BERYLLIUM											5 10.81 B BORON	6 12.01 C CARBON	7 14.006 N NITROGEN	8 15.99 O OXYGEN	9 18.99 F FLUORINE	10 20.18 Ne NEON
11 22.99 Na SODIUM	12 24.30 Mg MAGNESIUM											13 26.98 Al ALUMINUM	14 28.08 Si SILICON	15 30.97 P PHOSPHORUS	16 32.06 S SULFUR	17 35.45 Cl CHLORINE	18 39.94 Ar ARGON
19 39.09 K POTASSIUM	20 40.08 Ca CALCIUM	21 44.95 Sc SCANDIUM	22 47.9 Ti TITANIUM	23 50.94 V VANADIUM	24 51.99 Cr CHROMIUM	25 54.93 Mn MANGANESE	26 55.84 Fe IRON	27 58.93 Co COBALT	28 58.7 Ni NICKEL	29 63.54 Cu COPPER	30 65.38 Zn ZINC	31 69.72 Ga GALLIUM	32 72.59 Ge GERMANIUM	33 74.92 As ARSENIC	34 78.96 Se SELENIUM	35 79.90 Br BROMINE	36 83.8 Kr KRYPTON
37 85.46 Rb RUBIDIUM	38 87.62 Sr STRONTIUM	39 88.9 Y VITRIUM	40 91.22 Zr ZIRCONIUM	41 92.9 Nb NIOBIUM	42 95.9 Mo MOLYBDENUM	43 98 Tc TECHNETIUM	44 101 Ru RUTHENIUM	45 102.9 Rh RHODIUM	46 106.4 Pd PALLADIUM	47 107.8 Ag SILVER	48 112.4 Cd CADIUM	49 114.8 In INDIUM	50 118.6 Sn TIN	51 121.7 Sb ANTIMONY	52 127.6 Te TELLURIUM	53 126.9 I IODINE	54 131.3 Xe XENON
55 132.90 Cs CESIUM	56 137.33 Ba BARLUM	57 138.9 La LANTHANUM	72 178.4 Hf HAFNIUM	73 180.9 Ta TANTALUM	74 183 W TUNGSTEN	75 186 Re RHENIUM	76 190 Os OSMIUM	77 192 Ir IRIDIUM	78 195 Pt PLATINUM	79 196.9 Au GOLD	80 200.6 Hg MERCURY	81 204.3 Tl THALLIUM	82 207 Pb LEAD	83 208.9 Bi BISMUTH	84 209 Po POLONIUM	85 210 At ASTATINE	86 222 Rn RADON
87 223 Fr FRANCIUM	88 226 Ra RADIUM	89 227 Ac ACTINIUM	104 Unq	105 Unp	106 Unh												

Figure 11.3-2 *Periodic Table*

In a chemical equation, the raw materials or reactants are placed on the left side. As the reactants are mixed together, they yield predictable products. A yield sign or arrow immediately follows the reactants. The products are placed on the right side of the equation. Because atoms cannot be created or destroyed, a common rule of thumb is "what goes into a chemical equation must come out." *The sum of the reactants must equal the sum of the products.*

Example:

$$C + O_2 \xrightarrow[\text{(yields)}]{} CO_2$$

1 Carbon	=	1 Carbon
2 Oxygen	=	2 Oxygen
(reactants)	⟶	(products)

Note: What goes in must come out!

$$Cu + H_2SO_4 \longrightarrow CuSO_4 + H_2$$

1 Copper	1 Copper
2 Hydrogens	2 Hydrogens
1 Sulfur	1 Sulfur
4 Oxygens	4 Oxygens

Process Chemistry ● Chapter 11

$$4NH_3 + 3O_2 \longrightarrow 2N_2 + 6H_2O$$

4 Nitrogen	4 Nitrogen
12 Hydrogen	12 Hydrogen
6 Oxygen	6 Oxygen

Mass Relationships—Chemical Equations
When working out mass relationships, you need to have a good understanding of the chemical element chart. Certain elements combine to form chemicals that you will recognize easily. An example of this is water (H_2O) or carbon dioxide (CO_2). The elements and atomic mass units are listed on the periodic table.

Example:

$$H_3PO_4 + 3\, NaOH \longrightarrow Na_3PO_4 + 3H_2O$$

Phosphoric acid and sodium hydroxide react to form sodium phosphate and water. What is the product's total molecular weight?

Phosphoric acid (H_3PO_4) = 1 molecule (molecular weight?)
Sodium hydroxide (NaOH) = 1 molecule (molecular weight?)
Sodium phosphate (Na_3PO_4) = 1 molecule (molecular weight?)
Water (H_2O) = 1 molecule (molecular weight?)

H_3PO_4 - Reactant

    3 hydrogens = 3 × 1.008 AMU = 3.024
    1 phosphorus = 1 × 30.98 AMU = 30.98
    4 oxygens = 4 × 16.00 AMU = <u>64.00</u>
                                           98.00 grams, pounds, or tons

3NaOH - Reactant

- 3 sodium = 3 × 23.00 AMU = 69.00
- 3 oxygen = 3 × 16.00 AMU = 48.00
- 3 hydrogen = 3 × 1.008 AMU = <u>3.024</u>
                                           120.02 grams, pounds, or tons

98.00 + 120.02 = 218.02
Reactant's total molecular weight = 218 grams, pound, or tons

Na_3PO_4 - Product

    3 sodiums = 3 × 23.00 AMU = 69.00
    1 phosphorus = 1 × 30.98 AMU = 30.98
    4 oxygens = 4 × 16.00 AMU = <u>64.00</u>
                                           163.98 grams, pounds, or tons

3H$_2$O- Product

    6 hydrogens   = 6 × 1.008 AMU = 6.048
    3 oxygens   = 3 × 16.00 AMU = 48.000
                     54.050 grams, pounds, or tons

163.98 + 54.05 = 218.03
Product's total molecular weight = 218 grams, pound, or tons

Example:
$$4H_2 + 2O_2 \longrightarrow 4H_2O$$
Four volumes of hydrogen reacts with two volumes of oxygen to produce four volumes of water vapor.
What is the product's total molecular weight?

    Hydrogen (2H$_2$) = 1 molecule (molecular weight?)
    Oxygen (O$_2$)   = 1 molecule (molecular weight?)

2H$_2$ Reactant
    8 hydrogens   = 8 × 1.008 AMU   = 8.064 grams, pounds, or tons

O$_2$ Reactant
    4 oxygens   = 4 × 16 AMU   = 64.00 grams, pounds, or tons

8.064 + 64.00 = 72.064
Reactant's total molecular weight = 72.06

4H$_2$O Product
    8 hydrogens   = 8 × 1.008 AMU = 8.064
    4 oxygens   = 4 × 16.00 AMU = 64.00
                     72.06 grams, pounds, or tons

Product's total molecular weight = 72.06 grams, pounds, or tons

Solve: Given the chemical equation:

$$H_3PO_4 + 3\ NaOH \longrightarrow Na_3PO_4 + 3H_2O$$
    18 tons 10 tons   (28 total tons)

 a. Change H$_3$PO$_4$ (18 tons) to 1,800 tons.
 b. What does the (3NaOH) weight need to be to balance the equation?

211

Process Chemistry • Chapter 11

Solution: The first thing to remember in this type of problem is to identify the relative weight and the actual weight. The relative weight in this problem is 1,800 tons. The actual weight is 18 tons. Now, divide the relative weight by the actual weight.

$$1,800 \div 18 = 100$$

Use this new factor to adjust the 10 tons of 3NaOH.
    10 tons × 100 = 1,000 tons
    1,000 tons balances the 3NaOH equation.

Solve: Given the chemical equation:

$$H_3PO_4 + 3\ NaOH \longrightarrow Na_3PO_4 + 3H_2O$$

If you are told to add 25 lb of phosphoric acid (H_3PO_4) to the previous equation, how many pounds do you need to add to the sodium hydroxide (NaOH) to keep the equation balanced?

Solution: The first thing to remember in this type of problem is to identify the relative weight and the actual weight. The relative weight in this problem is 25 lb. The actual weight is the total AMU of H_3PO_4, which is 98 AMUs. (This can be in pounds or tons.)

Now, divide the relative weight by the actual weight.

$$25 \div 98 = 0.255$$

H_3PO_4 Phosphoric Acid					3NaOH Sodium Hydroxide				
H_3	=	3 × 1.0079	=	3.0237	3Na	=	3 × 23	=	69
P	=	1 × 31	=	31	3O	=	3 × 16	=	48
O_4	=	4 × 16	=	64	3H	=	3 × 1	=	3
		TOTAL		98			TOTAL		120

Use this new factor to adjust the 120 AMUs of 3NaOH.
    0.255 × 120 = 30.6 lb
    30.6 lb balances the 3NaOH equation.

Solve:

N_2	+	$3H_2$	$\longrightarrow$	$2NH_3$
120 lb		39 lb		159 lb
? lb		? lb		556 lb
420 lb		136.5 lb		

212

Solution: The relative weight is 556 lb. The actual is 159 lb.
556 ÷ 159 = 3.5

3.5 × 120 = 420
3.5 × 39 = 136.5

Solve:

$$CH_4 + 2O_2 \longrightarrow CO_2 + 2H_2O$$
1,600 lb   ?   ?   ?

Solution: The relative weight is 1,600 lb. The actual weight is CH_4 (16 AMUs).

1 C 1 × 12 = 12	$4O_2$ 4 × 16 = 64	1 C 1 × 12 = 12	4 H 4 × 1 = 4
4 H 4 × 1 = 4		$2O_2$ 2 × 16 = 32	$2O_2$ 2 × 16 = 32
16		44	36

1,600 ÷ 16 = 100 *atomic weight*

CH_4	100 × 16	= 1,600 lb
$2O_2$	100 × 64	= 6,400 lb
CO_2	100 × 44	= 4,400 lb
$2H_2O$	100 × 36	= 3,600 lb

11.4 Chemical Reactions

Exothermic
Exothermic reactions are characterized by a chemical reaction accompanied by the liberation of heat. As the reaction rate increases, the evolution of heat energy increases. Exothermic reactions can be controlled by controlling reactant flow rates, removing heat, or providing cooling.

Endothermic
Endothermic reactions must absorb energy in order to proceed.

Replacement
Industrial manufacturers use replacement reactions to remove dissolved mineral ions from process water. A number of dissolved minerals can be found in process fluids. A common compound found in process water is calcium chloride. Calcium chloride ($CaCl_2$) forms positive calcium (Ca^+) ions and negative chloride (Cl_2^-) ions when it is dissolved in water.

A replacement reaction can remove the Ca^+ ions and the Cl_2^- ions with synthetic resins. Resins are plastic strands rolled into balls and charged with ions. An H^+ ion on a resin ball is replaced by the Ca^+ ion as the process fluid moves through the resin bed. The replacement reaction will take place until all of the Ca^+ ions are removed from the fluid or the H^+ ions are used up on the resin balls.

Resin balls can be treated with positively or negatively charged ions and used for replacement reactions. For example, resin balls charged with hydroxyl ions (OH^-) can be used to replace the chloride ion (Cl^-).

Neutralization
Neutralization reactions remove hydrogen ions (acid) or hydroxyl ions (base) from a liquid. Neutralization reactions are designed to neutralize the acidity or alkalinity of a solution. Hydrogen ions (acid) and hydroxyl ions (base) neutralize each other.

Combustion
Combustion reactions are exothermic reactions that require fuel, oxygen, and heat to occur. In this type of reaction, oxygen reacts with another material so rapidly that fire is created. A fired furnace or a boiler is an example of a combustion reaction. Natural methane gas is pumped to the burner, mixed with oxygen, and ignited. This type of reaction can be represented by the following chemical equation:

$$CH_4 + 2O_2 \longrightarrow CO_2 + 2H_2O$$

In this equation, one molecule of methane (CH_4) chemically reacts with two molecules of oxygen to produce one molecule of carbon dioxide and two molecules of water.

Heat and Pressure
For a chemical reaction to occur, the atoms of the reactants must collide with each other. The addition of heat to a process increases reactant molecular activity. This increased activity ensures a much higher rate of energy transfer between molecules as they collide into each other. Reaction rates double every 10 degrees.

The addition of heat energy impacts a process by increasing molecular activity, increasing atomic collisions, and enhancing the formation of chemical bonds.

Note: High temperatures can cause undesirable products to form. Process temperatures are closely monitored during operation to ensure smooth and efficient reaction rates. Another important factor in a chemical reaction is pressure. Pressure has its greatest impact on gases. Gases are much easier to compress than liquids. Pressure can change the boiling point of a liquid and slow down molecular activity. As pressure builds, it pushes the gas molecules closer together and back into the liquid. More heat is required to boil the liquid, which wastes time and money.

Chapter 11 • *Process Chemistry*

Heat speeds up molecular activity.
Pressure pushes molecules closer together.

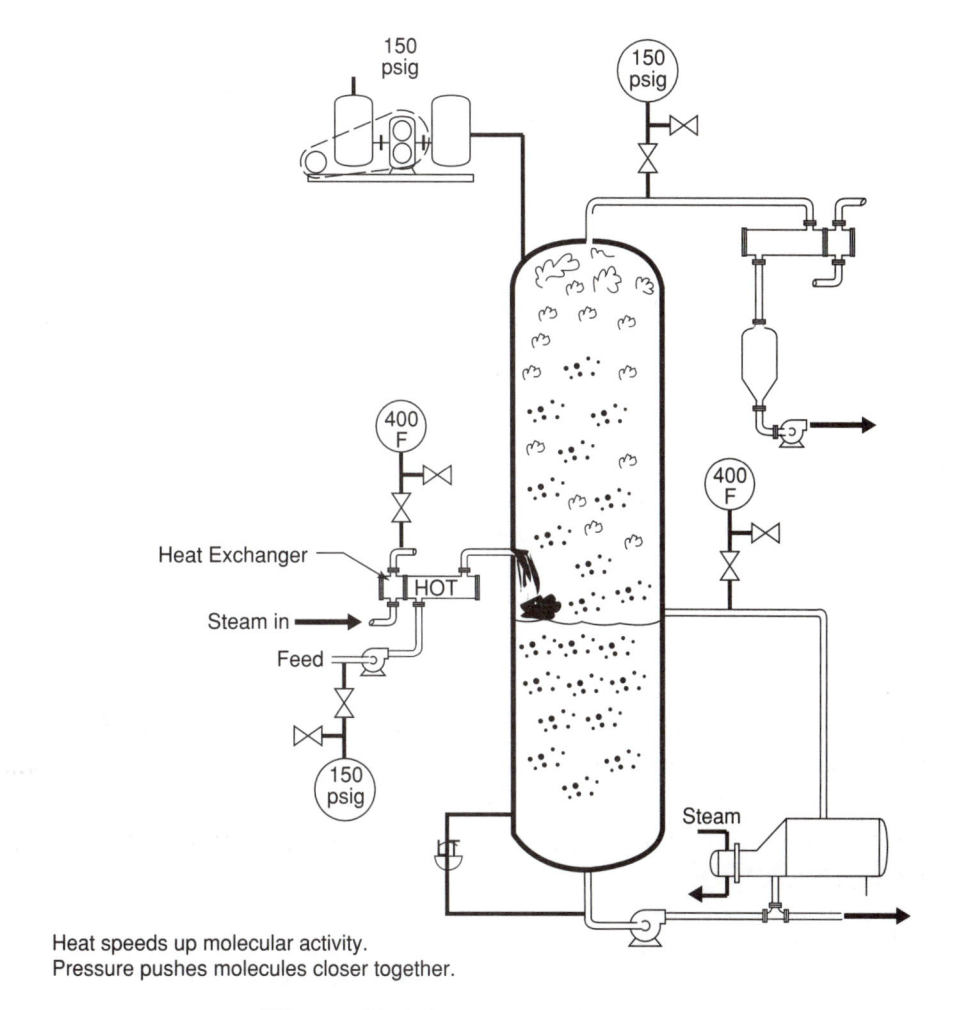

Figure 11.4-1 *Heat and Pressure*

Reaction rates are impacted by:

- Heat—molecular activity increases
- Heat—atomic collisions increase
- Heat—the formation of chemical bonds is enhanced
- Surface area—solids
- Concentration—liquid and gas reactants
- Pressure
- Flow rates—reactants and products
- Catalyst

Catalyst
A catalyst is a chemical that can increase or decrease the reaction rate without becoming part of the product.

Types of catalysts:
- Adsorption-type catalyst—a solid that attracts and holds reactant molecules so a higher number of collisions can occur. It also stretches the bonds of the reactants it is holding, weakening the bonds, requiring less energy to break and rebond.
- Intermediate-type catalyst—forms an intermediate product by attaching to the reactant and slowing it down so collisions can occur. This type of catalyst does not become part of the final product.
- Inhibitor-type catalyst—decreases reaction rate.
- Poisoned catalyst—no longer functions, used up.

11.5 Material Balance

Material balancing is a method used by technicians to determine the *exact* amount of reactants needed to produce the specified products. This method is used where two or more substances are combined in a chemical process. Reactants must be mixed in the proper proportions to avoid waste. Material balancing provides an operator with the correct reactant ratio.

The steps in checking a material balance are (1) determine the weight of each molecule, (2) insure reactant total weight is equal to product total weight, and (3) determine relative numbers of reactant atoms or ions.

Relative and Actual Weights

Step 1 $H^+ + OH^- \longrightarrow H_2O$

Step 2 H^+ (1 AMU) + OH^- (17 AMU) $\longrightarrow$ H_2O (18 AMU)

Note: The relationship between AMUs and other units is 1 AMU = 1 gram, pound, or ton.

Step 3 H^+ (1 g) + OH^- (17 g) $\longrightarrow$ H_2O (18 g)

         Add 10 grams of hydrogen ions and balance the equation.

Step 4 H^+ (10 × 1 g) + OH^- (10 × 17 g) $\longrightarrow$ H_2O (? g)

Step 5 H^+ (10 g) + OH^- (170 g) $\longrightarrow$ H_2O (180 g)

Example 1
Solve:
     $Na_2O + 2HOCl \longrightarrow 2NaOCl + H_2O$.

List the reactant elements. List the product elements. Is this chemical equation balanced?

$$Na_2O + 2HOCl \longrightarrow 2NaOCl + H_2O$$

2Na	2Na
3O	3O
2Cl	2Cl
2H	2H

Yes, the chemical equation is balanced.

Example 2
Solve:
$$2H_3PO_4 \longrightarrow H_2O + H_4P_2O_8.$$

List the reactant elements. Is this chemical equation balanced?

$$2H_3PO_4 \longrightarrow H_2O + H_4P_2O_8$$

6H	6H
2P	2P
8O	9O

No, the chemical equation is not balanced.

11.6 Percent-by-Weight Solutions

Percent-by-weight solutions are expressed as a percentage of the weight of the solution or, in other words: the weight of the solute (material being dissolved) is taken in relationship to the weight of the entire solution.

In a weight percent problem the amount of the solute and solvent can be calculated. For example, a 400 lb barrel has a 6 percent catalyst solution. The weight of the catalyst can be determined by multiplying the weight of the solution by the percent of the solute.

Weight of solution	×	percent of solute	=	Weight of Solute
400 pounds	×	6% or 0.06	=	24 lb

Figure 11.6–1 is an example of a solution.

Process Chemistry • Chapter 11

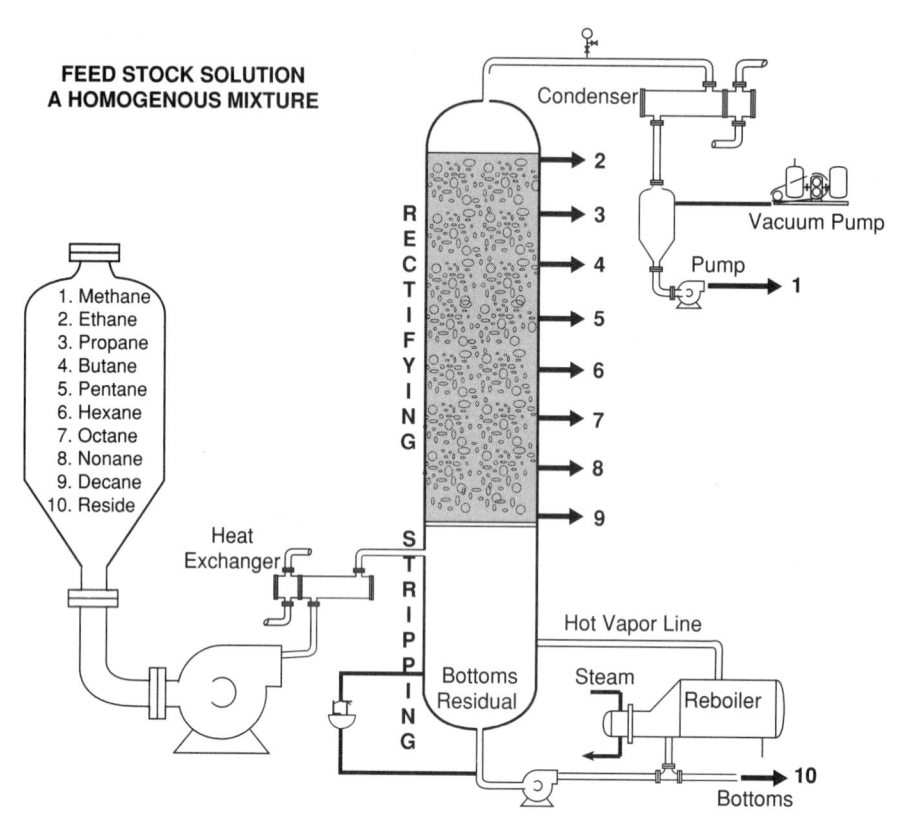

Figure 11.6-1 *Solution*

11.7 pH Measurements

The term *pH* is a measurement system used to determine the acidity or alkalinity of a solution. An acid is a bitter-tasting chemical compound that has a pH value below 7.0. It changes blue litmus to red and yields hydrogen ions in water. pH has a high concentration of hydrogen ions.

A base is a bitter-tasting chemical compound that has a soapy feel and a pH value above 7.0. It turns red litmus paper blue and yields hydroxyl ions.

The methods for determining pH include:
- the pH comparator—an indicator solution is added to the fluid to be checked. The color of the solution can be compared to the pH comparator standards.
- pH paper—red or blue litmus is impregnated with an indicator that causes a color change to occur in the presence of an acid or base.
- pH meter—measures the concentration of hydrogen ions in a solution.

11.8 Hydrocarbons

A hydrocarbon is a chemical compound that contains hydrogen and carbon. One of the best known hydrocarbons is crude oil. Crude oil is a mixture of hydrocarbons that vary from simple to complex. Industrial manufacturers separate the various components of crude oil by boiling or distilling it. The lighter carbon molecules have different boiling points than the heavier molecules.

The simplest hydrocarbon is methane or natural gas. Methane has one carbon atom and four hydrogen atoms. Close examination of the atomic structure of methane indicates that the outer valence electrons tend to couple in pairs. Compounds made up of carbon atoms have four possible bonds on each atom. The four arms (valence electrons) on each carbon atom bond with a hydrogen or another carbon atom. Each slot on the carbon atom must be filled. There are millions of possible combinations for these carbon atoms. Chemists have divided these hydrocarbons into two very large families.

Alkanes

Figure 11.8–1 shows the composition of some alkanes.

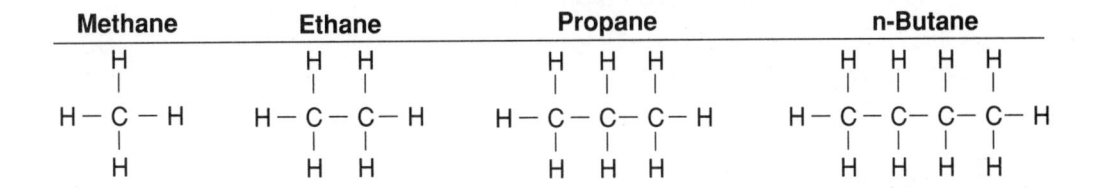

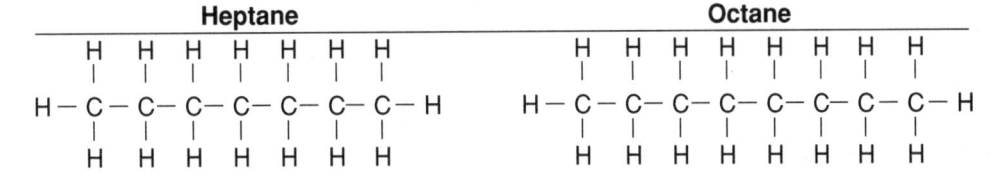

Figure 11.8-1 *Alkanes*

Process Chemistry • Chapter 11

The First 10 Alkanes

Name	Molecular Formula
1. Methane	CH_4
2. Ethane	C_2H_6
3. Propane	C_3H_8
4. Butane	C_4H_{10}
5. Pentane	C_5H_{12}
6. Hexane	C_6H_{12}
7. Heptane	C_7H_{16}
8. Octane	C_8H_{18}
9. Nonane	C_9H_{20}
10. Decane	$C_{10}H_{22}$

Chain Length Effects on Boiling Point

Alkane	Structure	Mole. Wt.	Boiling Point
Methane	CH_4	16	-164C
Ethane	CH_3—CH_3	30	-89C
Propane	CH_3—CH_2—CH_3	44	-42C
Butane	CH_3—CH_2—CH_2—CH_3	58	-0.5C
Pentane	CH_3—CH_2—CH_2—CH_2—CH_3	72	36C
Hexane	CH_3—CH_2—CH_2—CH_2—CH_2—CH_3	86	69C
Heptane	CH_3—CH_2—CH_2—CH_2—CH_2—CH_2—CH_3	100	98C
Octane	CH_3—CH_2—CH_2—CH_2—CH_2—CH_2—CH_2—CH_3	114	126C
Nonane	CH_3—CH_2—CH_2—CH_2—CH_2—CH_2—CH_2—CH_2—CH_3	128	151C
Decane	CH_3—CH_2—CH_2—CH_2—CH_2—CH_2—CH_2—CH_2—CH_2—CH_3	142	174C

Ethylene C_2H_4

```
H   H
|   |
C = C
|   |
H   H
```
BP = -155 °F

Propylene C_3H_6

```
    H   H   H
    |   |   |
H – C – C = C
    |       |
    H       H
```
BP = -54 °F

Butylene C_4H_8

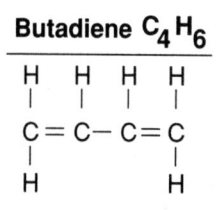

```
H   H   H   H
|   |   |   |
C = C – C – C – H
|       |   |
H       H   H
```
BP = 21 °F

Butadiene C_4H_6

```
H   H   H   H
|   |   |   |
C = C – C = C
|           |
H           H
```

Figure 11.8-2 *Olefins*

Olefins

Olefins, Figure 11.8–2, do not occur naturally in crude oil. Olefins are created by a manmade process called cracking. Each molecule of an olefin has at least one double bond.

11.9 Applied Concepts to Chemical Processing

Distillation

There are a number of fractions obtained from the distillation of petroleum. Distillation is defined as the separation of the various fractions in a mixture by individual boiling points. Hydrocarbon fractions obtained from petroleum include straight run gasoline, kerosene, heating oil, diesel, jet fuel, lubricating oil, paraffin wax, asphalt, and tar, Figure 11.9–1. Additional processes can be applied to these different fractions to create other products.

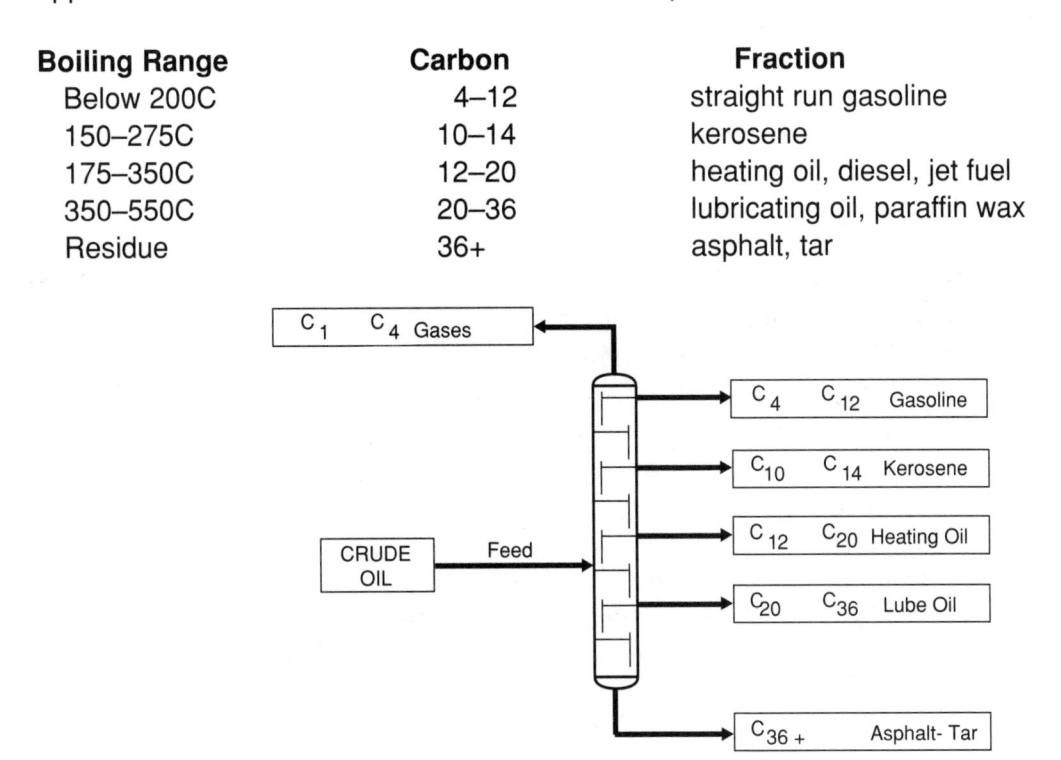

Boiling Range	Carbon	Fraction
Below 200C	4–12	straight run gasoline
150–275C	10–14	kerosene
175–350C	12–20	heating oil, diesel, jet fuel
350–550C	20–36	lubricating oil, paraffin wax
Residue	36+	asphalt, tar

Figure 11.9-1 *Hydrocarbon Fractions*

A distillation tower is a series of stills placed one on top of the other. As vaporization occurs the lighter components of the mixture move up the tower and are distributed on the various trays. The lightest component goes out the top of the tower in a vapor state and is passed over the cooling coils of a shell and tube condenser. As the hot vapor comes in contact with the coils it condenses and is collected in the overhead accumulator. Part of this product is sent to storage while the other is returned to the tower as reflux.

Heat balance on the tower is maintained by a device known as a reboiler. Reboilers take suction off the bottom of the tower. The heaviest components of the tower are pulled into the reboiler and stripped of smaller molecules. The stripped vapors are returned to the column and allowed to separate in the tower.

Reactors

Process technicians combine raw materials and modern reaction technology to form new products. This is the foundation upon which modern chemistry is based. Chemistry is the study of the characteristics or structure of elements and the changes that take place when they combine to form other substances. A reactor is designed to make or break chemical bonds, changing the molecular structure of the raw materials. A reactor is a device used to convert raw materials into useful products through chemical reactions. Process operators are responsible for the safe and efficient operation of the reactor and its associated equipment. Process technicians play a major role in the production and manufacturing of chemicals. Modern chemistry is an essential element in the initial training plans of process technicians.

Catalytic cracking. Crude oil comes into a refinery and is processed in a fractionating tower. The side-stream of the column is rich with light gas oil. Fluid catalytic cracking units split this gas oil into smaller, more useful molecules. Generally, only twenty percent of a barrel of crude oil can be used to produce gasoline. Fluid catalytic cracking is a process that uses a reactor to split large covalent gas oil molecules into smaller more useful ones. Cracking a C_{12} kerosene molecule yields two C_6 molecules (hexane and hexene). Figure 11.9–2 illustrates this process. The catalytic process increases yields from 20% to 50% by splitting the kerosene and heating oil fractions of the crude.

A typical fluid catalytic cracking unit includes a catalyst regenerator, reactor, and fractionating tower. During operation gas oil enters the reactor and is mixed with a superheated powdered catalyst. The term cracking is applied to the process because during vaporization the molecules literally split and are sent to a fractionation tower for further processing. The chemical reaction between the catalyst and light gas oil produces a solid carbon deposit. This deposit forms on the powdered catalyst and deactivates it. The spent catalyst is drawn off and sent to the regenerator where the coke is burned off. Catalyst regeneration is a continuous process during operation.

In the fractionation tower the light gas oil is separated into five different cuts: cat cracked gas, cat cracked naphtha, cat cracked heating oil, light gas oil, and reside.

Hydrocracking. Hydrocracking is a process that industrial manufacturers use to boost gasoline yields. During this process heavy gas oil molecules are split into smaller, lighter molecules called hydrocrackate. Heavy gas oil feed is mixed with hydrogen before being sent to the first-stage reactor. The reactor is filled with a fixed bed of catalyst. As process flow moves from the top of the reactor to the bottom the cracking reaction takes place. First-stage hydrocrackate is sent to a separator drum where the hydrogen is reclaimed and the hydrocrackate is moved on to a fractionating tower.

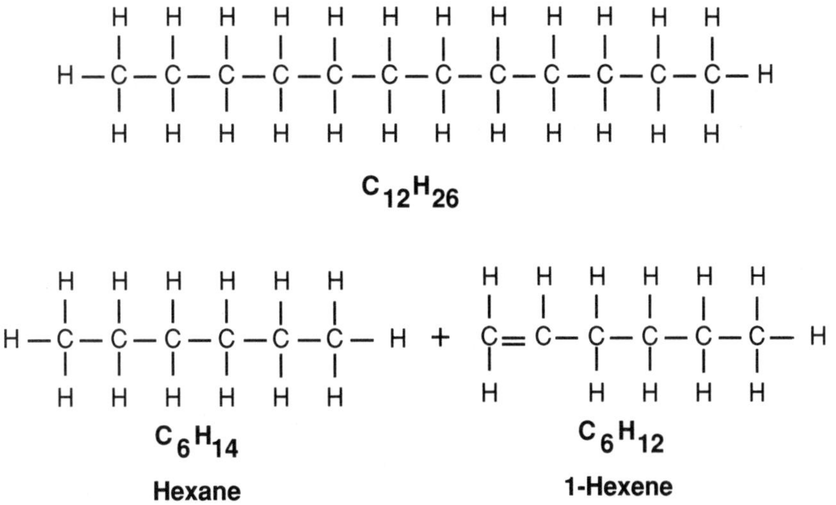

Chapter 11 • *Process Chemistry*

$$C_{12}H_{26}$$

$$C_6H_{14} \text{ Hexane} + C_6H_{12} \text{ 1-Hexene}$$

Figure 11.9-2 *Cracking*

In the fractionation tower the hydrocrackate is separated into five different cuts: butane, light hydrocrackate, heavy hydrocrackate, heating oil, and heavy bottoms. The heavy bottoms is mixed with hydrogen and sent to the second-stage reactor for further processing. The second-stage reactor reclaims as much of the hydrocrackate as possible before sending it to the separator and tower.

Alkylation. Alkylation uses a reactor to make one large molecule out of two small molecules. Alkylation units take two small molecules of isobutane and olefin (propylene, butylenes, or pentylenes) and combine them into one large molecule of high-octane liquid called alkylate. This combining process takes place inside a reactor filled with an acid catalyst. Alkylate is a superior anti-knock product that is used in blending unleaded gasoline.

After the reaction a number of products are formed which require further processing to separate and clean the desired chemical streams. A separator and an alkaline substance are used to remove (strip) the acid. The stripped acid is sent back to the reactor while the remaining reactor products are sent to a distillation tower. Alkylate, isobutane, and propane gas are fractionally separated in the tower. Isobutane is returned to the alkylation reactor for further processing. Alkylate is sent on to the gasoline blending unit.

Summary

Chemistry is the study of the characteristics or structure of elements and the changes that take place when they combine to form other substances. Process operators play a major role in the production and manufacturing of raw materials that have been converted into finished products. Modern chemistry is an essential part of the process environment and for this reason a vital part in the initial training plans of most technicians.

The four physical states of matter are solid, liquid, gas, and plasma.

The purest form of matter is an element. Elements cannot be broken down or changed by chemical or physical means. The chemical structure of an element is composed of identical components called atoms. A list of all of the known elements, natural and man made can be found on the chemical element chart (periodic table). The periodic table provides information about all known elements, including atomic mass, symbol, atomic number, and boiling point.

An atom is the smallest particle of an element that still retains the characteristics of an element. Atoms are composed of positively charged particles called protons and an equal number of neutral particles called neutrons. Protons and neutrons make up the majority of the mass in an atom and reside in an area referred to as the nucleus. The sum of the masses in the nucleus (protons and neutrons) is called the atomic mass unit (AMU).

The atomic number of an element is determined by the number of protons in its nucleus. The atomic number is used to place the element in its proper place on the chemical element chart.

Orbiting the nucleus are negatively charged particles known as electrons. Electrons and protons are equally balanced in an atom. This is important because it ensures that the atom is electrically neutral. The electrons that reside in the outermost shell of an atom are referred to as valence electrons. Valence electrons are important to chemistry because they provide the links in virtually every chemical reaction. Atoms share their valence electrons to form chemical bonds.

The two most common models for chemical bonding are covalent and ionic. Covalent bonds occur when elements react with each other by sharing electrons. This forms an electrically neutral molecule because the protons and electrons electrically balance each other.

If an atom has unequal numbers of protons or electrons, it is called an ion. Ions are electrically charged atoms (positive or negative). Ionic bonds occur when positively charged elements react with negatively charged elements to form ionic bonds through the sharing or lending of electrons. Ionic bonds have higher melting points and are held together by electrostatic attraction.

Compounds are the products of chemical reactions. A compound is a substance formed by the chemical combination of two or more substances in definite proportions by weight. A molecule is the smallest particle that retains the properties of the compound.

Solutions are a type of homogenous mixture. The term *homogenous* refers to the evenly mixed composition of the solution.

Mixtures do not have a definite composition. A mixture is composed of two or more substances that are only physically mixed. Because a mixture is not chemically combined, it can be separated through physical means, such as boiling or magnetic processes. Crude oil is a simple example of a mixture. It is composed of hundreds of different hydrocarbons. Process operators separate the different components in the crude oil by boiling in a distillation column.

Chemists describe elements with letters from the alphabet. A list of all known chemical symbols can be found on a periodic table or chemical element chart. A chemical reaction can be described by associated numbers and symbols.

Chapter 11 • Process Chemistry

In a chemical equation, the raw materials or reactants are placed on the left side. As the reactants are mixed together, they yield predictable products. A yield sign or arrow immediately follows the reactants. The products are placed on the right side of the equation. Because atoms cannot be created or destroyed, a common rule of thumb is "what goes into a chemical equation must come out." *The sum of the reactants must equal the sum of the products.* When working out mass relationships, you need to have a good understanding of the chemical element chart.

Exothermic reactions are characterized by a chemical reaction accompanied by the liberation of heat. As the reaction rate increases, the evolution of heat energy increases. Exothermic reactions can be controlled by controlling reactant flow rates, removing heat, and providing cooling.

Endothermic reactions must absorb energy to proceed.

Replacement reactions can be used to remove undesired products and replace them with desired ones.

Neutralization reactions remove hydrogen ions (acid) or hydroxyl ions (base) from a liquid. Neutralization reactions are designed to neutralize the acidity or alkalinity of a solution. Hydrogen ions (acid) and hydroxyl ions (base) neutralize each other.

Combustion reactions are exothermic reactions that require fuel, oxygen, and heat to occur. In this type of reaction, oxygen reacts with another material so rapidly that fire is created. A fired furnace or a boiler is an example of a combustion reaction. Natural methane gas is pumped to the burner, mixed with oxygen, and ignited. This type of reaction can be represented by the following chemical equation:

$$CH_4 + 2O_2 \longrightarrow CO_2 + 2H_2O$$

In this equation, one molecule of methane CH_4 chemically reacts with two molecules of oxygen to produce one molecule of carbon dioxide and two molecules of water.

For a chemical reaction to occur, the atoms of the reactants must collide with each other. The addition of heat to a process will increase reactant molecular activity. This increased activity ensures a much higher rate of energy transfer between molecules as they collide into each other. Reaction rates double every 10 degrees C. The addition of heat energy affects a process by increasing molecular activity, increasing atomic collisions, and enhancing the formation of chemical bonds.

Reaction rates are affected by heat, surface area, concentration, pressure, flow rates, and catalysts.

A hydrocarbon is a chemical compound that contains hydrocarbon and carbon. Crude oil is a mixture of hydrocarbons that vary in size from very small to large. Much modern manufacturing is based on separating the various components in crude oil by their individual boiling points. Important applied concepts of chemical processing include distillation, catalytic cracking, hydrocracking, and alkylation.

Process Chemistry • Chapter 11

Chapter 11
Review Questions

1. What is chemistry? Why is it important to a process technician?

2. What is matter? List the four states of matter.
   gas, solid, liquid, matter

3. Describe an atom. What is a proton, electron, valence, neutron, AMU, and atomic number?

4. What is an element?

5. What is the function of the periodic table? Define atomic number, atomic mass units, and element symbols. *groups all elements to weight*

6. Define the terms ions and atoms.

7. Define the terms covalent bonds and ionic bonds.

8. Describe the differences between mixtures and compounds.
   2 things mixed together   *Chemical Reaction*

9. What is a chemical equation? Describe reactants and products. What does the yield sign mean?

10. $H_2 + O \longrightarrow H_2O$. Is this chemical equation balanced? List reactant elements. List the product elements.

11. $8NH_3 + 6O_2 \longrightarrow 4N_2 + 12H_2O$. Is this chemical equation balanced? List reactant elements. List the product elements.

12. Given the chemical equation: $H_3PO_4 + 3\ NaOH \longrightarrow Na_3PO_4 + 3H_2O$, determine if it is balanced. List reactant elements and AMUs. List the product elements and AMUs.

Chapter 11 • *Process Chemistry*

13. You are given the chemical equation H₃PO₄ + 3 NaOH ⟶ Na₃PO₄ + 3H₂O. If you are told to add 15 lb of phosphoric acid (H₃PO₄) to this equation, how many pounds will you need to add to the sodium hydroxide (NaOH) to keep the equation balanced?

14. What is an exothermic reaction? How do you control it?
    cut heat, stop flow, add coolant

15. Describe the different types of chemical reactions.

16. How do heat and pressure affect a chemical reaction?

17. What affects reaction rates?

18. List the different types of catalysts.

19. A 500-lb barrel has a 10 percent catalyst solution. What is the weight of the catalyst?   500 × 10% = 50 lbs

20. Contrast an acid and a base.
    Ph Balance   acid low   base high Ph   7 neutral

21. List the first four hydrocarbons in the paraffin and olefin families.

22. Describe crude oil distillation.
    Putting oil in a distillation tower and separate by boiling point

227

Chapter 12
Applied Physics

OBJECTIVES

After studying this chapter, the student will be able to:

- *Review the key terms used in process physics.*
- *Describe the fundamental concepts of physics.*
- *Contrast and compare density and specific gravity.*
- *Describe the principle of pressure in fluids.*
- *Describe complex and simple machines.*

Physics • Chapter 12

KEY TERMS

API gravity—standards to measure the heaviness of a hydrocarbon. A specially designed hydrometer marked in units API; used to determine the heaviness or density of a hydrocarbon.

Baume gravity—the standard used by industrial manufacturers to measure nonhydrocarbon heaviness.

Density—the heaviness of a substance.

Energy—anything that causes matter to change and does not have the properties of matter.

Inertia—a principle used to explain a body's ability to resist motion.

Kinetic energy— the energy of motion or velocity.

Mass—the mass of an object is identified as the quantity of matter.

Matter—anything that occupies space and has mass or volume.

Potential energy—stored energy.

Specific gravity—a method for determining the heaviness of a fluid. The specific gravity of gasoline is 6.15 lb/gal. ÷ 8.33 = 0.738.

12.2 Fundamental Concepts

Matter and Energy

Physics is the study of matter and energy. Matter is anything that occupies space and has mass or volume. Energy is anything that causes matter to change and does not have the properties of matter. Energy takes the form of heat (Btu, causes matter to expand), electricity (Kilowatt hour), potential (height, foot-pound), kinetic (moving, foot-pound), light (produces chemical changes in film), magnetism (creates motion in certain materials), and mechanical work (horsepower hour).

There are two basic states of matter: potential and kinetic. Potential energy is stored energy. Kinetic energy is the energy of motion or velocity.

Specific Properties of Matter

Weight is closely related to mass. If the weight of two bodies is the same, the mass of these bodies is the same. One of the key principles associated with matter is that of attraction or gravitation. All molecules are attracted to each other. The force of molecular gravitation is called weight. Gravitational force between two objects is dependent upon the weight of the bodies and the distance between them. The larger the body, the greater the attraction. Force is inversely proportional to the square of the distance. When the distance between two attracted objects is doubled, the force is only one-fourth as great.

The mass of an object is identified as the quantity of matter. The measure of a body's mass is often identified by its weight.

Inertia is a principle used to explain a body's ability to resist motion. A force must be exerted to move a body that is at rest. To change the speed or direction of a moving object, a force must be applied. All matter has inertia. The total amount of inertia a body contains depends on the total mass in the body.

Volume is the space occupied by a body. See Figures 12.2–1, 12.2–2, and 12.2–3 for volume formulas for a sphere, cylinder, and rectangular solid.

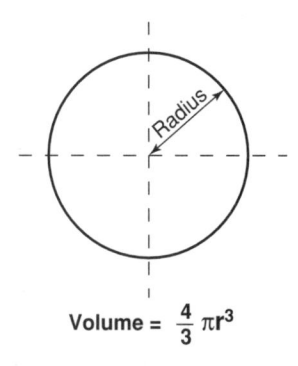

$$\text{Volume} = \frac{4}{3}\pi r^3$$

Figure 12.2-1 *Volume Formula: Sphere*

Volume = πr²h

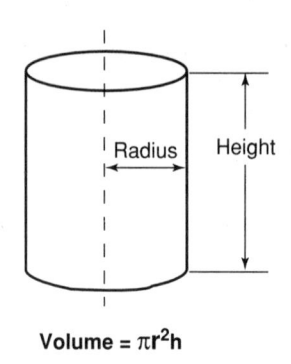

Figure 12.2-2 *Volume Formula: Cylinder*

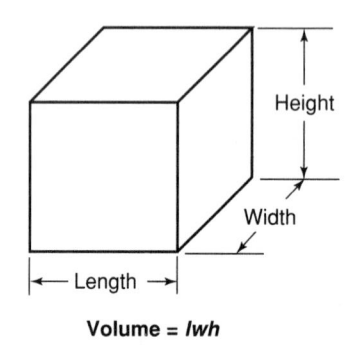

Volume = *lwh*

Figure 12.2-3 *Volume Formula: Rectangular Solid*

A fundamental principle of matter is that it cannot be created or destroyed, only changed from one state to another. This principle of indestructibility holds true for energy. It can only be transformed from one form to another.

Porosity or particle structure is a principle of matter that deals with the vast amounts of space that exist between molecules. This principle helps explain why mixtures of gases, liquids, or solids can occupy a smaller volume than the original materials.

Archimedes' Principle

Archimedes' principle applies to specific weight of solids denser than water. The principle states that:

- A submerged object will displace its own volume of water. (See Figure 12.2–4)
- Weight loss of the object equals the weight of the water displaced.

Example:
A chunk of rock weighs 1,000 g in air and 650 g in water. What is its specific weight?

Solution:

Specific weight of metal $= \dfrac{\text{Wt. of metal in air}}{\text{Loss of weight in water}} = \dfrac{1,000}{(1,000 - 650)} = 2.85$

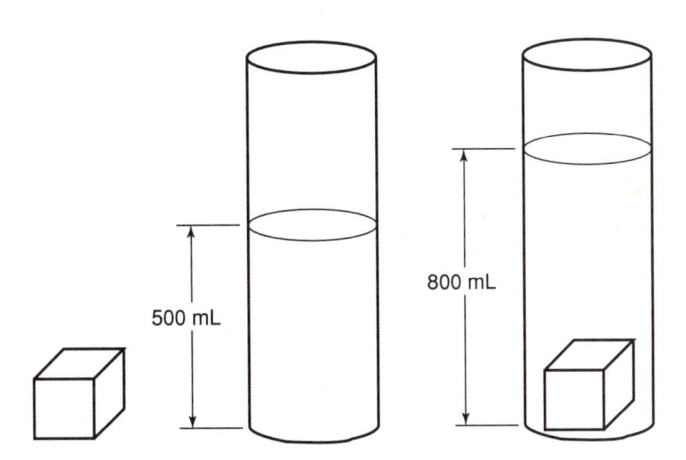

Figure 12.2-4 *Displacement*

12.3 Density and Specific Gravity

Because the density of liquids and solids varies so much, it is convenient to have a standard to compare them to. The standard used to compare densities is water. Water weighs 62.5 lb per cubic foot or one gram per cubic centimeter. The terms specific gravity and specific weight are used to compare the density of water to another substance. Hydrocarbons are typically lighter than water. Their specific gravities will be less than one, whereas substances above are listed as greater than one.

Specific gravity is defined as the comparison of a fluid (liquid or gas) to the density of water or air. It is a common mistake for operators to confuse specific gravity with density. This is easy to understand because specific gravity is a method for determining the heaviness of a fluid. Density is the heaviness of a substance whereas specific gravity compares this heaviness to a standard and then calculates a new ratio. Most hydrocarbons have specific gravities below 1.0.

Key points to remember:

- The specific gravity of water is 8.33 lb. gal. ÷ 8.33 = 1.0
- The specific gravity of gasoline is 6.15 lb/gal ÷ 8.33 = 0.738
- The density of water is 8.33 lb/gal
- The density of air is 0.08 lb per cubic foot
- Density is calculated by weighing unit volumes of a fluid at 60 degrees F (15.5 degrees C)

Physics • Chapter 12

Determining Specific Gravity
Specific gravity is determined by comparing the weight of a volume of material with the weight of the same volume of water. There are two methods for determining specific gravity.

$$\text{Specific gravity} = \frac{\text{Weight of definite volume of given material}}{\text{Weight of the same volume of water}}$$

$$\text{Specific gravity of a sinking solid} = \frac{\text{Weight of the object in air}}{\text{Loss of weight in water}}$$

Density
Industry uses four different ways to express a fluid's heaviness:

- Density—(D = Weight ÷ Volume) The density of water is 8.33 lb/gal. The density of a fluid is defined as the mass of a substance per unit volume. Density measurements are used to determine heaviness.

- Specific gravity—the specific gravity of water is 8.33 ÷ 8.33 = 1. The specific gravity of gasoline is 6.15 lb/gal ÷ 8.33 = 0.738.

- Baume—Baume gravity is the standard used by industrial manufacturers to measure nonhydrocarbon heaviness.

- API—The American Petroleum Institute applies API gravity standards to measure the heaviness of a hydrocarbon. A specially designed hydrometer marked in units API is used to determine the heaviness or density of a hydrocarbon. High API readings indicate low fluid gravity.

The density of a fluid is defined as the mass of a substance per unit volume. Density measurements are used to determine heaviness. For example, ~~one pound~~ *gallon* of:

    water   = 8.33 lb
    crude oil = 7.20 lb
    gasoline   = 6.15 lb

Viscosity
Another common term used by industry to describe the flow characteristics of a substance is viscosity. *Viscosity* is defined as a fluid's resistance to flow.

Weight = Volume X Density

Density of water   = 1 g per cubic centimeter
                          = 62.5 lb per cubic foot
                          = 1687.5 lb per cubic yard
                          = 16.41 g per cubic inch

Example 1:
Find the density of helium gas: twenty liters of the gas weighs 3.4 grams.

Solution:

$$D = \frac{W}{V}$$

$$D = \frac{3.4 \text{ g}}{20 L} = 0.17 \text{ g/l}$$

Example 2:
Find the volume of a granite object with a density of 2.6 g per cubic centimeter and a weight of 1,280 g.

Solution:

$$V = \frac{W}{D}$$

$$V = \frac{1{,}280 \text{ g}}{2.6 \text{ g/cm}^3} = \frac{1{,}280 \text{ (g} \times \text{cm}^3)}{2.6 \text{ g}} = 492 \text{ cm}^3$$

Elasticity

Elasticity is a principle that refers to the tendency of a substance to return to its original shape after a distorting force is removed. Substances like iron and steel have a high degree of elasticity whereas substances like clay or putty have a very low elasticity rating. Steel can be subjected to thousands of pounds per square inch and will return to its original shape when the distorting force is removed.

The term *strain* is used to describe the total distortion that occurs after the distorting force is removed. Robert Hooke's law states that strain is proportional to stress if the stress remains within the elastic limit of the material. A spring is a device that is an example of Hooke's law. Distorting forces take the form of compressing, stretching, tearing, twisting, and bending.

The elastic limit of a substance is the maximum force a substance can withstand without breaking or becoming permanently deformed.

Hardness

The hardness of a substance is determined by its ability to scratch or mark another substance. Diamond is the hardest substance known; gold is very soft.

Tenacity

Tenacity is the ability of a substance to resist being pulled or torn apart. Tenacity per unit area is called the tensile strength. Tensile strength is measured in pounds per square inch.

Ductility
Ductility is a material's ability to be drawn into fine threads. Copper and aluminum wire is a good example of two materials with high ductility.

Malleability
Malleability is a characteristic of a substance that allows it to be beaten or rolled into thin sheets. Examples of this principle include aluminum and gold. Gold can be rolled to a thickness of about 1/300,000 inch thick.

Adhesion
Dissimilar molecules have very powerful attractive forces referred to as adhesion. Examples of materials with great adhesion include concrete and glue.

Surface Tension
Surface tension is the result of molecular attraction, which is stronger along the outer perimeter and weaker toward the middle. The liquid acts like a stretched sheet of rubber. Surface forces vary from those found deeper in the liquid, because there are no upward forces.

Capillary Action
When a liquid comes in contact with the outside of its container, it experiences two forces: cohesive force and adhesion. The adhesive force is the result of the attractive forces between the walls of the container and the fluid; the cohesive force is related to the internal characteristics of the liquid. When the adhesive forces of the system are greater than the cohesive internal forces of the liquid, the liquid tends to cling to the sides of the container. This tendency of a liquid to cling to and climb up the walls of the container is called "wetting."

Mercury is so dense it has the opposite reaction as most liquids: Wetting does not occur and the mercury adhesive forces dome up the mercury near the wall of the container. The density of the liquid and the size of the tube determine how high or low a liquid will move inside a container.

Temperature and Cohesive Force
When the temperature inside a process system is increased, the cohesive forces between molecules is reduced.

Example 1:
What is the density of a cube of iron, 15 cm on an edge, that weighs 9.6 kg?

Solution:

$$D = \frac{W}{V}$$

$$D = \frac{9.6 \text{ kg}}{15 \text{ cm}^3} = 2.84 \text{ g/cm}^3$$

9,600 ÷ 3,375 = 2.84 g/cm^3

Answer: 2.84 g/cm^3

Example 2:
How many liters of alcohol will weigh 30 kg? (Density of alcohol = 0.8 g/cm^3)

Solution:

$$V = \frac{W}{D}$$

$$V = \frac{30 \text{ kg}}{0.8 \text{ g/cm}^3} = 37,500$$

Liters of alcohol = 37,500 grams ÷ 1,000 grams = 37.5 L

Answer: 37.5 L

Note: 1 kg = 1000 g

Practical Exercises
1. A beam of cedar wood 20 ft long, 2 ft wide, and 6 in. thick weighs 150 lb. Calculate its density. Show your work!

2. A cylinder 4 cm in diameter and 40 cm long is made of brass. (Density = 8.5 g/cm^3) Calculate its weight. Show your work!

3. What is the weight of a rectangular steel bar 15 ft long, 1 ft wide, and 2 in. thick? (Density of steel = 461 lb/ft^3) Show your work!

12.4 Pressure in Fluids

Force and Pressure
Force is a push or a pull that is used to change the direction, speed, or shape of a body. Gravitational force in liquids and pressure in fluids share a unique relationship. Pressure is the total force divided by the area. Force is measured in units of weight.

Example 1:
A rectangular tank 10 ft square and 8 ft deep is filled with water. The volume of the tank is 800 ft^3. Water weighs 62.5 lb/ft . Calculate the total force exerted by the water against the bottom of the tank.

Physics • Chapter 12

Solution:

Total Force	=	Area	×	Height	×	Density		
F	=	A	×	H	×	D		
F	=	10 ft²	×	8 ft	×	62.5 lb/ ft³	=	50,000 lb

Solution:
The total force exerted by the water against the bottom of the tank is 50,000 lb.
Pressure (P) = Force (weight) ÷ Area

Example 2:
Calculate the pressure produced by a 2,000 lb stone block,
40 in. length × 20 in. width × 60 ft height.

Solution:
The area occupied by the stone = 800 in.²
40 in. length × 20 in. wide × 60 ft height = 800 in.²
P = 2,000 ÷ 800 = 2.5 psi

Answer:
The psi at the base of the stone = 2.5 psi

Example 3:
Calculate the pressure produced by a 456-lb granite block,
10 in. length × 15 in. width × 72 in. height.

Example 4:
Calculate the pressure produced by a 10 ft onion tank filled with a hydrocarbon fluid (0.72 sg). Vapor pressure is 200. Add 45 psi N_2 to the total. What is the final pressure?

Solution:
Height × 0.433 × specific gravity = Pressure
10 ft × 0.433 × 0.72 sg = 3.1 psi
3.1 psi + 200 + 45 psi = 248.1 psi

Answer:
248.1 psi

Practice Exercises
1. Calculate the pressure produced by a 40 ft onion tank filled with 17.5 ft of a hydrocarbon fluid (0.54 sg). Vapor pressure is 300 psi. Add 15 psi N_2 to the total. What is the final pressure?

2. Calculate the pressure produced by water in a 12.5 ft high vessel.

3. Calculate the pressure exerted on the bottom of a 69-ft distillation column by a 10-ft hydrocarbon level. Specific gravity is 0.67. Vapor pressure at 240 degrees F (115.5 degrees C) is 236. One hundred psi is added to the column giving a top gauge reading of _____ psi and a bottom gauge reading of _____ psi.

4. What pressure, in pounds per square inch, is a scuba diver subjected to when descending to an ocean depth of 125 ft?

5. A rectangular vessel, 10 ft wide, 20 ft long, and 12 ft deep is filled with mercury (density = 13.6 g/cm). Using the following equations:
   (a) What is the pressure on the bottom of the tank?
   (b) Identify the average pressure on one side of the vessel.
   (c) What is the total force on the bottom of the tank?
   (d) Identify the total force against one end of the tank.

   a. $P = HD$
   b. $P_{ave} = H_{ave} \times D =$
   c. $F = AHD$
   d. $F = AHD \div 2 =$

12.5 Complex and Simple Machines

Work
Work is the process of overcoming the downward pull of gravity and moving a body that has been at rest. When you attempt to lift a body at rest, work is not accomplished unless the object is lifted. Work is equal to force × distance.

$W = FD$

Example 1:
Calculate how much work is done when a force of 50 lb is applied to push a wagon 20 ft.

Solution:
To solve this problem, multiply the applied force [50 lb] by the distance through which the force acts [20 ft].

Work = Force × Distance
     = F × D
     = 50 lb × 20 ft   = 1,000 ft-lb

Physics • Chapter 12

Answer:
1,000 ft-lb

Example 2:
Calculate the amount of work accomplished by a 3-ft table holding up a 2-lb book.

Example 3:
Calculate the amount of work accomplished when a weight is lifted 240 ft by a force of 600 lb. (Work = Force × Distance)

Example 4:
A man weighing 190 lb climbs a 35-ft staircase in 25 sec. Calculate how much work was performed.

Mechanical Advantage

Mechanical advantage is the ratio between resistance overcome and effort applied. When determining the mechanical advantage of a system, the resistance is divided by the effort. For example, when a 100-lb force moves a resistance force of 400 lb, the machine has a mechanical advantage of 4. Actual MA is calculated using the following equation.

$$MA = \frac{resistance}{effort} = \frac{R}{E}$$

Inclined Plane

When an object is rolled or slid up a ramp, it is using the scientific principle of the inclined plane. Inclined planes are very useful when large, heavy objects need to be moved from one level to the next. Other examples of this principle include stairways, ramps, inclined roads, and inclined tracks.

In the inclined plane principle, ideal MA is calculated by using the resistance force (gravity) that overcomes the vertical height of the plane and the effort force (length) that acts through the entire length of the plane. For example:

$$\text{Ideal } MA = \frac{D_e}{D_r} = \frac{\text{Length of plane}}{\text{Height of plane}}$$

Example:
A 24-ft long inclined ramp extends from the ground to a height of 8 ft. A force of 180 lb is required to roll a 420-lb cart up the ramp.
    (a) Calculate the actual MA.
    (b) Calculate the ideal MA.
    (c) Calculate the efficiency.

(d) Calculate how much work is accomplished against gravity.
(e) Calculate how much work is done in overcoming friction.

Solutions:

a. Actual MA $= \frac{R}{E} = \frac{420 \text{ lb}}{180 \text{ lb}} = 2.33$

b. Ideal MA $= \frac{De}{Dr} = \frac{24 \text{ ft}}{8 \text{ ft}} = 3.0$

c. Efficiency $= \frac{\text{actual MA}}{\text{ideal MA}} = \frac{2.33}{3} = 77\%$

d. Work output $= R \times Dr = 420 \text{ lb} \times 8 \text{ ft} = 3360 \text{ ft-lb}$

e. Work to overcome friction
   $=$ Work input − Work output
   $= (E \times De) - (R \times Dr)$
   $= (180 \text{ lb} \times 24 \text{ ft}) - (420 \text{ lb} \times 8 \text{ ft})$
   $= 960 \text{ ft-lb}$

The Principle of Moments and Levers

The principle of moments and levers can be illustrated using an ordinary playground seesaw. A seesaw is designed to operate like a balanced lever. Two arms of equal length extend across the fulcrum. When a force acts upon the lever arm, it causes a reaction. The lever will remain balanced only if the two forces acting on the seesaw are distributed equally. The point along the lever where the force is applied is important to this distribution concept.

The moment of a force is equal to the product of the force and the perpendicular distance from the fulcrum.

Example 1:
Do the total clockwise moments balance the total counterclockwise moments in Figure 12.5–1? (Page 242)

Solution:

Counterclockwise					*Clockwise*				
Force	×	Distance	=	Moment	Force	×	Distance	=	Moment
50 lb	×	6 ft	=	300 ft-lb	60 lb	×	5 ft	=	300 ft-lb
12 lb	×	4 (ft)	=	48 ft-lb	4 lb	×	12 ft	=	48 ft-lb
CCW Moment			=	348 ft-lb	CW Moments			=	348 ft-lb

If the force and distance are perpendicular to each other, no work is accomplished. When a lever is in equilibrium, the total counterclockwise and clockwise forces are equal.

Physics • Chapter 12

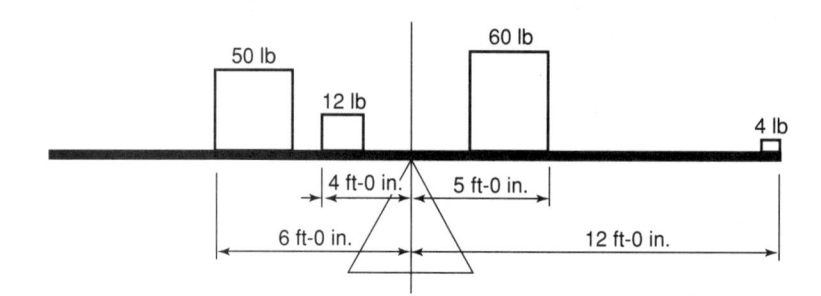

Figure 12.5-1 *Law of Moments*

Example 2:
A balanced lever arm rests on a fulcrum at its center. A 200-lb force is applied 5 ft from the fulcrum. To maintain equilibrium, how far from the fulcrum, on the other lever arm, should a 100-lb force be applied?

Solution:

CCW moments = CW moments
200 lb × 5 ft = 100 lb × x ft
x = 10 ft

Answer: 10 ft

Example 3:
A 200-lb weight and a 100-lb weight rest 8 ft and 6 ft from the fulcrum. A 200-lb weight rests on the opposite side. How many feet from the fulcrum must the weight be placed to establish equilibrium?

Solution:

(200 lb × 8 ft) + (100 lb × 6 ft) = 200 lb × x ft

$$x = \frac{1600 + 600}{200} = 11 \text{ ft}$$

Answer: 11 ft

Summary

Physics is the study of matter and energy. Matter is anything that occupies space and has mass or volume. Energy is anything that causes matter to change and does not have the properties of matter. Energy takes the form of heat, electricity, potential, kinetic, light, magnetism, and mechanical energy.

Chapter 12 • *Physics*

There are two basic states of energy: potential and kinetic. Potential energy is stored energy and kinetic energy is the energy of motion or velocity.

The specific properties of matter are weight, mass, inertia, volume, indestructibility, and porosity.

Because the density of liquids and solids varies so much, it is convenient to have a standard to compare them to. The standard used to compare densities is water. Water weighs 62.5 lb per cubic foot or one gram per cubic centimeter. The terms specific gravity or specific weight are used to compare the density of water to another substance. Hydrocarbons are typically lighter than water. Their specific gravities will be less than one, while substances above are listed as greater than one.

Specific gravity is the comparison of a fluid (liquid or gas) to the density of water or air. It is a common mistake for operators to confuse specific gravity with density. This is easy to understand because specific gravity is a method for determining the heaviness of a fluid. Density is the heaviness of a substance whereas specific gravity compares this heaviness to a standard and then calculates a new ratio. Most hydrocarbons have specific gravities below 1.0.

Key points to remember when contrasting specific gravity and density:
- The specific gravity of water is 8.33 lb/gal ÷ 8.33 = 1.0
- The specific gravity of gasoline is 6.15 lb/gal ÷ 8.33 = 0.738
- The density of one gal of water is 8.33 lb/gal
- The density of air is 0.08 pound per cubic foot
- Density is calculated by weighing unit volumes of a fluid at 60 degrees F

Specific weight is determined by comparing the weight of a volume of material with the weight of the same volume of water.

Industry uses four different ways to express a fluid's heaviness: density, specific gravity, baume, and API.

The density of a fluid is the mass of a substance per unit volume. Density measurements are used to determine heaviness.

Another common term used by industry to describe the flow characteristics of a substance is viscosity. Viscosity is a fluid's resistance to flow.

Elasticity is a principle that refers to the tendency of a substance to return to its original shape after a distorting force is removed.

The term *strain* is used to describe the total distortion that occurs after the distorting force is removed. Robert Hooke's law states that strain is proportional to stress if the stress remains within the elastic limit of the material. A spring is a device that is an example of Hooke's law.

within the elastic limit of the material. A spring is a device that is an example of Hooke's law. Distorting forces take the form of compressing, stretching, tearing, twisting, and bending.

The elastic limit of a substance is the maximum force a substance can withstand without breaking or becoming permanently deformed.

The hardness of a substance is determined by its ability to scratch or mark another substance. Diamond is the hardest substance known, gold is very soft.

Tenacity is the ability of a substance to resist being pulled or torn apart. Tenacity per unit area is called the tensile strength. Tensile strength is measured in pounds per square inch.

Ductility is a material's ability to be drawn into fine threads. Copper and aluminum wire are good examples of two materials with high ductility.

Malleability is a characteristic of a substance that allows it to be beaten or rolled into thin sheets. Examples of this principle include aluminum and gold. Gold can be rolled to a thickness of about 1/300,000 in.

Dissimilar molecules have very powerful attractive forces referred to as adhesion. Examples of adhesion include concrete and glue.

Surface tension is the result of molecular attraction, which is stronger along the outer perimeter and weaker toward the middle. The liquid acts like a stretched sheet of rubber. Surface forces vary from those found deeper in the liquid because there are no upward forces.

When a liquid comes in contact with the outside of its container, it experiences two forces: cohesive force and adhesion. The adhesive force is the result of the attractive forces between the walls of the container and the fluid. Cohesive force is related to the internal characteristics of the liquid. When the adhesive forces of the system are greater than the cohesive internal forces of the liquid, the liquid tends to cling to the sides of the container. This tendency of a liquid to cling to and climb up the walls of the container is called "wetting."

Mercury is so dense that it has the opposite reaction from most liquids: Wetting does not occur, and the mercury adhesive forces dome up the mercury near the wall of the container. The density of the liquid and the size of the tube determines how high or low a liquid will move inside a container.

When the temperature inside a process system is increased, the cohesive forces between molecules is reduced.

Force is defined as a push or a pull that is used to change the direction, speed, or shape of a body. Gravitational force in liquids and pressure in fluids share a unique relationship. Pressure is described as the total force divided by the area. Force is measured in units of weight.

Work is defined as the process of overcoming the downward pull of gravity and moving a body that has been at rest. When you attempt to lift a body at rest, work is not accomplished unless the object is lifted. Work is equal to force × distance. (W = FD)

The moment of a force is equal to the product of the force and the perpendicular distance from the fulcrum.

Mechanical advantage is defined as the ratio between resistance overcome and effort applied. When determining the mechanical advantage of a system, the resistance is divided by the effort.

Physics • Chapter 12

Chapter 12
Review Questions

1. What is the volume of a rectangular object 4 ft long, 3 ft wide, and 10 ft high?

2. What is the volume of a cylinder that has a 15-ft diameter and stands 22 ft tall?

3. Find the density of hydrogen gas: Forty liters of the gas weighs 3.3 g.

4. Find the volume of a metal object with a density of 4.6 g per cubic centimeter and a weight of 4,280 grams.

5. What is the density of a cube of iron, 25 cm on an edge, that weighs 16.5 kg?

6. How many liters of alcohol will weigh 80 kg? Density of alcohol = 0.8 g/cm^3

7. A beam of cedar wood 43 ft long, 3 ft wide, and 6 in. thick weighs 350 lb. Calculate its density.

8. A cylinder 6 cm in diameter and 30 cm long is made of brass. (Density = 8.5 g/cm^3) Calculate its weight.

9. What is the weight of a rectangular steel bar 10 ft long, 2 ft wide, and 2 in. thick? (Density of steel = 461 lb/ft^3)

10. Calculate the pressure produced by a 3,500-lb stone block, 60 in. length × 20 in. width × 72 in. height.

11. Calculate the pressure produced by a 956-lb granite block, 110 in. length × 15 in. width × 72 in. height.

12. Calculate the pressure produced by a 22-ft onion tank filled with a hydrocarbon fluid (0.72 sg). Vapor pressure is 430. Add 55 psi N$_2$ to the total. What is the final pressure?

13. Calculate the pressure produced by water in a 19.5 ft high vessel.

14. Contrast density and specific gravity.

Chapter 12 • *Physics*

15. Calculate the pressure exerted on the bottom of a 79-ft distillation column by a 10-ft hydrocarbon level. Specific gravity is 0.67. Vapor pressure at 240 degrees F (115.5 degrees C) is 236. One hundred psi is added to the column giving a top gauge reading of _____ psi and a bottom gauge reading of _____ psi.

16. What pressure, in pounds per square inch, is a scuba diver subjected to when descending to an ocean depth of 105 ft?

17. A chunk of rock weighs 1,000 g in air and 350 g in water. What is its specific weight?

18. Calculate how much work is done when a force of 50 lb is applied to push a wagon 20 ft. To solve this problem, multiply the applied force (50 lb) by the distance through which the force acts (20 ft).   50 × 20 = 1000 ft-lbs   W = F × D

19. Calculate the amount of work accomplished by a 2-ft table holding up a 2-lb book.
    Zero

20. Calculate the amount of work accomplished when a weight is lifted 140 ft by a force of 300 lb.

21. A man weighing 150 lb climbs a 35-ft staircase in 15 seconds. Calculate how much work was performed.

22. A 20 ft long inclined ramp extends from the ground to a height of 6 ft. A force of 190 lb is required to roll a 520-lb cart up the ramp.
    (a) Calculate the actual MA.
    (b) Calculate the ideal MA.
    (c) Calculate the efficiency.
    (d) Calculate how much work is accomplished against gravity.
    (e) Calculate how much work is done in overcoming friction.

23. A 1,600-lb weight and a 400-lb weight rest 8 ft and 6 ft from the fulcrum. A 700-lb weight rests on the opposite side. How many feet from the fulcrum will the weight need to be placed to establish equilibrium?

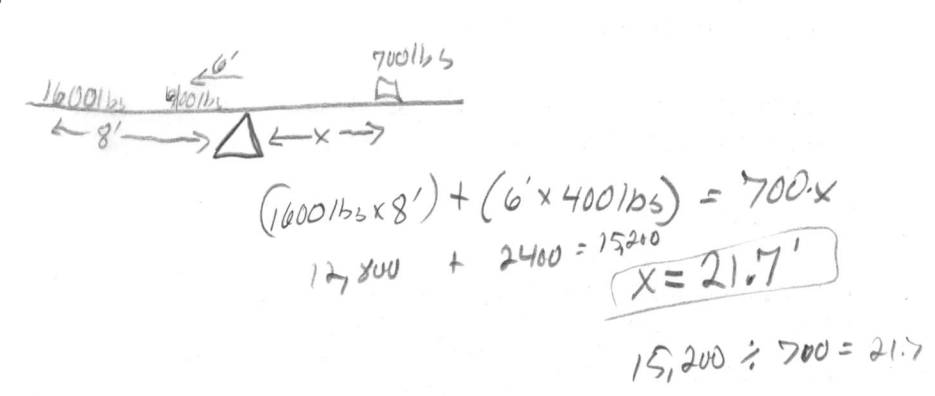

(1600 lbs × 8') + (6' × 400 lbs) = 700·x
12,800 + 2400 = 15,200
x = 21.7'
15,200 ÷ 700 = 21.7

247

Chapter 13

Environmental Control

OBJECTIVES

After studying this chapter, the student will be able to:

- *Define the key terms associated with environmental awareness training.*

- *Describe air pollution control.*

- *Discuss water pollution control.*

- *Explain solid waste control.*

- *Describe toxic substances control.*

- *Explain emergency response.*

- *Discuss community right-to-know.*

Environmental Control • Chapter 13

KEY TERMS

Air permits—permits must be obtained for any project that has the possibility of producing air pollutants.

Air pollution—the contamination of the air especially by industrial waste gases, fuel exhausts, or smoke.

CAER—Community Awareness and Emergency Response. CAER is a program designed to inform the community of potentially hazardous situations, inform the community of hazardous chemicals found in the plant, work with the community to develop emergency response programs, and open the lines of communication between industry and the community.

Clean Air Act—the purpose of the Clean Air Act is to enhance the quality of the nation's air; accelerate a national research and development program to prevent air pollution; provide technical and financial assistance to state and local government; and to develop a regional air pollution control program.

Community right-to-know—a principle that increases community awareness of the chemicals manufactured or used by local chemical plants and business; involves community in emergency response plans, improves communication and understanding; improves local emergency response planning, and identifies potential hazards.

Emergency response—an industrial environment responding to an emergency follows a specific set of standards. "Emergency response" drills are carefully planned and include preparations for worst-case scenarios. Examples: vapor releases, chemical spills, explosions, fires, equipment failure, hurricane, high winds, loss of power, and bomb threats.

EPA—Environmental Protection Agency. The EPA is a government-approved agency with authority to make and enforce environmental policy.

RCRA—the Resource Conservation and Recovery Act was enacted as public law in 1976. The purpose of RCRA is to protect human health and the environment. A secondary goal is to conserve natural resources. It completes this goal by regulating all aspects of hazardous waste management including generation, storage, treatment, and disposal. This concept is referred to as "cradle to grave."

Solid waste—a by-product of modern technology; technically defined as a discarded solid, liquid, or containerized gas. This definition includes materials that have been recycled or abandoned through disposal, burning or incineration, accumulation, storage, or treatment.

Chapter 13 • *Environmental Control*

The Clean Water Act of 1972—legislation adopting the best available technology (BAT) strategy for all cleanups.

The Toxic Substances Control Act of 1976 (TSCA)—intended to protect human health and the environment, and to regulate commerce by requiring testing and necessary restrictions on certain chemical substances. TSCA imposes requirements on all manufacturers, exporters, importers, processors, distributors, and disposers of chemical substances in the United States.

Water pollution—contamination of the water especially by industrial wastes.

13.2 Air Pollution Control

Modern technology produces a variety of useful products. This same technology produces by-products that can harm the environment. Because of the potential hazards that exist with new technology, environmental laws and regulations have been passed to protect our future. The purpose of the Clean Air Act is to enhance the quality of the nation's air; accelerate a national research and development (R&D) program to prevent air pollution; provide technical and financial assistance to state and local government; and develop a regional air pollution control program.

Air Pollution Control

In 1955, the original Clean Air Act was passed. Over the years, a number of modifications have been made.

    1960 amendment—directed Surgeon General to study vehicle pollution
    1963 amendment—directed research into fuel desulfurization and development of air quality criteria
    1965 amendment—to study new sources of pollution
    1967—Quality Air Act
    1970—Clean Air Amendment
    1977 amendment—the Clean Air Act for emission standards
    1990—reauthorization of Federal Clean Air Act

- Air toxins
- Acid deposition
- Job training for laid-off workers due to Clean Air Act
- Air quality standards
- Permits
- Stratospheric ozone and global climate protection
- Provisions for enforcement
- Acid rain and air monitoring research
- Provision to improve air quality and visibility near national parks
- Provisions relating to mobile sources

Agencies

The Environmental Protection Agency (EPA) was established in 1970, the EPA is an independent agency of the United States government whose primary purpose is to protect the environment from pollution. The EPA has authority to develop and enforce environmental policy.

The Air Pollution Control Board maintains numerous regional offices throughout each state. Each location receives public complaints, coordinates investigations, documents violations, and recommends enforcement actions.

Air Permitting

Permits must be obtained for any project that has the possibility of producing air pollutants. The Air Control Board places limits on emissions and requires about 3 to 8 months to complete the permit process. After the ACB issues the permit, a yearly inspection is scheduled. Penalties for civil and criminal abuses of the Clean Air Act range from $25,000 a day to $250,000 and 2 to 15 years in jail. For example, smoking flares in excess of 5 minutes should be reported. Failure to report results in severe penalties.

13.3 Water Pollution Control

The Federal Clean Water Act was passed in 1898. Fifty years later, Congress provided funds for the construction of municipal waste water treatment facilities. The Water Control Act of 1965 took a "water quality" approach and initiated close examination of receiving waters. States were required to establish standards for water quality.

The Clean Water Act of 1972 adopted the best available technology (BAT) strategy for all cleanups. Under the 1987 amendments, states are required to identify waters that are not expected to meet quality standards.

Water Pollution Controls

The Clean Water Act regulates waste water. Waste water standards are applied to:
- Process waste water—process contact water, contaminated water from vessels and equipment, tanks, slab cleanup, and so on.
- Rainwater—sewer system releases
- Once-through cooling water—cooling tower blow-down or boiler blow-down

The Federal Clean Water Act is designed to protect United States water quality. The EPA, State Water Commissions, Army Corps of Engineers, State Parks and Wildlife, and U.S. Fisheries and Wildlife help enforce the Clean Water Act.

National Water Quality Standards

National Water Quality Standards state that:
- All U.S. waters shall be fishable and swimmable
- No discharge of toxic pollutants in toxic quantities will be allowed
- Technology must be developed to eliminate pollutant discharge

Water Permitting
The Clean Water Act requires a company to have a water permit. In some states, a two-permit system exists.

13.4 Solid Waste Control

Solid waste is a by-product of modern technology. Solid waste is technically defined as a discarded solid, liquid, or containerized gas. This definition includes materials that have been recycled or abandoned through disposal, burning or incineration, accumulation, storage, or treatment.

The Resource Conservation and Recovery Act (RCRA) was enacted as public law in 1976. The purpose of RCRA is to protect human health and the environment. A secondary goal is to conserve our natural resources. RCRA completes this goal by regulating all aspects of hazardous waste management: generation, storage, treatment, and disposal. This concept is referred to as "cradle to grave."

Solid waste is categorized as:
    Class One   Hazardous: ignitable, reactive, corrosive, toxic
    Class One   Non-hazardous: RCRA regulations do not apply
    Class Two Examples include garbage, cured epoxy resin,
                              biopond filter solid
    Class Three   Examples include uncontaminated or inert material, "wood"

Laws
The RCRA states these penalties: civil penalty of $25,000 a day; criminal penalty from knowing endangerment $250,000 and 15 years in jail ($1,000,000 for a company); liability includes any person involved in breaking the law.

State water commissions are organized to regulate and control the solid waste generated within their boundaries.

Permitting
More than 90-day storage of a hazardous chemical requires a permit.

An ideal facility includes:
- A covered facility to prevent rain water contamination
- No contact with soil
- Containment for all equipment
- Raised equipment to permit inspection for leaks

Environmental Control • Chapter 13

13.5 Toxic Substances Control

The Toxic Substances Control Act (TSCA) is a federal law enacted in 1976. TSCA was intended to protect human health and the environment. TSCA was also designed to regulate commerce by requiring testing and necessary restrictions on certain chemical substances. TSCA imposes requirements on all manufacturers, exporters, importers, processors, distributors, and disposers of chemical substances in the United States.

Controls

The TSCA inventory (70,000 toxic chemicals) was established to record all products manufactured, imported, sold, processed, or used for commercial purposes. Exemptions include R&D chemicals and by-products without commercial purpose. The TSCA also controls premanufacture review of new chemical substances, risk assessment by testing and information gathering, recordkeeping and reporting on health and environmental effects associated with chemical substances, and restrictions on known hazardous chemicals.

The Toxic Substances Control Act has severe penalties for those who break the law. Yearly penalties for current violations are estimated at over $40,000,000. The EPA is the primary agency charged with enforcing toxic substance control.

13.6 Emergency Response

In an industrial environment, responding to an emergency follows a specific set of standards. "Emergency response" drills are carefully planned and include preparations for worst-case scenarios; e.g., vapor releases, chemical spills, explosions, fires, equipment failure, hurricane, high winds, loss of power, and bomb threats.

Programs like CAER are designed to work with the community while industry utilizes:
- Action Plans
- Emergency Response Coordinators & Teams
- "Site-Specific" Drills
- Incident Reports

13.7 Community Right-to-Know

The community right-to-know principle increases community awareness of the chemicals manufactured or used by local chemical plants and business, involves the community in emergency response plans, improves communication and understanding, improves local emergency response planning, and identifies potential hazards.

The Comprehensive Environmental Response Compensation and Liability Act (CERCLA) holds generators and disposers of hazardous waste liable for past practices, and established the "Superfund" of $1.6 billion to pay for clean-up operations of abandoned hazardous waste sites. It also informs the public of these sites and the known hazards. Community right-to-know and Community Awareness and Emergency Response (CAER) programs work with CERCLA to protect the community.

Under the community right-to-know principle
- CERCLA
- Emergency Planning and Community Right-To-Know
- Superfund Amendments and Reauthorization Act "SARA"
- Hazard Communication Act "HAZCOM" (See MSDS)
- OSHA Hazard Communication Act (OSHA HAZCOM) (See MSDS)

work with agencies like the
- Department of Health
- State Water Commission
- US Environmental Protection Agency

to protect citizens.

Quality Standards

Industry believes that reducing and recycling wastes at their source is the first priority of responsible waste management. Industry has put in place environmental management systems to make, use, handle, and dispose of its products safely. Industry is committed to making major expenditures in environmental technology to reduce emissions and protect the environment.

Summary

Air pollution is defined as the contamination of the air especially by industrial waste gases, fuel exhausts, or smoke.

Water pollution is defined as the contamination of the water especially by industrial wastes.

Solid waste is technically defined as a discarded solid, liquid, or containerized gas. Solid waste is a by-product of modern technology. This definition includes materials that have been recycled or abandoned through disposal, burning or incineration, accumulation, storage, or treatment.

In an industrial environment responding to an emergency follows a specific set of standards. Emergency response drills are carefully planned and include preparations for worst-case scenarios, such as vapor releases, chemical spills, explosions, fires, equipment failure, hurricane, high winds, loss of power, and bomb threats.

Environmental Control • Chapter 13

The community right-to-know principle: increases community awareness of the chemicals manufactured or used by local chemical plants and business, involves the community in emergency response plans, improves communication and understanding, improves local emergency response planning, and identifies potential hazards.

The Resource Conservation and Recovery Act (RCRA) was enacted as public law in 1976. The purpose of RCRA is to protect human health and the environment. A secondary goal is to conserve our natural resources. It completes this goal by regulating all aspects of hazardous waste management, including generation, storage, treatment, and disposal. This concept is referred to as "cradle to grave."

The Toxic Substances Control Act is a federal law enacted in 1976. It is intended to protect human health and the environment. TSCA was also designed to regulate commerce by requiring testing and necessary restrictions on certain chemical substances. TSCA imposes requirements on all manufacturers, exporters, importers, processors, distributors, and disposers of chemical substances in the United States.

Chapter 13 Review Questions

Select the correct choice to complete the statements.

1. Solid waste is technically defined as
   a. a discarded solid, liquid, or containerized gas.
   b. recycled or abandoned materials through disposal, burning or incineration, accumulation, storage, or treatment.
   c. a material composed of 2 percent hydrocarbons.
   d. a and b.

2. The community right-to-know principle: (select all correct responses)
   a. increases community awareness of the chemicals manufactured or used by local chemical plants and business.
   b. identifies potential hazards.
   c. improves communication and understanding.
   d. improves local emergency response planning.
   e. involves the community in emergency response plans.

3. The purpose of RCRA is to protect
   a. human health and the environment.
   b. industrial equipment.
   c. industrial manufacturers from liability cases.
   d. a and c.

4. "Cradle to grave" is part of which act?
   a. Resource Conservation and Recovery Act
   b. Clean Air Act
   c. Clean Water Act
   d. Toxic Substance Control Act

5. Smoking flares in excess of _____ minutes should be reported. Failure to report results in severe penalties.
   a. 3
   b. 10
   c. 5
   d. none of these

Environmental Control • Chapter 13

6. Vapor releases, chemical spills, explosions, fires, equipment failure, hurricane, high winds, loss of power, and bomb threats fall under which main program?

7. What must be obtained for any project that has the possibility of producing air pollutants?

Chapter 14

Quality Control

OBJECTIVES

After studying this chapter, the student will be able to:

- *Define quality control principles and terms.*
- *Describe the principles of continuous quality improvement.*
- *Describe the four phases of the qualities improvement cycle: plan, observe and analyze, learn, and act.*
- *Describe the supplier-customer relationship.*
- *Identify and describe quality tools used in the industry.*
- *Describe statistical process control.*
- *Use flow charts.*
- *Use run charts.*
- *Use cause-and-effect diagrams (fishbone).*
- *Use pareto charts.*
- *Describe planned experimentation.*
- *Describe histograms or frequency plots.*
- *Describe forms for collecting data.*
- *Describe scatter plots.*

Quality Control • Chapter 14

KEY TERMS

Cause-and-effect diagrams (fishbone)—a method for summarizing available knowledge about the causes of process variation.

Control charts—a statistical tool used to determine and control process variations.

Dr. W. E. Deming and Joseph M. Juran—pioneers in the development of statistical and quality methodology.

Flow chart—a picture of the activities that take place in a process.

Forms for collecting data—can vary from notes jotted down on a napkin to complex forms.

Histogram or frequency plot—a graphical tool used to understand variability. The chart is constructed with a block of data separated into five to twelve bars or sections from low number to high number. The vertical axis is the frequency and the horizontal axis is the "scale of characteristics." The finished chart resembles a bell if the data is in control.

Improvement cycle—a four-phase system for quality improvement: plan, observe and analyze, learn, and act.

Pareto chart—a simple bar graph with classifications along the horizontal and vertical axes. The vertical axis is usually the number of occurrences, cost, or time. The horizontal axis orders the bars from the most frequent to the least frequent.

Planned experimentation—a tool used to test and implement changes to a process (aimed at reducing variation) and to understand the causes of variation (process problems).

Run chart—a graphical record of a process variable measured over time.

Scatter plot—used to indicate relationships between two variables or pairs of data.

Statistical process control—statistical control methodology applied to a process.

14.2 Principles of Continuous Quality Improvement

This section discusses the technology that provides the foundation for quality improvement. Process technicians use this technology and are a valuable component of the continuous quality improvement team.

The principles of continuous quality improvement include:
- Innovate and improve services and products
- Innovate and improve processes
- Make a commitment to quality
- Integrate suppliers and customers into the quality process
- Use quality tools
    - statistical process control
    - flow charts
    - cause-and-effect diagrams (fishbone)
    - pareto charts
    - run charts
    - control charts
    - planned experimentation
    - histograms or frequency plots
    - forms for collecting data
    - scatter plots
- Audit and evaluate
- Provide continuous quality improvement training to all employees
- Make an unrelenting commitment to quality and involve all levels in the organization
- Document what you do, and do what you say

14.3 Quality Improvement Cycle

Phase 1: Plan
The first step in the improvement cycle is to increase current knowledge of the process. The more the team knows about the process, the more likely the changes submitted by the team affect quality. At the conclusion of Phase 1, a plan should be developed that addresses specific questions, and considers methods, resources, schedules, and people. Phase 1 takes a significant amount of time for the team to complete. The planning phase should address specific objectives and questions, predictions, and a plan for test.

Phase 2: Observe and Analyze
Phase 2 implements the data collection process. The data collected is used to address the questions from Phase 1.

Phase 3: Learn
This phase combines Phase 1 and Phase 2 activities. The results of the data analysis can be compared to current knowledge to see if contradictions exist.

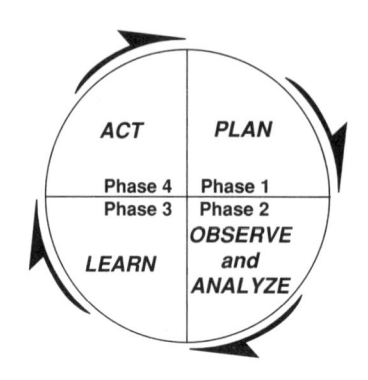

Figure 14.3-1 *Improvement Cycle*

Phase 4: Act
The results from Phase 3 are used to decide whether or not a change to the process is required. If a change is required, a modified brainstorming session should be conducted to determine what changes to the process would result in improvement. These changes should be stated clearly.

Figure 14.3–1 illustrates the improvement cycle.

14.4 Supplier-Customer Relationship

Industrial manufacturers buy raw materials from suppliers to produce products for their customers. Companies depend on suppliers to provide them with quality raw materials. Customers depend on companies to provide them with quality products. In today's global economy, a new relationship exists between suppliers, companies, and customers. Each is dependent on the other for financial success.

Companies are becoming more and more involved with customers and suppliers. Products and raw materials are tracked from inception. Documentation, quality charts, and external audits follow products and raw materials from cradle to grave. Customers are providing more information about their needs to companies.

14.5 Quality Tools

Process operators use a number of analytical tools to perform their jobs.

A list of quality tools includes:
- Flow charts
- Cause-and-effect diagrams (fishbone)
- Pareto charts

Chapter 14 • *Quality Control*

- Run charts
- Control charts
- Planned experimentation
- Histograms or frequency plots
- Forms for collecting data
- Scatter plots

14.6 Statistical Process Control

Statistical process control (SPC) is a quality tool based on the principles of statistical mathematics and applied to a process to control product quality. The theory behind SPC is based on some complex mathematics, but you do not need to be a mathematician to understand how to use the system.

In any process, a certain amount of variability can be found. Variability is defined as the tendency to vary. Process equipment does not heat up to 450.5 degrees F and stay at 450.5 degrees F. Instead, it tends to move a few degrees low and high. These variations can be found with temperature, pressure, flows, and levels. Each process has its own variability. SPC identifies this variability and enhances an operator's ability to control the process.

In the past, SPC operators would adjust their equipment based on current readings. If the reading is taken when the process's normal variation is in a low cycle, the operator's adjustment will swing the process variable high. Adjustments made without SPC methodology tend to make the process swing high and low.

Customers identify the key target setpoints for the products they require. The other process variables are identified in-house that support the ability of the process to produce the desired products: temperatures, flows, level, and pressure.

SPC allows the normal fluctuations on the equipment to be considered over a longer period of time. Sample results must demonstrate a downward or upward trend above the key target setpoints.

Product Directives
Most companies use a product directive or recipe for each product they produce. Product directives detail operating parameters and control points:
- Correct feed
- Temperature and pressure profile
- Equipment speed
- Additive setpoint
- Level and flow setpoint

Sample Types
Raw materials normally are sampled before and during a process run. However, when troubleshooting problems, it is advisable to catch a sample frequently.

Additives
Additives are sampled for purity and conformation to specifications when received at the warehouse. When troubleshooting control problems, additional samples are caught from containers being used at the time problems occur.

Products
Products are sampled for customer specifications. Sampling is performed to confirm that the process is in control, but sample results also can be a flag for unseen problems.

Typical Sample Tag
A blue tag is used for samples to be processed by the laboratory according to the usual procedures. A green tag indicates to the laboratory that this sample should be given high priority. Figure 14.6–1 shows sample tags.

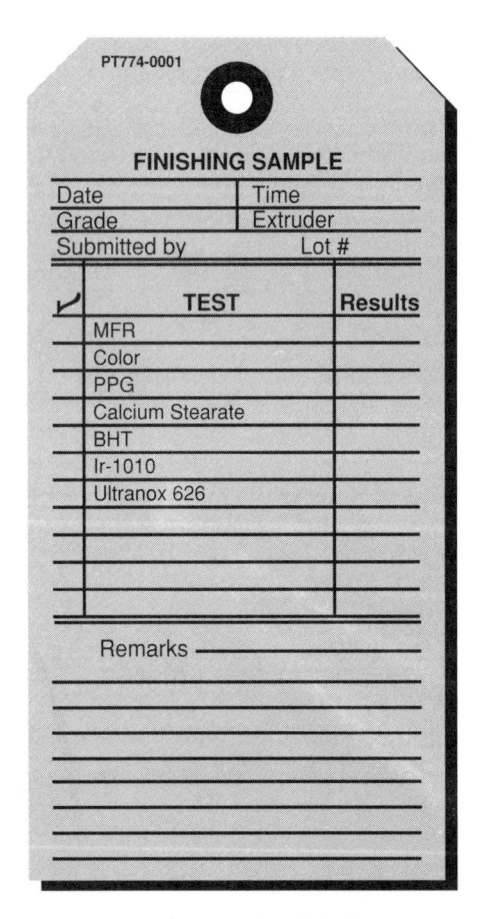

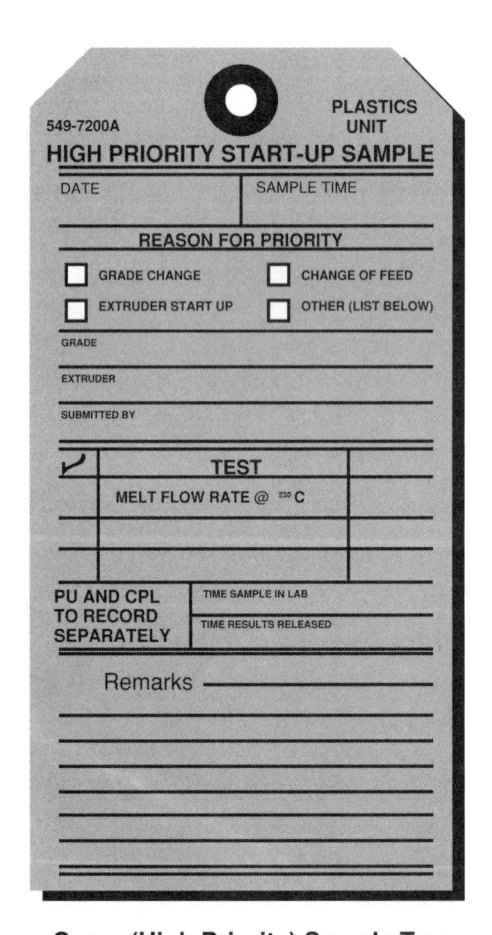

Blue (Normal) Sample Tag   Green (High Priority) Sample Tag

Figure 14.6-1 *Sample Tags*

Figure 14.6-2 description

EXTRUDER

GRADE	PP0001
DATE	2/20/98
LOT	99546789

TIME	16	18	20	22	0	2	4	6	8	10
VALUE	2.2	2.11	2.2	2.17						

QUALITY LIMITS: 2.5 / 1.8

- $3^r = 2.36$
- $2^r = 2.29$
- $r = 2.22$
- $\bar{x} = 2.15$
- $-r = 2.08$
- $-2^r = 2.01$
- $-3^r = 1.94$

Ⓡ 0.156 **KEY ADDITIVE- IR-1010**

VALUE	0.120	0.120	0.116	0.156	0.134

QUALITY LIMITS: 0.167 / 0.083

- $3^r = 0.143$
- $2^r = 0.137$
- $r = 0.131$
- $\bar{x} = 0.125$
- $-r = 0.119$
- $-2^r = 0.113$
- $-3^r = 0.107$

#1

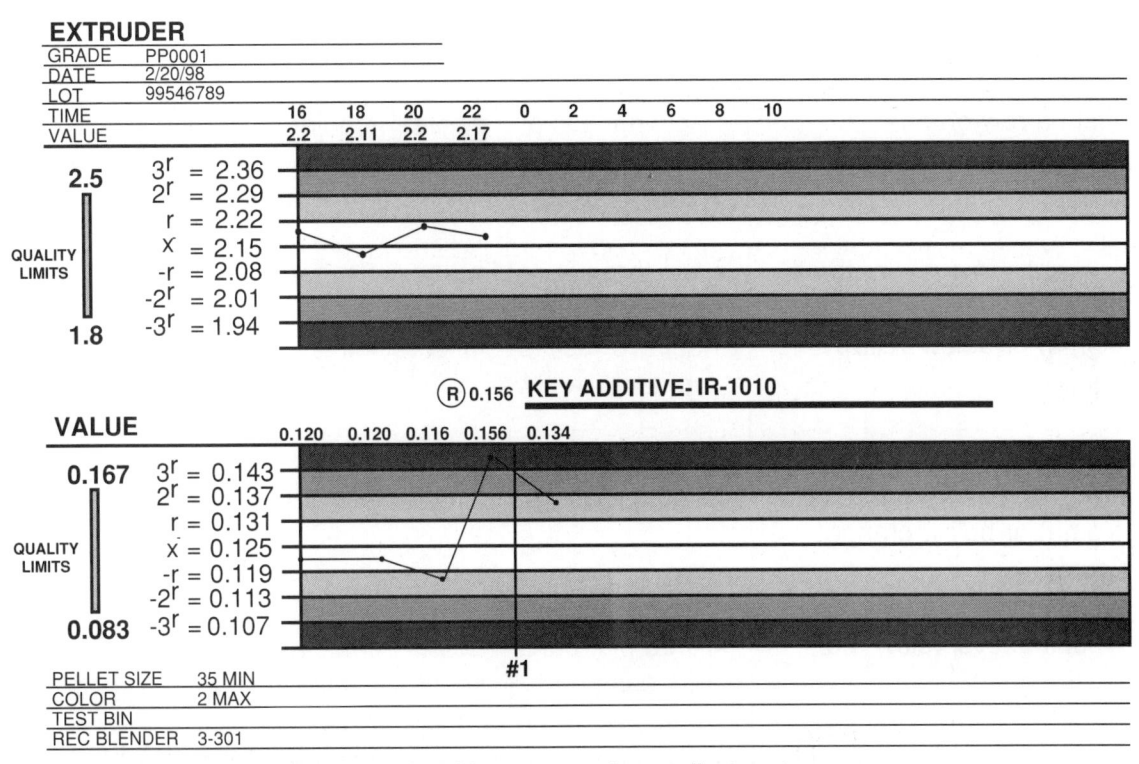

PELLET SIZE	35 MIN
COLOR	2 MAX
TEST BIN	
REC BLENDER	3-301

COMMENTS: #1 No apparent assignable cause; made no adjustment (possibly end of mix-higher concentration of IR-1010?)

COMPOSITES

DATE/TIME	RESULTS	
	MFR	ADDITIVE

Figure 14.6-2 *Quality Control Run Chart (Melt Flow Rate)*

Control Charts

Statistical process control (SPC) charts are used to plot quality parameter points from samples taken at different times during a run. All of the points may be on specification, yet when plotted on a graph, you may see quite clearly that there is a trend, which in time will result in off-specification material unless an adjustment is made. An upset or out-of-control situation is documented vividly on the chart in Figure 14.6–2.

SPC Guidelines

SPC guidelines account for normal process deviations and process upsets.

Evaluation	Action
Seven in a row on one side of the target.	Action is usually a small setpoint change. Usually indicates a process shift. Using the average of the last two results, adjust according to the product directive.
Three results in a row or three out of four results above yellow (+1 sigma) or below yellow (-1 sigma).	The three out of four case scenario keeps one isolated plot point from upsetting the process. Using the average of the last two results, make adjustments according to the product directive.
Two results in a row or two out of three results above orange (+2 sigma) or below orange (-2 sigma).	Using the average of the last two results, make adjustments according to the product directive.
One result above/below red (+3 or -3 sigma).	A red plot point requires immediate evaluation. Check process trends to see if a significant step change occurred. If a shift in the normal process is identified, make adjustments according to the product directive. If the change does not look reasonable (no visible change in operating conditions), resample and send a green tag to the laboratory. If the green tag result is in control, disregard the questionable result and follow normal procedures. If the resample confirms the previous result, take the appropriate action and resample after 30 minutes. If the green tag result is above quality limits, divert to off test until the process is back in control.
One result crosses 4 sigma lines.	A four sigma jump requires immediate investigation. Check process trends to see if a significant step change occurred. If a shift in the normal process is identified, make (+3 or -3 sigma) product directive adjustments. If the change does not look reasonable (no visible change in operating conditions) resample and send a green tag to the laboratory. If the green tag result is in control, disregard the questionable result and follow normal procedures. If the resample confirms the previous result, take the appropriate action and resample after 30 minutes. If the green tag result is above quality limits, divert to off test until the process is back in control.

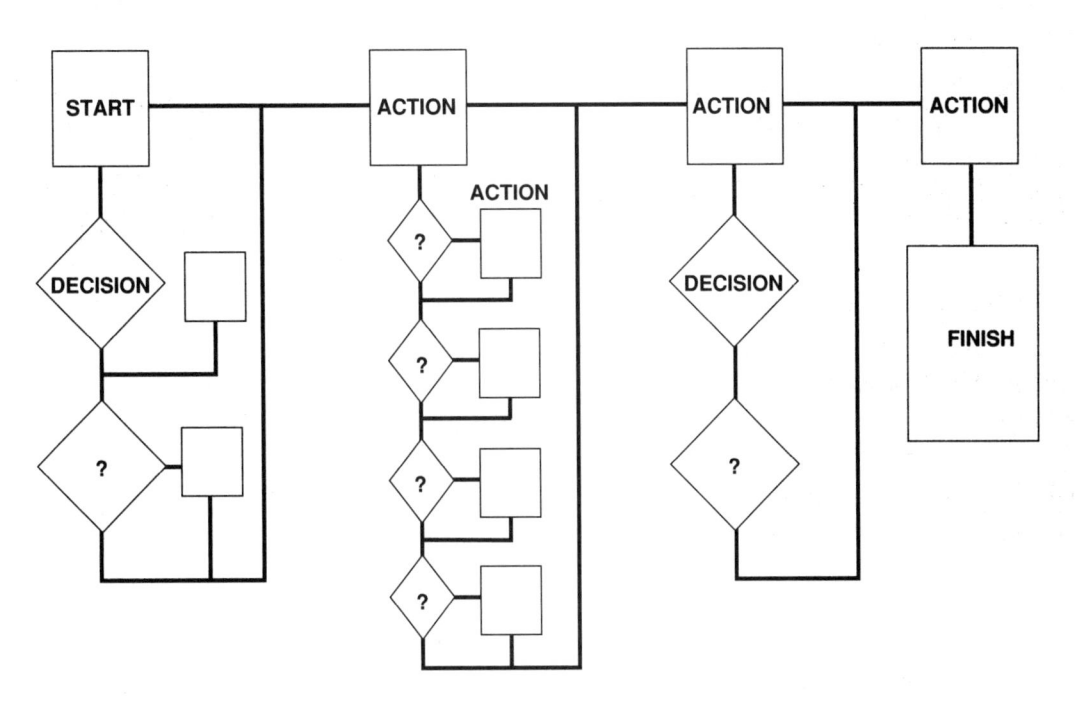

Figure 14.7-1 *Flow Chart*

14.7 Flow Charts

A flow chart is a picture of the key activities that take place in a process, Figure 14.7–1. Flow charts describe how the process actually works today. One of the common mistakes people make when flow charting is to add too much information to the chart. Flow charts should include action or step boxes and yes/no decision diamonds.

14.8 Run Charts

One of the most common tools used in industry is a run chart, Figure 14.8–1. Run charts are very powerful tools that show a graphical record of a process variable measured over time. The following steps should be used when building a run chart:
- Estimate the expected range of data points.
- Develop a vertical scale for the data that uses 50 to 70 percent of the overall range so the chart is not too narrow or too wide.
- Plot the data over time.

14.9 Cause-and-Effect Diagrams (Fishbone)

Another important quality tool is a cause-and-effect (C&E) diagram, also called fishbone, Figure 14.9–1. Cause and effect diagrams organize the causes of variation into general categories:

Quality Control • Chapter 14

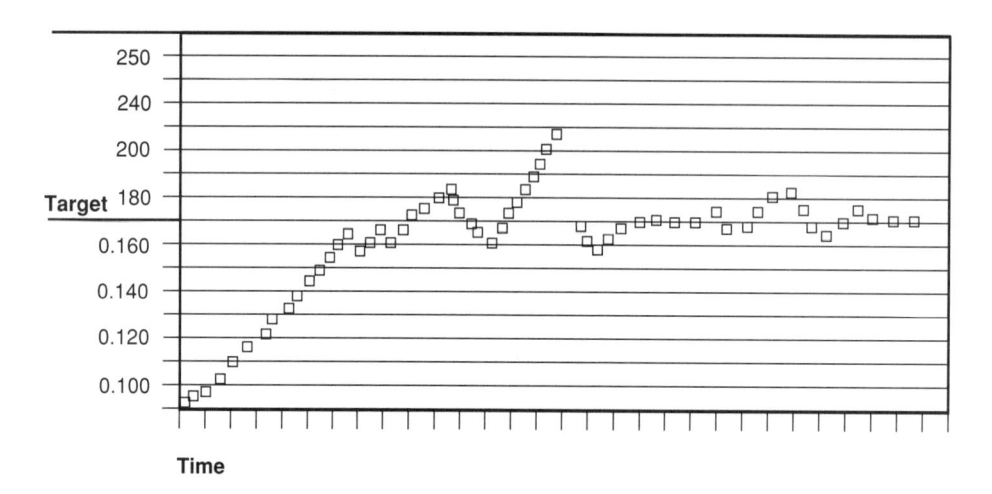

Figure 14.8-1 *Run Chart*

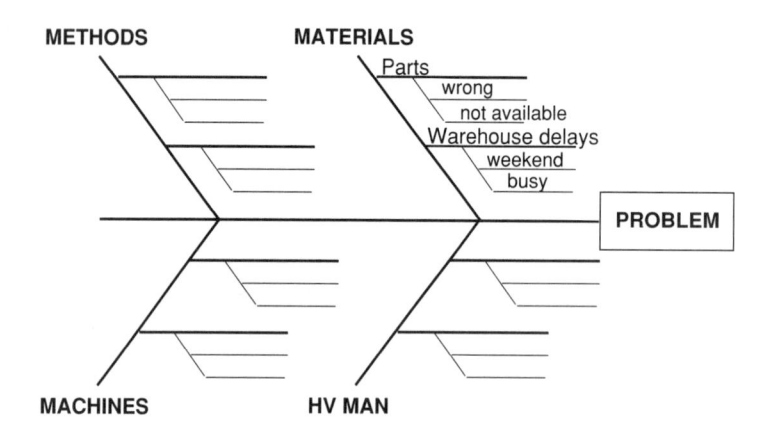

Figure 14.9-1 *Fishbone*

(1) method, (2) materials, (3) equipment, and (4) human. Each of these four sections summarize available knowledge about the causes of process variation. C&E diagrams were developed by Kaoru Ishikawa in 1943.

14.10 Pareto Charts

A pareto chart is a simple bar graph with classifications along the horizontal and vertical axes, Figure 14.10–1. The vertical axis is usually the number of occurrences, cost, or time. The horizontal axis orders the bars from the most frequent to the least frequent. The term *pareto* takes its name from a man named Vilfredo Pareto who pioneered income distribution studies.

Chapter 14 • *Quality Control*

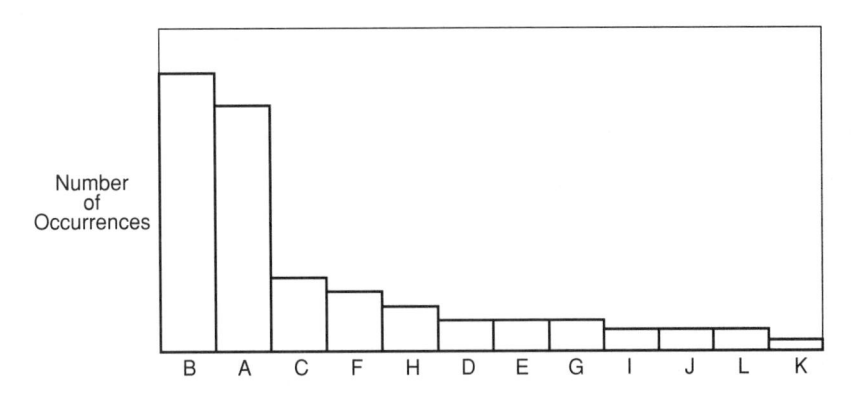

Figure 14.10-1 *Pareto Chart*

14.11 Planned Experimentation

Planned experimentation is a tool used to test and implement changes to a process (aimed at reducing variation) and understand the causes of variation (process problems). See Figure 14.11–1.

14.12 Histograms or Frequency Plots

Histograms or frequency plots are graphical tools used to understand variability, Figure 14.12–1. The chart is constructed with a block of data separated into five to twelve bars

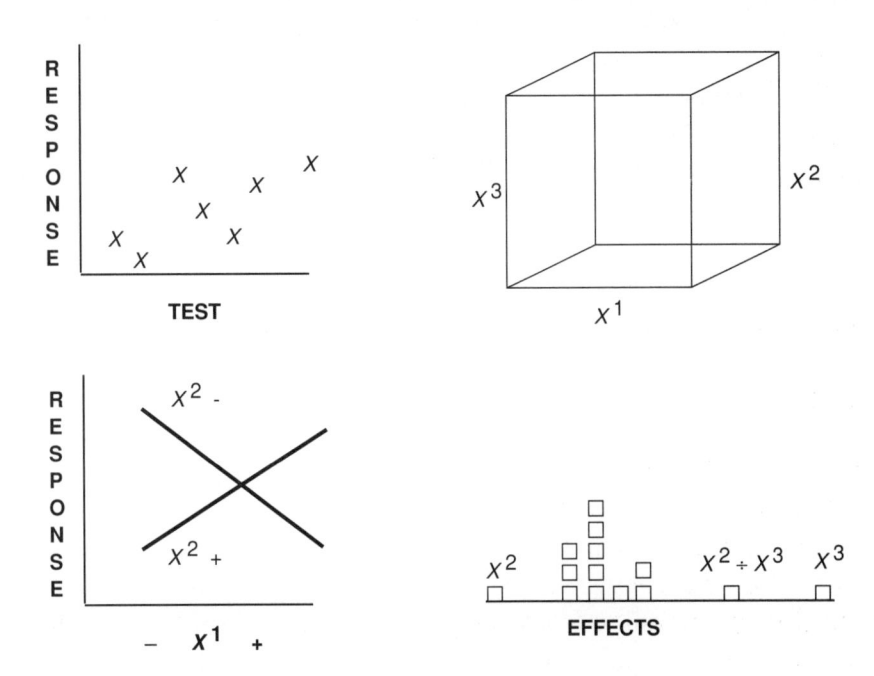

Figure 14.11-1 *Planned Experimentation*

269

Quality Control ● Chapter 14

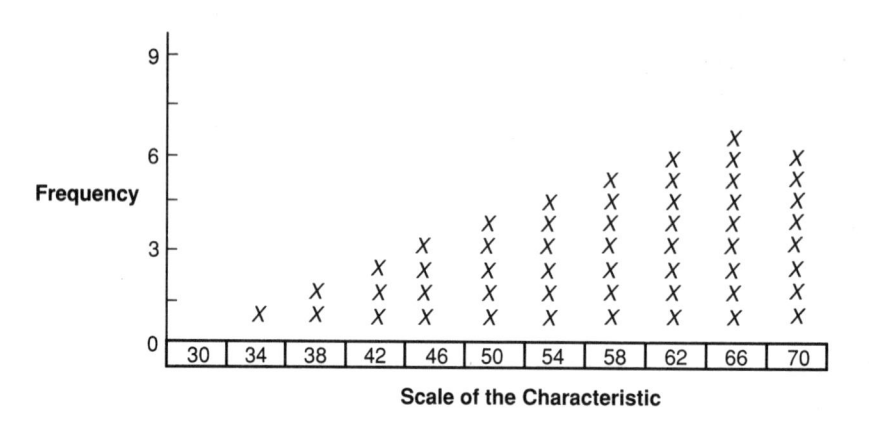

Figure 14.12-1 *Histograms*

or sections from low number to high number. The vertical axis is the frequency and the horizontal axis is the "scale of characteristics." The finished chart resembles a bell if the data is in control.

14.13 Forms for Collecting Data

The forms for collecting data can range from notes jotted down on a napkin to complex checklists. Forms are very helpful in collecting and organizing raw data. Most operators carry around small notebooks to record information collected during routine rounds. Figure 14.13–1 shows a sample form for collecting data

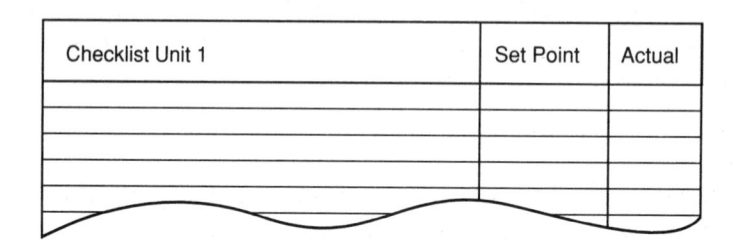

Figure 14.13-1 *Form for Collecting Data*

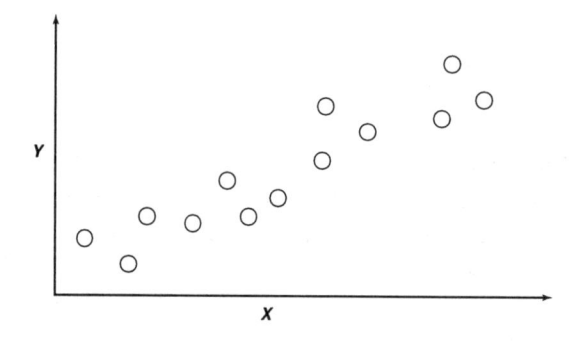

Figure 14.14-1 *Scatter Plots*

270

14.14 Scatter Plots

Scatter plots are used to indicate relationships between two variables or pairs of data, Figure 14.14–1.

Summary

The principles of continuous quality improvement include innovation and improvement of services, products, and processes; integration of suppliers and customers into the quality process; use of quality tool; audit and evaluation; unrelenting commitment and involvement of *all* levels in the organization; documentation of what you do; and do what you say.

Companies are becoming more and more involved with customers and suppliers. Products and raw materials are tracked from inception. Documentation, quality charts, and external audits follow products and raw materials from cradle to grave. Customers are providing more information about their needs to companies.

The four phases of quality improvement are plan, observe and analyze, learn, and act. The first step in the improvement cycle is to increase current knowledge of the process. The more the team knows about the process the more likely the changes submitted by the team will affect quality. At the conclusion of Phase 1, a plan should be developed that will address specific questions, and consider methods, resources, schedules, and people. Phase 1 takes a significant amount of time for the team to complete. Phase 2 implements the data collection process. In Phase 3, the results of the data analysis can be compared to current knowledge to see if contradictions occur. In Phase 4, the results are used to decide whether or not a change to the process is required.

Statistical process control (SPC) is a quality tool based on the principles of statistical mathematics. Process technicians use quality tools during normal operations.

Statistical process control charts are used to plot quality parameter points from samples taken at different times during a run.

A flow chart is a picture of the key activities that take place in a process. Flow charts describe how the process actually works today.

Run charts are very powerful tools that show a graphical record of a process variable measured over time. To build a run chart, estimate the expected range of data points, develop a vertical scale for the data that uses 50 to 70 percent of the overall range so the chart is not too narrow or too wide, and plot the data over time.

Chapter 14

Review Questions

1. List five quality tools.

2. Describe the improvement cycle.

3. What type of quality chart is a picture of activities that take place in a process?

4. The principles of quality improvement include all of the following except
   a. use quality tools
   b. contingency perspective
   c. audits and evaluations
   d. innovation and improvement of services and products

5. Which of the following is *not* a quality chart?
   a. control
   b. flow
   c. pareto
   d. gant

6. What do the initials SPC stand for?

7. Which chart uses quality methodology to control a process?
   a. flow
   b. pareto
   c. run
   d. SPC

8. Name four charts that work with process variation.

9. What is the other name for a fishbone chart?

10. Planned experimentation is
    a. a tool used to test and implement changes in a process.
    b. a tool designed to reduce variation.
    c. a tool designed to help operators understand process variability.
    d. all of the above.

Chapter 15

Self-Directed Job Search

OBJECTIVES

After studying this chapter, the student will be able to:

- *Explain how to conduct a successful job search.*
- *Write an effective cover letter.*
- *Write an effective process technology resume.*
- *Obtain job lists from local chamber of commerce.*

KEY TERMS

Job lists—these lists contain contact name, address, phone number, and size of company. Can be obtained from local chamber of commerce for a small fee.

Pre-employment tests—Examples include: Bennett Mechanical Comprehension Test (BMCT) by George K. Bennett, S & T version; The Richardson, Bellows, Henry & Company "Test of Chemical Comprehension" S & T version, 1970; and California Math Test. Types include reading comprehension, checking for accuracy, block counting, and in-house developed tests.

Resume—a one-page document designed to sum up a job applicant's skills, work history, hobbies, and education.

Successful job search—requires 4 to 6 months, a good resume and cover letter, a certificate or degree, good investigative skills to identify who is hiring and who to contact, application method, interest cards, tests, etc. Job searches are very difficult and require serious dedication, time, and a "thick skin."

15.2 The Job Search

Understanding the job market is very important for conducting a successful job search. Although statistics indicate that a large number of process technician positions will become available over the next ten years, the chemical processing industry is very cynical in nature. Employment opportunities may surface only once a year. Each company has different hiring needs and different mechanisms for granting an interview. It is the responsibility of each job applicant to become familiar with the hiring practices of the companies in their area. A list of these companies can be obtained at the chamber of commerce for a small fee. A phone call to the company can also provide an applicant with invaluable information.

As a new job applicant you will be competing against a large number of people for a select few jobs. Only a small fraction of process technician job seekers are qualified to work in the chemical processing industry. A student that has a one-year certificate in process technology is a much better prepared job applicant. Graduates from these programs have statistically proven that they complete mandatory training and job post training quicker and have a very low drop-out rate after starting a new job. The truth is that it is very difficult to compete with a new graduate if he or she can find their way through the unqualified candidates and into the interview.

Job Market Facts

Job lists can be picked up at your local chamber of commerce.
Hiring practices are cyclical.
A job search can take 4 to 6 months.
The majority of process technician job applicants are not qualified for the job.
Each company has its own prospective employee hiring practice.
Pre-employment testing procedures vary between companies.
The Process Technician degree is valuable.
Finding a job is your responsibility.
Networking finds more job opportunities.
Newspaper ads have poor placement rates.
A good resume and cover letter are important.

Since you have enrolled in the process technology degree program there are a number of things you should know in order to position yourself as the top candidate. To be a top candidate you will need to be better at searching for a job than other people. High GPAs and test scores do not indicate how successful you will be in a job search. The truth is, most people are not very good at looking for work. There are a number of critical elements that determine job search success:
- Job searching is your responsibility.
    - Do not believe that someone else will find you a job.
- Develop a job search plan
- Narrow the search
    - Houston area, Salt Lake City, near your home

- Interview
    - This is a numbers game. The more interviews the better your chance.
- Networking
    - 75% of all jobs come from networking.
- Newspaper ads (poor)
- Placement agencies
    - Certified Personnel; Manpower, Inc; Skillmaster; Kelly Scientific Services; Allstates Personnel; Staffing Professionals; Job Placement-College, etc.
- Become a private investigator
    - Contact employers
    - Call hot lines
    - Write sales letters
    - Visit HR departments
    - Find out hiring procedure
- Write a good resume
- Develop good interview skills
    - Research the company
    - Make a positive first impression: solid handshake, dress appropriately
    - State your strengths, develop chemistry
    - Practice answering hard questions
- Follow-up

Cover Letter and Resume

A successful job search requires a cover letter and a resume. The resume is a summary of your life experiences and should be presented in a positive manner. A variety of formats are available and should be selected based upon your individual preference. It is important for you to write your own resume. Do not have a friend or relative write things about you that may be difficult to explain in an interview. The resume is designed before the cover letter. The cover letter introduces you to the company and sums up your resume.

Elements of a cover letter include:
- Your address
- Date
- Name and address of the person you are contacting.
- Greeting
- Paragraph #1: Briefly explain why you are writing and how you found out about the company.
- Paragraph #2-3: Describe how your education and skills benefit the organization.
- Last Paragraph: Request a reply; tell them how they can reach you.
- Complimentary closing

Elements of a resume include:
- Your name
- Address
- Phone number(s)
- PTEC education
- Work experience
- Reference

During the selection process a job placement officer sees hundreds of resumes. Since recruiting is such an important feature of good business, most companies take it very seriously. Job interviewers carefully screen for the best applicants. In a job search, aggressive, professional marketing goes a long way. The next step can be summed up as, "it's how you package the product." A well-written, clean resume and cover letter can go a long way.

During the selection process, resumes and job applications are typically separated into three stacks: AAS degree—process technology, one-year certificate, and uneducated. A company may use a pre-employment test to select which resumes will be retained. Process experience is another important variable used to determine second phase recruiting or who will get an interview.

Companies select employees based upon a wide array of needs. Attrition rates and new plant expansions create opportunities for process technicians. Companies will attempt to hire the best possible candidate based upon company needs, EEO requirements, and personal relationships.

Typical questions asked by employers include:
- Tell us about yourself. Why should we hire you?
- Why do you want to leave your present position?
- Why did you leave your last job?
- Tell me what you have learned in your college classes.
- Tell me about the different types of pumps you have studied.
- What is distillation? What is reflux and what is its purpose?
- List the elements of a control loop.
- Do you have any hands-on experience?
- Tell me how to put a distillation system on-line.
- What do you know about our company?
- What are your personal goals in this job for the next year? The next five years? The next ten years?
- How do you handle stress?
- If you were asked to perform an unsafe act by your supervisor, how would you respond?
- Why did you choose to be a process technician?
- Do you have plans to continue your education?

Questions asked by job applicants include:
- Can you tell me about your safety program.
- What specific responsibilities of the position do you consider the most important?
- How are process technicians evaluated at your company?
- What would you expect me to accomplish during my first six months? Year? Two years?
- What long and short term problems will I face as a new technician at your company?

Suggestions for job search by process technology students
- Get one-year certificate in process technology.
- Get two-year degree in process technology.
- Prepare for pre-employment tests, sign up for placement agency tests.
- Improve skills: math, reading comprehension, communications.
- Develop a network: friends, instructors; get references; gather documents; prepare for interviews; research companies; develop job search plan.
- Prepare a resume, target companies you want to work for, mail out resumes, follow-up on job leads, get jobline numbers.
- Dress appropriately: jeans, long sleeve shirt, work boots, little make-up, tone down jewelry.
- Be accessible—allocate time, get an answering machine.
- Sign up with your state's work force commission, placement agencies.
- Identify strengths and weaknesses.
- Don't get discouraged; deal with stress.
- Go to chamber of commerce and get job lists.
- Go to companies and develop relationships.
- Check out WWW.TWC.STATE.US (Process internet for Texas).
- Improve your appearance.

15.3 Pre-Employment Testing

At the present time a number of tests are given to prospective job applicants. The most common type of tests include mechanical aptitude, chemical comprehension, reading comprehension, basic math, psychological, and block-counting. Some companies spend thousands of dollars developing their own test while others use standardized exams. Mechanical aptitude tests are administered frequently to individuals wishing to work in the chemical processing industry. Mechanical aptitude deals with which way an object will move when influenced by an outside set of forces. The most common form is the Bennett Mechanical Comprehension Test (BMCT) by George K. Bennett. The BMCT has an S and T format that includes sixty-eight questions. Another mechanical aptitude test that can be purchased at local bookstores is the ARCO Mechanical Aptitude and Spatial Relationship book. There are a number of good texts that will help new technicians improve their ability to work out mechanical aptitude problems.

In 1970, Richardson, Bellows, Henry & Company developed the "Test of Chemical Comprehension." This test comes in a 50 question, S and T format. Questions on the test were developed from information that should be learned in a high school science class. Most pre-employment tests have a math section that covers addition, subtraction, multiplication, division of fractions and whole numbers, decimals, averaging, percentages, and low level algebra. The block-counting tests ask the job applicant to look at a three dimensional drawing and identify the total number of blocks. Since some of the blocks are hidden, this test can be tricky. Carelessly rushing through this section of the test is a mistake. The block counting test is designed to screen for observational accuracy. Reading comprehension is a common testing practice that screens for how well a technician can read a paragraph or two and then answer or respond to specific instructions or questions. People who read and comprehend quickly should not be concerned about this type of testing. If you like to read instructions carefully before you respond you will need to develop a system that increases your speed since all of these tests are timed.

15.4 Sample Cover Letter and Resume

Cover Letter Example

75 East Payton St.
Houston, TX 77409
May 25, 2000

Mr. S. Bigg Jr.
Human Resource Manager
Bigg Chemical Company
Pasadena, TX 77409

Dear Mr. Bigg:

I am writing in response to your classified ad for a process technician placed in the Houston Chronicle on May 4. Enclosed you will find my resume which describes my educational background and work experience.

On May 15, 1999 I graduated with an Associate of Applied Science degree in process technology from Process Technical College. As my enclosed resume indicates I have taken courses that have prepared me to take an entry level position as a process technician at your company.

I look forward to meeting with you for an interview. If you have any additional questions please call 555-456-1234 after 3:00 p.m.

Sincerely,

Louis Hicks

Self-Directed Job Search • Chapter 15

Resume Example

LOUIS HICKS   75 East Payton St., Houston, TX 77409   555-456-1234

OBJECTIVE   To obtain an entry level position as a process technician.

EDUCATION   In 1999 I graduated from technical college with an Associate of Applied Science (AAS) degree in Process Technology. My course of studies included the operation and maintenance of a full scale pilot plant, modern process control, process equipment and systems instrumentation, chemistry, math and physics. Additional topics of study include safety, quality control, troubleshooting and the academic core.

EMPLOYMENT   True Value Hardware, Baytown, Texas
06/98-Present   Supervisor: Jane Johnson 555-425-1234
Performed general maintenance, stocking, sales.

REFERENCES   My Favorite Instructor—Process Technology
Process Technical College
P.O.Box 848, Baytown, TX 77522-0818

15.5 Work Experience

For many years industrial employers have valued prospective employees with industrial experience. Some companies require five years experience before initiating the interview process. Industrial experience provides a track record for a person's stability, ability to work rotating shifts, and exposure to industrial processes and the environment.

At this point the argument can be made that experience or exposure to all industrial processes is impossible to obtain since over 40 petrochemical processes and 19 refinery processes can be identified. These lists do not include the gas processes. The only common thread between these facilities is the equipment and technology used in different arrangements. Prospective employees with strong science and math backgrounds, good mechanical aptitude, troubleshooting skills, and a process technology degree provide prospective employers with an informed trainee prepared to start site specific training.

Experience does not carry the weight that it used to. Since 1989 the government and industrial manufacturers have been raising the bar for process technicians. Displaced process operators are being required to go back to school and obtain a certificate or degree before they can return to their occupation. This added level of protection is designed to protect the process technician,

community, and industrial manufacturers from the rapid advances in technology. Experienced technicians have a number of negative issues to address. Some of these concerns include:
- Why did you leave your last job?
- Bad habits?
- Trained incorrectly?
- Do you have a two-year degree in process technology?
- What industrial processes have you been exposed to?
- Did you complete a Department of Labor approved apprentice training program?

Summary

A successful job search requires 4 to 6 months, a good resume and cover letter, a certificate or degree, good investigative skills to identify who is hiring, who to contact, application method, interest cards, tests, etc. Job searches are very difficult and require serious dedication, time, and a "thick skin."

At the present time a variety of pre-employment tests are being offered by the chemical processing industry including the Bennett Mechanical Comprehension Test (BMCT) by George K. Bennett. The BMCT has an S and T format that includes 68 questions. Another mechanical aptitude test that can be purchased at local bookstores is the ARCO Mechanical Aptitude and Spatial Relationship book. In 1970, Richardson, Bellows, Henry & Company developed the "Test of Chemical Comprehension." This test comes in a 50 question, S & T format.

Chapter 15

Review Questions

1. List the key elements of a resume.

2. Write a resume.

3. List the key elements of a cover letter.

4. Write a cover letter.

5. Tell us about yourself. Why should we hire you?

6. Why do you want to leave your present position?

7. Why did you leave your last job?

8. Tell me what you have learned at Process Technical College.

9. Tell me about the different types of pumps you have studied.

10. What is distillation? What is reflux and what is its purpose?

11. List the elements of a control loop.

12. Do you have any hands-on experience?

13. Tell me how to put a distillation system on-line.

14. How do you find information about a company?

15. What are your personal goals in this job for the next year? The next five years? The next ten years?

16. How do you handle stress?

17. If you were asked to perform an unsafe act by your supervisor, how would you respond?

18. Why did you choose to be a process technician?

19. Do you have plans to continue your education?

20. Select an ad from the paper and apply for a process job.

Glossary

Absorber—a device used to remove selected components from a gas stream by contacting it with a gas or liquid.

Adsorber—a device, such as a reactor or dryer, filled with a porous solid designed to remove gases and liquids from a mixture.

Aerator—a device used to stir up and add oxygen to a microbiological system.

Atom—the smallest particle of a chemical element that still retains the properties of an element. An atom is composed of protons and neutrons in a central nucleus surrounded by electrons. Nearly all of an atom's mass is located in the nucleus.

Boilers—provide steam for industrial applications.

Career portfolio—a collection of college notes, tests, articles, technical assignments, textbooks, graphics, transcripts, certificates, letters, and awards. This information can be used to explain to a prospective employer the curriculum you have completed and to refresh your memory on key topics asked during job interviews.

Chemistry—the science and laws that deal with the characteristics or structure of elements and the changes that take place when they combine to form other substances.

Compound—a substance formed by the chemical combination of two or more substances in definite proportions by weight.

Compressor—similar to a pump in design and operation. The primary difference between them is that pumps move liquids while compressors move gases. Compressors come in two basic designs: (1) positive displacement (rotary and reciprocating) and (2) dynamic (axial and centrifugal). A compressor is designed to accelerate or compress gases.

Cooling towers—consist of a box-shaped collection of multilayered wooden slats and louvers that direct airflow and break up water as it falls from the top of the tower or water distribution header.

Cyclone—a device used to remove solids from a gas stream.

Demineralizer—a filtering-type device that removes dissolved substances from a fluid.

Density—equals the mass of any substance in grams divided by the volume of the substance in milliliters.

Distillation towers—a series of stills arranged so the vapor and liquid products from each tray flow counter-currently to each other.

Electron—a negatively charged particle that orbits the nucleus of an atom.

Element—composed of identical atoms.

Equipment and technology—a term used to describe the basic equipment found in the chemical processing industry and the technology associated with the operation of that equipment.

Extruder—a complex piece of equipment composed of a heated jacket, a set of screws or a screw, a heated die, large motor, gear box, and a pelletizer. An extruder converts raw plastic material into pelletized plastics ready for further processing into finished products.

Filter—device that removes solids from fluids.

Glossary

Fired heaters—consist of a battery of tubes that pass through a firebox. Fired heaters or furnaces are commercially used to heat large volumes of crude oil or hydrocarbons.

Flare—safely burns excess hydrocarbons. A flare system is composed of a flare, knock-out drum, flare header, fan optional, steam line and steam ring, fuel line, and burner.

Fluid—anything (gas or a liquid) that flows from one area to another.

Fluid flow—characterized by fluid particle movements such as laminar and turbulent. Fluids assume the shape of the container they occupy. A fluid can be classified as a liquid or a gas. When a liquid is in motion it will remain in motion until it reaches its own level or is stopped.

Fluid pressure—the pressure exerted by a confined fluid. Fluid pressure is exerted equally and perpendicularly to all surfaces confining it.

Heat—a form of energy caused by increased molecular activity. A basic principle of heat states that it cannot be created or destroyed, only transferred from one substance to another.

Heat exchangers—transfers energy in the form of heat between two fluids without physically coming into contact with each other. A typical heat exchanger is composed of a battery of tubes surrounded by a shell. A heat exchanger can heat or cool a fluid (liquid or gas). Common names include condenser, intercooler, aftercooler, reboiler, and preheater.

Heat transfer—heat is transmitted through conduction (heat energy is transferred through a solid object; e.g., a heat exchanger); convection requires fluid currents to transfer heat from a heat source; e.g., the convection section of furnace or economizer section of boiler; radiation (the transfer of energy through space by the means of electromagnetic waves; e.g., the sun).

Hydrocarbons—a class of chemical compounds that contain hydrogen and carbon.

Incinerator—a permitted device used to burn industrial wastes.

Inside operator—(board operator or console operator)—typically senior technicians who have a good understanding of the outside unit. The inside operator is responsible for controlling and monitoring unit functions from a computer console and working with an outside operator.

Matter—anything that occupies space and has mass.

Mixture—composed of two or more substances that are only physically mixed. Mixtures can be separated through physical means, such as boiling or magnetic attraction.

Neutron—a neutral particle in the nucleus of an atom.

Organic chemistry—the scientific study of substances that contain carbon.

Outside operator—works with the inside operator to control plant functions. Outside operators make rounds, monitor equipment, make line-ups, and report all activities to the inside person. The term "outside" indicates that most of this person's responsibilities are outside the control room area.

Periodic table—provides information about all known elements: atomic mass, symbol, atomic number, boiling point, etc.

pH—a measurement system used to determine the acidity or alkalinity of a solution.

Pressure—force or weight per unit area. (Force ÷ Area = Pressure) Pressure is measured in pounds per square inch. Pressure is directly proportional to height; the higher the atmosphere, gas or liquid, the greater the pressure. Atmospheric pressure is produced by the weight of the atmosphere as it presses down on an object resting on the surface of the Earth. Atmospheric pressure at sea level is 14.7 psi.

Process instruments—devices that control processes and provide information about pressure, temperature, levels, and flows.

Process technician responsibilities—inside operators; work mainly inside a control room monitoring and controlling process variables, carefully watching and interfacing with computer screens and control instruments, filling out unit log books and quality charts, and working with the outside operator. Outside operators: make rounds, inspect equipment, make relief, perform shift tasks, perform unit start-ups and shutdowns, troubleshoot problems, perform routine housekeeping and maintenance, catch readings, samples, and respond to emergency situations.

Proton—a positively charged particle in the nucleus of an atom.

Pumps—used primarily to move liquids from one place to another. Pumps come in two basic designs: (1) positive displacement (rotary and reciprocating) and (2) centrifugal. Centrifugal pumps displace liquid by centrifugal force while positive displacement pumps displace fluids positively.

Reactor—a device used to combine raw materials, heat, pressure, and catalysts in the right proportions. Reactors are classified as either batch or continuous and as fixed bed or fluidized bed design. Industrial manufacturers use reactor technology in alkylation, fluid catalytic cracking, hydrodesulfurization, hydrocracking, VNB synthesis, coal gasification, isomerization, and fluid coking. The shape and design of a reactor will depend on the process it is to be used in.

Scrubber—device used to remove chemicals and solids from process gases.

Solution—homogenous mixture.

Specific gravity—equals the mass of a substance divided by the mass of an equal volume of water. When metric units are used, specific gravity has the same numerical value as density.

Temperature—the hotness or coldness of a substance.

Steam turbines—used as a driver to turn pumps, compressors, electric generators and propeller shafts on naval vessels. A steam turbine is an energy conversion device that converts steam energy (kinetic energy) to useful mechanical energy. Steam turbines come in two basic designs: (1) condensing and (2) noncondensing.

Steam trap—a device used to remove condensate from steam systems.

Strainer—a device used to remove solids from a process before they can enter a pump and damage it.

Tanks and pipes—equipment that stores and contains fluids. Tank designs vary depending upon the service. Piping designs are carefully applied to specific processes. Pipe size and design determine flow rates, pump and valve sizes, turbulent or laminar flow, instrument type, and automation. Tank and pipe designs must be suitable to the process that will run through them.

Valve—a device used to control (start, stop, or direct) the flow of fluids. Valves come in a variety of shapes, sizes, and designs that throttle, stop, or start flow. Modern manufacturing plants contain a variety of automatic and manually operated valves. Typical valves found in the workplace include gate, globe, ball, plug, butterfly, diaphragm, check, solenoid, pressure relief, three way, and automatic.

Index

A

Abbott, William, first oil refinery and, 5, 24
Above ground vessels, 94–95
Absolute (PSIA) gauge, 68
Absorbed heat, effects of, 69
Absorber,
    column, 118
    defined, 110
Absorption, defined, 202
AC (Alternating current)
    defined, 102
    motor, 104
Acid, defined, 202
ACS (American Chemical Society), 30–31
    work force development and,, 32
Actual weight, 216–217
Actuator
    design of, 145
    electrically operated, 145
Addition, defined, 62
Additives, described, 264
Adhesion, defined, 236, 244
Administrative support staff, 23
Adsorber, defined, 110
Adsorption, 119
    defined, 202
Aeroators, defined, 110
A-frame direct fired furnace, 115
Air, density of, 72, 233
Air permits, 252
    defined, 250
Air pollution
    control, 251–252
    defined, 250, 255
Air Pollution Control Board, 252
Air-purifying respirator, 52
    defined, 44
Air-supplying respirator, 52
    defined, 44
Algebra, defined, 62
Alkanes, 219–220
Alkylation, 223
    defined, 170
    described, 179
Allergic response, described, 51
Alternating current (AC)
    defined, 102
    motor, 104

American Chemical Society (ACS), 30–31
American Petroleum Institute (API), gravity standards and, 72
AMU (Atomic mass unit), defined, 202
Anesthetic, described, 51
Anion, defined, 202
API gravity, 72, 234
    defined, 81, 230
Applied chemistry, 39
Applied learning, described, 19
Archimedes' principle, 232
ARCO, 34
ARCO Mechanical Aptitude and Spatial Relationship book, 278
Asphalt, development of, 7
Asphyxiation, described, 50
Atmospheric pressure, defined, 64, 80
Atomic mass unit (AMU), defined, 202
Atomic number, 195
    defined, 206, 224
Atoms, described, 202, 205–206, 224
Attorney, patent, 24
Automated valves, categories of, 145
Automatic valves, 92–93
    control loops and, 145
Axial
    compressor, 99
    pumps, 96–97

B

Balanced equation, defined, 202
Ball valves, 89, 90
    valves, 90
Barnsdall, William, first oil refinery and, 5, 24
Base, defined, 202
Batch process, history of, 5–7
Baume gravity, 72, 234
    defined, 81, 230
Bearings, 105
Benchtop operations, 195
Bennett Mechanical Comprehension Test (BMCT), 278
Benzene process, 173, 174
Bernoulli's principle, 71
    defined, 62, 81
Bhopal, 34
Blending section, plastic, 160
BMCT (Bennett Mechanical

Comprehension Test), 278
Boilers, 113–115, 160–161
    defined, 110
    energy conversion and, 73
    symbols for, 136
Boiling point
    defined, 63
    impact of pressure on, 64
Box direct fired furnace, 115
Boyle, Robert, 64–65, 80
Boyle's law, 64–65
    defined, 62
BTX aromatics, 173–174
Burton, William, thermal cracking and, 7–8
Burton process, 7–8
Butterfly valves, 90

C

Cabin direct fired furnace, 115
CAER (Community Awareness and Emergency Response), defined, 250
Calendering, described, 160
Capillary action, defined, 236
Carcinogens, described, 50
Careers, in chemical processing industry, 22–24
Casting, described, 160
Catalyst, 215–216
    defined, 202
Catalytic cracking, 222
    defined, 2
    history of, 8–9
    see also Catcracking
Catalytic reforming, 183
    defined, 184
Catcracker, defined, 170
Catcracking, 180, 222
    defined, 202
    see also Catalytic cracking
Cations, defined, 202
Cause-and-effect diagrams (fishbone), 191, 260, 267–268
Celsius, heat measurement system, 70
Centrifugal compressors, 99
    symbols for, 134
Centrifugal pumps, 96–97
    symbols for, 133
Check valves, 90, 91
Chemical bond

Index

covalent, 202, 206
ionic, 202, 206
Chemical equations, 210–213
  defined, 203
  described, 207–210
Chemical hazard communication program, 48, 50
Chemical processing industry
  applied concepts to, 221–223
  careers in, 22–24
  defined, 28
  history of, 3–11
  preparation/basic skills needed, 11–14
Chemical reaction, 213–216
  combustion, 204, 214
  defined, 203
  endothermic, 204, 213
  exothermic, 213
  neutralization, 205, 214
  replacement, 205, 213–214
Chemicals
  health hazards associated with, 50
  safe handling of hazardous, 50
Chemistry, 203
  applied, 39
  described, 223
  principles of, 205–207
Chemists, 23
Class A fire, 54
Class B fire, 54
Class C fire, 54
Class D fire, 54
Classifier
  described, 121
  illustrated, 122
Clayton, John, 4
Clean Air Act, defined, 250, 251
Clean Water Act of 1972, defined, 251
Clerical assistant, 23
CO_2 extinguisher, 54
Cohesive force, temperature and, 236–237
College
  programs, 14–19
  succeeding in, 17
  system, understanding, 18
  transition to, 16–17
Combining processes, defined, 203
Combustible liquid, described, 50
Combustion
  defined, 204
  described, 214
  reactions, 225
Communication, as required skill, 12
Community Awareness and Community Response (CAER), defined, 250
Community right-to-know, 254–255
  defined, 250, 255–256
Compounds, defined, 203, 207, 224
Compressed gas, described, 50
Compressors, 99, 101
  defined, 86
  symbols for, 134
  system, 155
Computer science analyst, 24
Computer technology, as required skill, 12
Condensation, latent heat of, 69
Conduction, of heat, 69, 81
Contact engineer, 23
Continuous quality improvement, 261, 271
Control charts, 193, 265
  defined, 260
Controllers
  control modes and, 143–144
  defined, 140
  modes of, 143–144
  purpose of, 143
  transmitters and, 142
Controllers, proportional band of, 144
Control loop
  defined, 126, 139–140, 151
  elements of, 139–140
  process variables and, 140–141
  symbols for, 139
  transmitters and, 142–143
Convection, of heat, 69, 81
Conversion
  process, 10
  tables, defined, 62
Cooling towers, 112–113, 214
  defined, 110
  heat exchangers and, 113
  heat exchanger system, 154, 156
  symbols for, 136
  described, 134–135
Corrosive, described, 51
Cover letter, 279–280
Critical job functions, 33
Crosby, Phillip B., 190, 199
Crude oil, distillation, 183–184, 208
Crystallization, defined, 203
Cultural diversity, 20–21
Cyclone, defined, 110
Cylindrical direct fired furnace, 115

D

Dalton's law, 68
  defined, 62, 80
Data collection, forms for, 260, 270
DC (Direct current), defined, 103
Decimal point, defined, 62
Demineralizer, defined, 110
Deming, W. Edward, 35, 189, 199, 260
Denominator, defined, 62
Density
  calculating, 72, 233
  defined, 203, 230
  expressing fluid's heaviness with, 234
  fluid flow and, 71
  of fluids, 81, 243
Department of Transportation (DOT)
  defined, 44
  hazardous material shipment and, 56
Design engineer, 23
Diagrams
  flow, 146
  equipment relationships and, 155
  process, 127–128
  section flow, 128
Diaphragm valve, 91–92
Dimensional analysis, defined, 62
Diaphragm actuator, 145
Direct current (DC), defined, 103
Direct fired furnace, 115
Displacement, 233
Distillation, 116–120
  crude oil, 183–184, 208
  defined, 117, 161, 221
  described, 221–222
  system, 154, 161–163
Distillation column, 116–117, 133
  defined, 110
  symbols for, 137
Distillation tower, 221–222
  classification of, 118
  defined, 170
  energy conversion and, 73
Diversity, cultural, 20–21
Diversity training, defined, 2
Division, defined, 62
Divisor, defined, 62
DOT (Department of Transportation)
  defined, 44
  hazardous material shipment and, 56
  labels/signs/placards, 48, 57
Double-pipe heat exchanger, 111
Drake, Edwin L., 5, 24
Drawings
  electrical, 126, 148–149
  elevation, 126, 148–149
  equipment location, 126, 149
  foundation, 126, 148
Dry chemical extinguisher, 55

287

Index

Drying section, plastic, 160
Ductility, defined, 236, 244
Dynamic pumps, 96–97

E

Edison, Thomas A., 7
Elasticity, defined, 235
Elastic limit, defined, 243, 245
Electrical drawings, 126, 148
Electrical system, 154, 156, 157
   symbol for, 150
Electrician, 23
Electricity, motors and, 102–105
Electron, defined, 203, 206
Electronic transmitters, 141
Elements, 203
   described, 205, 224
Elevation drawings, 126, 148–149
Emergency response, 53–54, 254
   defined, 44, 250, 255
Endothermic reactions, described, 225
Energy, 231
   defined, 230, 242
   kinetic, 73, 231
   potential, 230, 231
Engineer, 23
Environment, 34, 35
Environmental awareness, as required skill, 14
Environmental Protection Agency (EPA), 252
   defined, 250
   Process Safety Management (PSA) and, 34–35
Equilibrium
   defined, 203
   reaction, defined, 204
Equipment
   complex/simple, 239–242
   plastics plant, 120–121
   process, 126
   symbol for, 147
   training programs and, 37–38
Equipment location drawings, 126, 149, 151
Ethylbenzene process, 175
Ethylene
   glycols, 175, 176
   process, 178
Exothermic reaction, defined, 204, 225
Explosive, described, 50
External gear pump, 97
Extruder
   defined, 110
   described, 120

   illustrated, 121
Extrusion
   described, 159
   section, plastic, 160
Exxon Valdez, 34

F

Faculty expectations, process safety management (PSM) standard, 2
Fahrenheit, heat measurement system, 70
Fan, on flare systems, 122
Feed and transfer section, plastic, 160
Filters, 95, 96
   defined, 86
Final control elements
   control loops and, 145
   defined, 145
Financial analyst, 24
Fired heaters, described, 110, 123
Fires
   classification of, 54
   extinguishers for, 54–55
Fire tube
   heaters, 113–114
   indirect furnace, 115
First-aid incident, defined, 44
First responder, defined, 44
Fishbone chart, 191, 260, 267–268
Fixed bed reactor, defined, 170
Fixed head, multi-pass heat exchanger, 111
Fixed heat, single pass heat exchanger, 111
Flammable
   gas, 50
   liquid, 50
Flare
   defined, 110, 122
   system, 122, 123
Floating head, multipass (u-tube) heat exchanger, 111
Floating roof tanks, 95
Flow charts, 260, 267, 271
   described, 150
Flow control loop, symbol for, 139
Flow diagrams, 126, 127–128, 146
   equipment relationships and, 155
Fluid catalytic cracking, 179, 180
   defined, 170
Fluid coking, 182, 183
   defined, 170, 184
Fluid energy conversion, 73
Fluid flow, 70–73
   defined, 62
   measuring flow rate, 73

Fluidized bed reactor, defined, 170
Fluid pressure, 237–239
   defined, 62
Fluids, density of, 81, 243
Foam fire extinguisher, 55
Force, defined, 237
Forced draft cooling tower, 112
Foundation drawings, 126, 148
   illustrated, 148
Foundations for Excellence in the Chemical Process Industries, 30
Fraction, defined, 62
Fractional distillation
   modern, 10-11
   defined, 203
Fractionating column
   defined, 2
   history of, 8
Frequency chart, 269–270
   defined, 260
Furnace, 115–116
   energy conversion and, 73
   symbols for, 136
   system, 158
Fusion, latent heat of, 69

G

Gas chromatography, 203
Gases
   laws of, 68–69
   pressure and, 68
   storage tanks for, 95
   transferring solids and, 73
Gasoline, specific gravity of, 72, 233
Gas processing, described, 184
Gate valves, 88
Gauge pressure, 68
Gesner, Abraham, 5
Globe valves, 88
Goal setting
   defined, 2
   described, 19
Gold collar, defined, 28
Gravity
API, 230, 234
Baume, 81, 230, 234
   specific, 230, 233
   calculating, 234
Grouping symbols, defined, 62

H

Halon fire extinguisher, 55
Hand tools
   defined, 86
   described, 87–88

Index

Hardness, defined, 235, 244
Hardwire, interlocks, 146
Harassment, sexual, 20–21
Hart, William Aaron, 4–5
Hazardous chemicals, safe handling of, 50
Hazardous materials
   permitting, 253
   regulation of shipment of, 56
   labels/signs/placards, 48, 57
HAZCOM, 35, 48–50
   defined, 28, 44
   delivery of standard to employees, 48, 50
   standard, 35
HAZWOPER, 56
   instrumentation and, 44
Health, 34
Hearing conservation, industrial noise and, 56
Heat
   chemical reactions and, 214–215
   defined, 63, 69, 80
   effects of absorbed, 80
   transmission of, 69
Heat exchangers, 111
   cooling towers and, 113, 154, 156
   described, 110, 122
   symbols for, 131, 135
Heat transfer, defined, 63
Helmont, Jan Baptista van, ver, 4
Highly toxic, described, 51
High school, transition to college, 16–17
Histogram, 269–270
   defined, 260
HMIS system, hazardous material shipment and, 57
Horizontal cylindrical tanks, 95
Houdry, Eugene J., catalytic cracking and, 8–9
Houdry process catalytic cracking, 8–9
Housekeeping, defined, 44
Human resources analyst, 24
Hydraulic
   actuator, 145
   system, 154
Hydraulics, 165–166
Hydrocarbons, 219–221
   defined, 203, 225
Hydrocracking, 181–182, 222–223
   defined, 170, 184
Hydrodesulfurization, 180–181, 184
Hydrogen ion, defined, 203
Hydrosulfurization, defined, 170
Hydroxyl ion, defined, 203

Hyperbolic cooling tower, 112

I

Ideal gas law, 68–69
Improvement cycle, defined, 260
Incinerator, defined, 110
Inclined plane principle, 240
Indirect fired furnace, 115
Induced draft cooling tower, 112
Induction motors, 105
Industrial hygienist, 24
Industrial processes, 40, 171–173
   defined, 154, 184
Inertia, defined, 230
Injection molding, described, 159
Inorganic chemistry, defined, 203
Instrumentation
   process, 126
   process control and, 39
   symbols, 129
Instruments, symbols for, 139, 147
Interlock, defined, 146, 151
Interpersonal skills, as required skill, 12
Ion, defined, 203, 206–207
Irritants, described, 51

J

Job lists, defined, 274
Job search, 275–278
Juran, Joseph M., 35, 190, 199, 260

K

Kelvin, heat measurement system, 70
Kettle reboiler heat exchanger, 111
Kier, Samuel, 5
Kinetic energy, 231
   defined, 230
   energy conversion and, 73

L

Labels, hazardous material shipment and, 48, 57
Laboratory, critical job functions, 33
Laminar flow, 72–73
Laminating, described, 160
Latent heat
   of condensation, 69
   defined, 69, 81
   of fusion, 69
   of vaporization, 69
Lawyer, patent, 24
LCD (Lowest common denominator), defined, 63
Legal assistant, 23
Legends, 126
   process, 146

   symbol for, 147
Level control loop, 141
   symbol for, 140
Levers, principle of, 241–242
Lifelong learning, defined, 2
Lift check valves, 90
Limiting factor, 204
Lines, symbol for, 147
Liquid energy, forms of, 73
Liquids
   determining pressure produced by, 65–67
   principles of pressure of, 68, 80
Listening, as required skill, 12
Lobe pumps, 98
Lock-out, tag-out, defined, 44
Lower control limit, defined, 188
Lowest common denominator (LCD), defined, 63
Lubrication
   equipment, 105
   system, 154, 156, 157

M

Machinist, 23
Malleability, defined, 236, 244
Mass
   defined, 230
   relationships, 210–213
Material balance, 216–217
Material balancing, defined, 204
Material Safety Data Sheets (MSDS), 52
   defined, 204
Math, as required skill, 12
Math skills, process technicians and, 73–79
Matter
   defined, 204, 230, 231
   properties of, 231–232, 243
Mean, defined, 188
Mechanical advantage, 240, 245
Mechanical crafts, 23
Mixed number, defined, 63
Mixtures, described, 204, 207, 224
Molding, 159
Molecules, described, 207
Moments, principle of, 241–242
Motors
   electricity and, 102–105
   symbols for, 134
   three-phase, 104–105
MSDS (Material Safety Data Sheets), 52
   defined, 204
Multiplication, defined, 63
Multitasking, as required skill, 13

289

Index

Murdock, William, 4
Mutagen, described, 50

N

National Institute for Occupational Safety and Health (NIOSH), 46
National Water Quality Standards, 252
Natural draft cooling tower, 112
Networks, people responsible for, 34
Neurotoxic, described, 51
Neutralization reaction, 205, 225
Neutron, defined, 204
NFPA diamond, hazardous material shipment and, 57
NIOSH (National Institute for Occupation Safety and Health, 46
Noise, hearing conservation and, 56
Nonhydrocarbon heaviness, measurement of, 72
Numerator, defined, 63

O

Occupational Safety and Health Administration (OSHA)
    process safety management (PSM) standard, 31
    purpose of, 46
Occupational Safety and Health Review Commission (OSHRC), 46
Olefins production, 177
Olekins, 221
Operations
    training programs and, 38–39
    troubleshooting, 193–195
Operators
    process training for, 37–39
    responsibilities of, 33–34
    team skills and, 189
Organic chemistry, defined, 204
Organic peroxide, described, 50
OSHA (Occupational Safety and Health Administration)
    described, 28
    process safety management (PSM) standard, 31
    purpose of, 46
OSHRC (Occupational Safety and Health Review Commission), 46
Oxidizer, described, 50

P

P&ID (Piping and instrument drawings), 126, 127–138, 149
    components, 146
    described, 28
    illustrated, 130
Paraxylene processes, 177
Pareto chart, 191, 268, 269
    defined, 260
Pascal, Blaise, 64
Pascal's law, 64
    defined, 63
Patent attorney, 24
Percent, defined, 63
Percent-by-weight solution, 217–218
    defined, 204
Periodic table, 208, 209
    defined, 204
    information box, 208
Permissive, interlocks and, 146, 151
Permit systems, 35, 54
    defined, 44
Personal protective equipment (PPE), 52
    defined, 44
    for emergency response, 53–54
Petrochemical processes, 172, 173
Petroleum
    described, 3
    products, 3–4
PFDs (Process flow diagrams), 28, 127–138, 149
pH
    defined, 204
    measurements, 218
Phillips, 34
Physical hazard, defined, 44
Physics, 39
    defined, 231, 242
Pilot plant operations, 196–198
Pipe fittings, 94
Pipes, defined, 86
Piping, 93
    symbols for, 132
Piping and instrument drawings (P&ID), 126, 127–133, 149
    components, 146
    described, 28
    illustrated, 130
Piston compressor, 98
Piston pump, 98
Placards, hazardous material shipment and, 57
Planned experimentation, 269
    defined, 260
Plant permit system, 54
Plastic
    initial process in creating, 160
    system, 154, 158–160
Plastics plant, equipment, 120–121
Plug valves, 89

Pneumatically (AIR) operated actuator, 144
Pneumatic transmitters, 142
Pollution control
    air, 251–252
    water, 252–253
Polyethylene processes, 177, 178
Polymerization section, 160
Positive displacement
    compressors, 99
    symbols for, 134
    pumps, symbols for, 133
Potential energy, 231
    defined, 230, 243
PPE (Personal protective equipment), 52
    defined, 44
    for emergency response, 53–54
Pre-employment tests, 278–279
    defined, 274
Pressure
    chemical reactions and, 214–215
    defined, 63, 79, 237
    in fluids, 237–239
    gases and, 68
    liquid, principles of, 68
    principles of, 63–69
    produced by liquids, determining, 65–67
Pressure control loops, 141
    symbol for, 140
Pressure (PSIG) gauges, 68
Pressure relief, equipment, 122
Primary elements, process instrumentation, 142
Problem solving, as required skill, 13
Process
    defined, 28
    equipment, 126
Process control, instrumentation and, 39
Process diagrams, 128
Process flow diagrams (PFDs), 127–138, 149
Process instrumentation, 126
    drawings (P&Is), 149
Process instruments, 139
    defined, 86
Process legends, 146
    symbols for, 147
Process reaction, defined, 204
Process Safety Management Standard (PSM), 47
    defined, 44
EPA (Environmental Protection Agency), 34–35
OSHA standard, described, 28

290

Index

training programs and, 34
Process technician, 23, 199
   basic math for, 73–79
   career as, 19–22
   chemical mixtures and, 39
   critical job functions, 33
   defined, 2
   diversity and, 20–21
   inside/outside operators, 22
   roles/responsibilities of, 29–34
   sexual harassment and, 20–21
Process technology
   college programs for, 14–19
   defined, 2, 15
Process training, for operators, 37–39
Process troubleshooting, 40
Process variables
   categories of, 140
   control loops and, 140–141
Product directives, 263
Products, sampling, 264
Progressive cavity pumps, 98
Proportional band, controller, 144
Proportional-integral-derivative mode, 144
Proportional plus
   derivative mode, 144
   integral mode, 144
Proton, defined, 204
PSM. See Process Safety Management Standard
Pump-around system, 154
Pumps, 96–98
   defined, 86
   symbols for, 133
Punctuality, as required skill, 13
Pyrophoric, described, 50

Q

Quality, defined, 188, 199
Quality awareness, as required skill, 14
Quality control, 35
   pilot plant production and, 198
Quality improvement
   continuous, 261, 271
   cycle, 261–262
   phases of, 271
Quality standards, 255
Quality tools, 189–190, 262–263
   defined, 188
   listed, 189

R

Radiation, of heat, 69
Range, defined, 188

Rankine, heat measurement system, 70
Rate mode, controller, 143
RCRA (Resource Conservation and Recovery Act), 253
   defined, 250, 256
Reactants and products, defined, 204
Reaction
   combustion, 204
   endothermic, 204
   equilibrium, 204
   exothermic, 204
   neutralization, 204
   replacement, 204
   time, defined, 205
Reactors, 116, 138, 222
   defined, 111, 123, 170
   symbols for, 138
   system, 154
Reactor system, 160, 161
Reading, as required skill, 12
Reboilers, 117–118
   defined, 170
Reciprocating pumps, 98
Refining processes, 172, 178
Reformer, defined, 170
Refrigeration system, 164, 165
   components of, 162–163
Regenerator, defined, 170
Relative weight, 216–217
Relief valves, 92
Replacement reaction, 205
Reproductive toxin, described, 50
Research technician, 23
Reset mode, controller, 144
Residuum, development of, 7
Resource Conservation and Recovery Act (RCRA), 253
   defined, 250, 256
Respirator protection, defined, 44
Respirators, 52
Respiratory protection, programs, 52
Resume, 279–280
   defined, 274
Rotary pumps, 97, 98
Run chart, 267, 268, 271
   defined, 260

S

Safety, 34, 35
   awareness, as required skill, 13–14
   basic, 45–46
   overview, 197
   rules, 45–46
   training programs and applied, 197

issues in, 35
   topics covered, 45
   valves, 92
Samples, tag, 264
Sample types, 264
Scatter plots, 191, 270, 271
   defined, 260
Science, as required skill, 12–13
Screw pumps, 97, 98
Scrubber, 119–120
   defined, 111
   illustrated, 120
Seals, 105
Secretary, 23
Section flow diagram, 128
Sensible heat, defined, 69, 81
Sensitizer, described, 51
Sensors, process instrumentation, 141
Separation process, 10
Sexual harassment
   defined, 2
   described, 20–21
Shell heat exchangers, 111
Shell nomenclature heat exchanger, 111
Shewhart, Walter, 189, 199
Shipping papers, hazardous material shipment and, 57
Signs, hazardous material shipment and, 57
Sillman, Benjamin, Jr., 5
Simon, Leslie E., 190, 199
Smart transmitters, 143
Softwire, interlocks, 146
Solids feeders
   described, 120
   illustrated, 121
Solids, flow of, 73
Solid waste
   control, 253
   defined, 250, 255
Solute, defined, 205
Solutions, described, 205, 207, 224
Solvent extraction, defined, 205
SPC (Statistical Process Control), 190–193, 263–266, 271
   chart, 191, 271
   defined, 260
   guidelines, 265–266
Specific gravity
   defined, 81, 205, 230, 233, 243
   determining, 234
   fluid flow and, 71–72
Specific heat, defined, 69, 81
Spherical storage tanks, 95
Spheroidal storage tanks, 95

Index

Spindletop oil field, 7
Squirrel-cage induction motors, 105
Statistical Process Control (SPC), 190–193, 263–266, 271
    chart, 271
    defined, 260
    guidelines, 265–266
Steam generation system, 154, 160–161, 162
Steam ring, flare, 122
Steam traps, 102, 103
    defined, 86
Steam turbines, 101–103
    defined, 86
    energy conversion and, 73
    symbols for, 134
Stop check valves, 90
Storage tanks, 93–95
    symbols for, 133
Straight-through diaphragm valve, 91–92
Strain, defined, 243–244
Strainer, defined, 86
Stripping columns, 118
Subtraction, defined, 63
Successful job search, defined, 274
Supplier-customer relations, 262
Surface tension, defined, 236, 244
Swing check valve, 90
Synchronous motors, 105
System, defined, 28
Systems, work force development and,, 32

T

Tag, samples, 264
Taguchi, Genich, 190, 199
Tanks
    defined, 86
    design of, 94
    storage, 93–95
    symbols for, 133
Target organ effects, described, 51
Team skills
    described, 184, 188
    operations and, 189
Technical Notebook, 198
Temperature
    cohesive force and, 236–237
    control loop, symbol for, 140
    defined, 63
    determining, 81
    systems of, 70
Tenacity, defined, 235, 244
Tension, defined, 236
Teratogen, described, 50

Test of Chemical Comprehension, 278
The Economic Control of Manufactured Product, 189
Thermal cracking
    defined, 2, 205
    history of, 7–8
Thermosyphon reboiler heat exchanger, 111
Three-phase motors, 104–105
Time management
    defined, 2
    described, 19
Toxic, described, 51
Toxicology, defined, 52
Toxic substance control, 254
Toxic Substances Control Act of 1976 (TSCA), 254
    defined, 251, 256
Training programs, 31
    PSM (Process Safety Management) and, 34
Transducers, defined, 140
Transformer, defined, 149
Transmitters
    controllers and, 142
    control loops and, 142–143
Traps, steam, 86, 102
Treatment process, 10
Troubleshooting, 40, 199
    operations, 193–195
TSCA (Toxic Substances Control Act of 1976, 254
    defined, 251, 256
Tube heat exchangers, 111
Turbines
    steam, 101–102
    described, 86
    symbols for, 134
Turbulent flow, 72–73
Turning controllers, 144

U

University, programs, 14–19
Unstable, described, 50
Upper control limit', defined, 188
Utilities, 166

V

Vacuum
    (PSIV) gauge, 68
    boiling point and, 64
    systems, 64
Valence electrons, defined, 206
Valves, 146
    automatic, 92–93

    ball, 89, 90
    butterfly, 90
    check, 90
    diaphragm, 91–92
    gate, 88
    globe, 88
    plug, 89
    relief, 92
    safety, 92
    symbol for, 147
    symbols for, 131
Vane pumps
    actuator, 145
    pumps, 98
Vaporization, latent heat of, 69
Vapor pressure, defined, 64, 79
Variation, normal, 192
Viscosity, 234–235
    defined, 81, 243
    fluid flow and, 71

W

Water
    density of, 233
    one gallon, 72
    permitting, 253
    pollution, 251
    specific gravity of, 72, 233
Water fire extinguisher, 55
Water pollution
    control, 252–253
    defined, 255
Water reactive, described, 50
Water treatment section, 164, 165
Water tube boilers, 114–115
Weight, 216–217
Weir diaphragm valve, 91
Work, defined, 239, 245
Workers right to know., See HAZCOM
    Workers Right to Know Act, 48
Work experience, 280–281
Work force development, 32–34
Wound rotor induction motors, 105
Writing, as required skill, 12

X

X-bar, defined, 188
Xylene isomerization, 177
Xylenes, mixed, 175–176

Y

Young, James, 5